MEGAKARYOCYTE BIOLOGY AND PRECURSORS: IN VITRO CLONING AND CELLULAR PROPERTIES

MEGAKARYOCYTE BIOLOGY AND PRECURSORS: IN VITRO CLONING AND CELLULAR PROPERTIES

Proceedings of the Symposium on Megakaryocytes *In Vitro* held at the Centers for Disease Control, Public Health Service, U.S. Department of Health and Human Services, Atlanta, Georgia, U.S.A., May 1–2, 1980

Editors:

BRUCE LEE EVATT, M.D.
Director, Division of Hematology
Center for Disease Control &
Associate Professor of Medicine
Emory University
Atlanta, Georgia, U.S.A.

RICHARD F. LEVINE, M.D.
Associate Chief of Staff for Research and Development
V.A. Medical Center
Washington D.C., and
Associate Professor of Medicine
George Washington University
Washington D.C., U.S.A.
and

NEIL T. WILLIAMS, Ph.D.
Staff Scientist
Sloan-Kettering Institute for Cancer Research
New York, New York, U.S.A.

ELSEVIER/NORTH-HOLLAND
NEW YORK • AMSTERDAM • OXFORD

Published by:

Elsevier North Holland, Inc.
52 Vanderbilt Avenue, New York, New York 10017

Sole distributors outside the USA and Canada:
Elsevier Science Publishers B.V.
P.O. Box 211, 1000 AE Amsterdam, The Netherlands

Library of Congress Cataloging in Publication Data

Symposium on Megakaryocytes In Vitro (1980 : Center for
Disease Control)
Megakaryocyte biology and precursors : in vitro
cloning and cellular properties.
Includes bibliographies and index.
1. Megakaryocytes--Congresses. 2. Cloning--Congresses.
I. Evatt, Bruce Lee. II. Levine, Richard Frank.
III. Williams, Neil Thomas. IV. Title.
QP97.5.S94 1980 599.01'13 81-5458
ISBN 0-444-00585-4 AACR2

Manufactured in the United States of America

Contents

Preface

Megakaryocytes are unique cells. They occur only in mammals and their main function is to produce platelets as first demonstrated by Wright in 1906 (*Boston Med. Surg. J.* 154:643–645). Each one gives rise to a few thousand enucleate platelets by a precisely controlled cytoplasmic division (Behnke, *J. Ultrastr. Res.* 24:412–433, 1968). Megakaryocytes are polyploid, resulting from repeated mitoses without cell divisions (Jolly, *Traité Technique d'hématologie*, Paris: Maloine, 1923). The nature of these mechanisms and their consequences are still poorly understood. The cells are of interest to cell biologists and physiologists for these unusual mechanisms, to hematologists for their involvement in various disease processes, and more recently to experimental pathologists and others for their perspective on platelet contributions to hemostasis and arteriosclerosis.

Study of the biology of megakaryocytes and the regulation of their development from uncommitted stem cells to platelets has been severely limited by the extremely low incidence of megakaryocytes in the bone marrow (0.01–0.3% of all nucleated cells, depending on the age and species). In the last five years, *in vitro* techniques have become available which have permitted megakaryocytes to be isolated in reasonable yields (Levine and Fedorko, *J. Cell Biol.* 69:159–172, 1976) and their precursor cells to be studied and monitored in cloning assays (Metcalf et al., *Proc. Nat. Acad. Sci. USA* 72:1744, 1948, 1975).

This book and the conference from which it was drawn were planned to provide a forum for presentation of new data derived from these *in vitro* techniques. Although specific topics such as morphology, thrombopoietin, etc., have been the subjects of review articles, no single comprehensive source on megakaryocytes has been available. We have brought together in this volume a wide spectrum of information and have included discussions in an attempt to reconcile possible differences of interpretation. More than that, we hoped to provide a synthesis of all that has been done in this field, to reach a common understanding not only of terms but also some of the mechanisms in which these cells are involved. The *in vitro* work of the last five or six years has been clarified and the remaining problems are better defined. The highlights below illustrate some of the new perspectives achieved, but it is left to the reader to survey the rich complexity contained herein.

In this book, the definitions of stages involved in the maturation sequence have been more clearly described than ever before. Megakaryocytes are found in sizes down to 5 μm in diameter, but are recognized by routine microscopy down to only 10 μm. The smaller cells are detectable by specific markers such as acetylcholinesterase activity in the mouse and rat or fluorescent antibody detection of platelet antigens in the human. These cells were previously thought to be a separate class of progenitors, but with the presence of these biochemical markers they are now thought to be newly differentiated megakaryocytes. Little is known of the length of this stage or of the events which encompass it, but it does appear to be important in ploidy amplification in response to platelet demand.

Several chapters are devoted to data on qualitative or biochemical contributions of megakaryocytes to platelets. It now seems likely that all of the substances found in platelets are synthesized in the megakaryocytes. The one exception is serotonin, for which megakaryocytes have the same uptake capacity as do platelets.

Analysis of megakaryocytes using objective criteria has permitted the conclusion that, contrary to earlier reports, 8N megakaryocytes are the most frequent megakaryocyte class in the bone marrow. Furthermore, it was reported that each *in vitro* colony contained megakaryocytes of different rather than uniform ploidy levels.

Megakaryocytopoiesis appears to be controlled at two levels. A compartment of precursor cells was shown to proliferate in response to various conditioned media, but was not influenced by platelet levels. These pre-megakaryocytes are identifiable only by measuring their progeny, with the above biochemical substances as useful identifying markers. Ploidy amplification in the differentiated megakaryocyte compartment occurs in a positive feedback control mechanism in response to platelet demand and is mediated by thrombopoietin. Thus, separate factors independently influence proliferation and maturation events. It is not known whether only a single factor is involved at each level.

No attempt has been made to recommend particular nomenclature. The authors in this text and elsewhere have used different terms to describe similar factors or cell populations: thrombopoietin, thrombocytopoiesis stimulating factor, or megakaryocyte potentiator; CFU-M or Meg-CFC; and Meg-CSF, Meg-CSA, or MK-CSF.

Many useful observations about the biology and physiology of megakaryocytes have been assembled in this book from *in vitro* studies. It should be stressed, however, that the information gained is meaningful only in an *in vivo* context. For example, following induction of thrombocytopenia in animals, the increase in differentiated megakaryocytes precedes, not follows, the increase in clonable precursor cells (Meg-CFC). Data derived from both these approaches further our understanding of the normal development of megakaryocytes and thus provide a basis for exploring abnormalities in states of perturbed thrombocytopoiesis.

Acknowledgments

The editors of this book are deeply indebted to Dr. Kathryn Kellar for her editorial assistance and for the countless hours she spent in effectively shepherding the manuscripts through the editorial process. In addition, we are grateful for the services of Claudia Lewis and Nancy Cooey who were responsible for the word processing which produced the copy in final form, and to our capable secretaries, Mrs. Evelyn DuVal and Mrs. Peggy Fett, for their efforts in the initial stages of the editorial process.

This volume is the edited proceedings of a symposium entitled ''Megakaryocytes *In Vitro*'' held in Atlanta on May 1 and 2, 1980. The symposium was sponsored by the Centers for Disease Control and by Hemophilia of Georgia, Inc. The symposium organizing committee was comprised of Drs. Bruce L. Evatt, Richard Levine, and Neil Williams.

The committee is grateful to the following companies whose participation greatly enhanced the success of the symposium: Beckman Instruments, Inc., Becton-Dickinson Co., Coulter Electronics, Inc., E. Leitz, Inc., Ortho Instruments, and Vickers Instruments, Inc. In addition, the committee is indebted to the members of the Hematology Division, CDC. They spent tireless hours assisting in the organization of the symposium and it was their efforts that made the meeting run smoothly. We extend our special thanks to Dr. Kathryn Kellar, Mrs. Neile McGrath, Rosemary Ramsey, Peggy Fett, Evelyn DuVal, and to Mr. Walter Scheffel of the Instructional Media Division.

Contributors

Janine Breton-Gorius, M.D.
Unité de Recherches sur les Anémies
INSERM U.91
Hôpital Henri Mondor
9401 Creteil, FRANCE

Paul A. Bunn, Jr., M.D.
NCI-VA Medical Oncology Branch
VA Medical Center
50 Irving Street, N.W.
Washington, D.C. 20422

Samuel A. Burstein, M.D.
Harbor View Medical Center
Harbor View Hall
Division of Hematology, 6th Floor
325 Ninth Street
Seattle, Washington 98104

Hugo Castro-Malaspina, M.D.
Sloan-Kettering Institute for Cancer Research
1250 First Avenue
New York, New York 10021

Peter P. Dukes, Ph.D.
Hematopoiesis Research Laboratory
Childrens Hospital of Los Angeles
P.O. Box 54700
Terminal Annex
Los Angeles, California 90054

Shirley N. Ebbe, M.D.
Lawrence Berkeley Laboratory, Bldg. 55
University of California
Berkeley, California 94720

Bruce L. Evatt, M.D.
Director, Hematology Division
Centers for Disease Control
1600 Clifton Road, N.E.
Atlanta, Georgia 30333

Jerome E. Groopman, M.D.
Division of Hematology-Oncology
UCLA Medical Center
Los Angeles, California 90024

Gulgun D. Kalmaz, M.D.
University of Tennessee Memorial Research Center
1924 Alcoa Highway
Knoxville, Tennessee 37920

Kathryn L. Kellar, Ph.D.
Division of Hematology
Centers for Disease Control
1600 Clifton Road, N.E.
Atlanta, Georgia 30333

Robert M. Leven
Department of Anatomy
School of Medicine
University of Pennsylvania
Philadelphia, Pennsylvania 19104

Jack Levin, M.D.
Hematology Division
Johns Hopkins Hospital
Baltimore, Maryland 21205

Richard F. Levine, M.D.
VA Medical Center
50 Irving Street, N. W.
Washington, D.C. 20422

Michael W. Long, Ph.D.
Sloan-Kettering Institute for Cancer Research
145 Boston Post Road
Rye, New York 10580

Dr. Med. Manfred Mayer
Facult. f. clin. Medicine, I. Med.
Postfach 23, 6800 Mannheim
GERMANY

Ted P. McDonald, Ph.D.
University of Tennessee Memorial Research Center
1924 Alcoa Highway
Knoxville, Tennessee 37920

Eric M. Mazur, M.D.
Yale University School of Medicine
333 Cedar Street
New Haven, Connecticut 06510

Hans A. Messner, M.D.
Ontario Cancer Institute
500 Sherbourne Street
Toronto, M4X 1K9, Ontario
CANADA

Hideaki Mizoguchi, M.D.
Division of Hematology
Department of Medicine
Tokyo Women's Medical College
Kawadacho 10
Shinjuku-ku, Tokyo, 162, Japan

Alexander Nakeff, Ph.D.
Section of Cancer Biology
4511 Forest Park Blvd., Ste. 401
St. Louis, Missouri 63108

Jean-Michel Paulus, M.D., Ph.D.
Institut de Pathologie Cellulaire
Hôpital de Bicêtre
Le Kremlin-Bicêtre, FRANCE

Sigurdur R. Petursson, M.D.
University of Pittsburgh
School of Medicine
931 Scaife Hall
Pittsburgh, Pennsylvania 15261

Enrique Rabellino, M.D.
Cornell University School of Medicine
1300 York Avenue
New York, New York 10021

Paul K. Schick
Department of Physiology and Biochemistry
Medical College of Pennsylvania
Philadelphia, Pennsylvania 19129

Barbara P. Schick, Ph.D.
Temple University Health Sciences Center
3400 North Broad Street
Philadelphia, Pennsylvania 19140

N. Raphael Shulman, M.D.
Bldg. 10, Rm. 9N-250
9000 Rockville Pike
Bethesda, Maryland 20205

Howard M. Steinberg, Ph.D.
Beth Israel Hospital
330 Brookline Avenue
Boston, Massachusetts 02215

Neil T. Williams, Ph.D.
Sloan-Kettering Institute for Cancer Research
145 Boston Post Road
Rye, New York 10580

The Following Includes Those Participating in the Discussions

Bentfeld-Barker, M. E., Berkeley, CA
Bussel, J. B., New York, NY
Chervenik, P. A., Pittsburgh, PA
Dombrose, F. A., Chapel Hill, NC
Dukes, P. P., Los Angeles, CA
Groopman, J. E., Los Angeles, CA
Hempling, H. G., Charleston, SC
Hoffman, R., New Haven, CT
Jackson, C. W., Memphis, TN
Kalmaz, G. D., Knoxville, TN
Inoue, S., Detroit, MI
Leven, R. M., Philadelphia, PA
Long, M. W., Rye, NY
Mayer, M., Mannheim, W. Germany
Mazur, E. M., New Haven, CT
Miller, J. L., Syracuse, NY
Mizoguchi, H., Tokyo, Japan
Paulus, J-M., Paris, France
Petursson, S. R., Pittsburgh, PA
Schick, P., Philadelphia, PA
Shulman, N. R., Bethesda, MD
Steinberg, R. M., Boston, MA
Vigneulle, R. M., Bethesda, MD
Walz, D. A., Detroit, MI
Warheit, D., Detroit, MI
Weil, S., Boston, MA
Zuckerman, K., Ann Arbor, MI

MEGAKARYOCYTE BIOLOGY AND PRECURSORS:
IN VITRO CLONING AND CELLULAR PROPERTIES

Introduction

Published 1981 by Elsevier North Holland, Inc.
Evatt, Levine, and Williams, Editors
MEGAKARYOCYTE BIOLOGY AND PRECURSORS:
IN VITRO CLONING AND CELLULAR PROPERTIES

Megakaryocytopoiesis *In Vivo*

Shirley Ebbe

Donner Laboratory, Lawrence Berkeley Laboratory, University of California, Berkeley and Department of Laboratory Medicine, University of California, San Francisco, California

Technological advances have profoundly influenced hemopoietic research in recent years by providing capabilities to isolate blood forming cells, study their physiology and chemistry, and grow clones of recognizable differentiated hemopoietic cells from morphologically unrecognizable stem cells. These new techniques should advance our understanding of hemopoiesis and its regulation, but it will be vital to determine how the findings in culture apply to hemopoiesis in intact organisms. Therefore, as a preamble to the first major conference on megakaryocytes *in vitro*, it is appropriate to review some of our knowledge and speculations about behavior of megakaryocytes *in vivo*.

Observations of platelet production and megakaryocytopoiesis in experimental animals and in human beings have led to certain concepts about the way that megakaryocytes develop from pluripotential stem cells (Fig. 1). The data are consistent with a model in which the cells become committed to megakaryocytopoiesis, but do not immediately differentiate further or enter into active proliferation. This compartment will be referred to as non-proliferating with the understanding that this terminology allows either for a small portion of cells to be in cycle or for all cells to be in cycle with a long generation time. Thereafter, the megakaryocyte precursors appear to undergo cellular proliferation and initiate nuclear replication without cell division before becoming recognizable as megakaryocytes. Nuclear replicative capacity persists in the most immature recognizable cells (Feinendegen et al., 1962), but the major activities in the recognizable compartment are related to cytoplasmic growth and development. Recognizable megakaryocytes are polyploid, and, in the normal steady state, the major classes of cells have 4, 8, and 16 times the normal diploid amount of DNA (Odell et al., 1965). Size of mega-

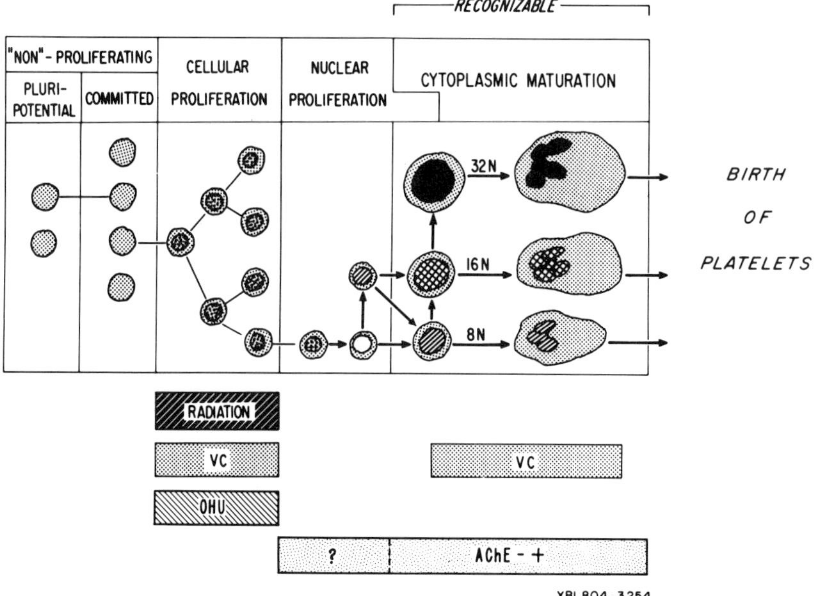

Fig. 1. Annotated model of megakaryocytopoiesis. Sites of major actions of cytotoxic agents are indicated (VC = vincristine; OHU = hydroxyurea). The area labeled AChE− + indicates cells that in rats or mice stain for the enzyme acetylcholinesterase; the ? indicates that some of the small cells undergoing nuclear proliferation may be small AChE− + cells.

karyocytes is proportional to ploidy (Odell et al., 1970), but the relationship of size and ploidy of a megakaryocyte to the number and size of platelets it produces is not clear (Paulus et al., 1979a).

It is convenient to think of these cells as compartmentalized, of the compartments as having well-defined boundaries, and to consider that the cells march through the compartments like soldiers. Such artificial notions are, in fact, consistent with some experimental observations. However, it is likely that the cells behave more in a civilian than military fashion and the lack of regimentation may become more apparent after perturbations than it is in the normal steady-state.

Platelet production has been found to be regulated, in part, by the number or mass of circulating platelets, but the relative importance of these two parameters of platelet concentration has not been established. It has been suggested that the total body mass of platelets may be a more important homeostatic parameter than is the number or mass of platelets in a given volume of circulating blood (Aster, 1966). In spite of these uncertainties, it is clear that deficiency of circulating platelets is associated with stimulation of megakaryocytopoiesis (Craddock et al., 1955) and an excess with its suppression (Cronkite, 1957). The homeostatic mechanisms affect megakaryocytic precursor cells, rather than recognizable megakaryocytes themselves, and this finding is consistent with the localization of proliferative activity largely to the precursor compartments. Cytotoxic agents that show their effect on proliferating cells also affect predominantly the precursor cells. By analyzing the ways in which homeostatic perturbations and cytotoxic agents, alone or in combination, ultimately affect recognizable megakaryocytes and platelet production, information about the precursor compartments can be gained.

It is known that pluripotential stem cells participate in the homeostatic response to thrombocytocytopenia. Their numbers increase in the spleens of thrombocytopenic mice, but not in the bone marrow (Ebbe et al., 1971). However, other observations suggest that homeostatic regulation of platelet production is mediated by committed precursor cells, thus implying that pluripotential stem cell changes may be secondary. For example, W/W^v mice are known to have a severe defect of pluripotential stem cells. Nevertheless, they respond to and recover from immunothrombocytopenia normally (Shreiner, 1976; Ebbe and Phalen, 1978). More convincing evidence that the homeostatic mechanisms do not affect the pluripotential stem cells directly comes from the experiments of Goldberg et al. (1977). They found that prevention of thrombocytopenia (by platelet transfusion) in lethally irradiated recipients of bone marrow transplants inhibited splenic megakaryocytopoiesis as measured 10 days later. This effect was found when platelet transfusions were given for the full 10-day period of observation or for only the final 4 days; it was not found when platelets were administered only on the first 2 days after irradiation and transplantation, a period when pluripotential stem cell seeding and proliferation occurs. Thus, suppression of megakaryocytopoiesis by high platelet counts, in these experiments, appeared to affect differentiation of precursor cells into recognizable megakaryocytes rather than pluripotential stem cell proliferation.

After exposure to sublethal doses of ionizing radiations, mouse platelet counts show characteristic changes (Ebbe and Stohlman, 1970). They are maintained at a normal level for 4 days, corresponding to continued production of platelets by radioresistant recognizable megakaryocytes and their immediate precursors. There is a subsequent 4-day period, corresponding to the mouse platelet survival time, during which the platelet count drops because production is reduced as a result of radiation-induced damage to a sensitive precursor compartment. Thereafter, recovery gradually occurs. It is noteworthy that onset of thrombocytopenia is preceded by megakaryocytopenia in the bone marrow, but that the recovery of platelet counts occurs while megakaryocytopenia persists (Ebbe and Phalen, 1979). The initial recovery may be mediated, in part, by the stimulatory effect of thrombocytopenia on the recovering megakaryocytes.

If megakaryocytopoiesis is acutely stimulated by induction of thrombocytopenia immediately before or after sublethal irradiation of the mice, the radiation-induced thrombocytopenia is reduced in severity and duration (Ebbe and Stohlman, 1970). Therefore, the stimulus of acute thrombocytopenia on day 0 must affect a population of precursor cells that was not heavily damaged by the radiation; the response of this population of cells appears to result in an earlier repopulation of the megakaryocytes. It seems likely that such a compartment of cells would be one that was not in active cell cycle at the time of irradiation and, thus, may be the committed, but non-proliferating stem cell.

The responses to radiation suggest that there are homeostatically responsive precursors that are not in active cell cycle in the normal steady state and that there are cycling precursors that are sensitive to radiation. Are the cycling and non-cycling cells both part of the same compartment? Does stimulation or suppression of the system alter the proportion of cells in cycle? To answer these questions, we looked at the effects of cytotoxic agents on megakaryocytopoiesis to determine if, by stimulating the system, we could increase the proportion of stem cells in cycle and make it more sensitive to cytotoxic agents. Stimulation was produced by injections of heterologous antiplatelet serum which destroyed essentially all circulating platelets; the duration of stimulation before exposure to cytotoxic agents was varied.

Sublethal doses of radiation were administered to determine if the degree or duration of radiation-induced thrombocytopenia would be enhanced (Ebbe and Stohlman, 1970). Radiation was administered after 1, 4 or 6 days of thrombocytopenia. In no case was radiation-induced thrombocytopenia more severe than in the controls. In some experiments, there was a suggestion that thrombocytopenia might be more prolonged in the mice that had been stimulated prior to irradiation, but this finding was neither impressive nor highly reproducible. Therefore, the stem cells responsible for platelet repopulation after irradiation did not become substantially more radiosensitive than normal when the marrow was stimulated by peripheral platelet depletion. This conclusion further minimizes the likelihood that pluripotential stem cells are directly affected by peripheral homeostatic mechanisms.

Similar studies were done in which the cytotoxic drugs, vincristine and hydroxyurea, were given to rats or mice. These drugs are considered to be cell cycle dependent in that vincristine is mainly toxic to cells during mitosis and hydroxyurea during DNA synthesis.

Administration of vincristine (0.3 mg/kg) to normal rats resulted in a biphasic reduction of recognizable megakaryocytes (Ebbe et al., 1975). The first occurred within 4 hours, so it appeared to be due to a direct toxic effect of the drug on megakaryocytes. It was followed by a recovery of the megakaryocytes which was, in turn, followed by a second reduction (to 60% of normal) on days 3 and 4. The second drop was attributed to vincristine toxicity to a proliferating precursor compartment. The first reduction in megakaryocytes was associated with a modest thrombocytopenia. The second was not, but, during the second reduction, megakaryocytes were larger than normal, and this may have explained why platelet production was not compromised.

Hydroxyurea (900 mg/kg) in normal mice produced a delayed megakaryocytopenia on the second and third days after drug administration, presumably as a result of toxicity to dividing precursors (Ebbe and Phalen, 1979). During this period, megakaryocytes were reduced to about 50% of normal values. The megakaryocytopenia was associated with macromegakaryocytosis and no thrombocytopenia, but subsequent studies with hydroxyurea have shown that mild thrombocytopenia may develop on days 4 and 5. With either vincristine or hydroxyurea, the decline in megakaryocytes that is attributable to interference with cell proliferation occurs 1–2 days later than the reduction in total marrow cellularity.

The finding of a 40–50% reduction in megakaryocytes with either vincristine or hydroxyurea indicated that there is an important precursor compartment in which virtually all the cells are in cycle. If it is valid to compare the results of these observations with studies in which the cycling status of the CFU-M (colony-forming unit-megakaryocyte) has been determined with *in vivo* suicide techniques, it could be concluded that the dividing cell population is an unlikely candidate to be the CFU-M. Only 7–20% of CFU-M have been found to be in DNA synthesis at any time as compared to the 50% indicated by our experiments (Nakeff et al., 1978; Burstein et al., 1979a; Paulus et al., 1979b). However, this must be considered to be an open question because of different experimental techniques resulting in probable differences in exposure of cells to toxic levels of drugs.

When vincristine or hydroxyurea was given to animals after 1 or 4 days of antiplatelet serum, there was a tendency for the subsequent reductions in megakaryocytes to be somewhat greater than in normal animals (Table 1). Interpretation of this finding is difficult, and the difficulty arises from the fact that virtually the entire drug-susceptible population of cells appeared to be in cycle in the normal steady state. Thus, the modest increases in drug sensitivity in stimulated animals probably reflected acceleration of maturation or shortening of generation time. It may be more profitable to determine if cell cycle status is changed by the suppressive effect of transfusion-induced thrombocytosis or if parameters of thrombocytopoiesis, other than the number of megakaryocytes, are changed when cytotoxic agents are administered to stimulated animals.

Can these data, in any way, be correlated with published results of CFU-M in perturbed animals? In one report, no change in cell cycle status of CFU-M was found in thrombocytopenic mice (Nakeff, 1979) but in another it was increased (Burstein et al., 1979b). Similar somewhat contradictory findings were present in our studies in which hydroxyurea was given to thrombocytopenic mice. In some otherwise identical experiments, the reduction in megakaryocytes was greater than normal, and in others it was not. These results indicate the importance of considering biological variability in comparative experiments and emphasize the need for large amounts of data.

A megakaryocyte precursor that can be identified in rodents is the small acetylcholinesterase (AChE)-positive cell. Jackson (1973) and Long and Henry (1979) have shown that perturbation of the system by alteration of circulating platelet levels

Table 1. Cytotoxic Megakaryocyte Reductions.

	Vincristine (rats)	Hydroxyurea (mice)
Normal	40%	50%
Thrombocytopenia, 1 day	75%	50-65%
Thrombocytopenia, 4 days	60%	50-65%

leads to appropriate changes in AChE-positive cells that precede comparable changes in recognizable megakaryocytes. Nakeff and Floeh (1976) proposed that AChE-positive cells may be the immediate precursor cells of recognizable megakaryocytes, because their density distribution is similar to that of megakaryocytes, thus suggesting that development of nuclear polyploidy may have commenced in these cells. It is not known if the polyploid cells are affected directly by homeostatic mechanisms to change the level of ploidy, but changes in their numbers would more logically occur by changes in rates of influx, enflux, or transit time.

When the platelet count is acutely perturbed in experimental animals, characteristic compensatory changes are observed in the megakaryocytes after a delay of several hours (Table 2). The size, ploidy, number, and mitotic index of megakaryocytes all increase in response to acute thrombocytopenia (Odell et al., 1962; Harker, 1968; Odell et al., 1969; Penington and Olsen, 1970). The rate of maturation is accelerated in the compartment of recognizable megakaryocytes as is the rate of influx into the compartment (Ebbe et al., 1968; Odell et al., 1971a). As a consequence of these changes in megakaryocytes, increased numbers of platelets are produced and rebound thrombocytosis develops (Craddock et al., 1955), the size of these "stress" platelets is increased (McDonald et al., 1964), and the increased platelet production can be detected by an increased incorporation of radioactive tracers (Harker, 1970). These findings all support the notion that megakaryocytopoiesis and platelet production are stimulated when the blood platelet count is low. Converse changes in most of these parameters are produced by transfusion-induced thrombocytosis, thus indicating suppression of megakaryocytopoieisis and platelet production (Odell et al., 1967; Evatt and Levin, 1969; Harker, 1970; Penington and Olsen, 1970).

When radiation, vincristine, or 5-fluorouracil is administered to mice or rats after only one day of thrombocytopenia (induced by antiplatelet serum), the development of rebound thrombocytosis is prevented (Ebbe et al., 1975; Radley et al., 1980). This observation may mean that cell proliferation is an essential intermediary of

Table 2. Megakaryocyte and Platelet Responses to Perturbations of Platelet Count.

	Thrombocytopenia	Thrombocytosis
Megakaryocyte		
Number	Increased	Decreased
Ploidy	Increased	Decreased
Size	Increased	Decreased
Maturation	Accelerated	Normal
Mitotic Index	Increased	
Platelet		
Production	Increased	Decreased
Rebound overcompensation	Present	Present
Size	Increased	

acute stimulation of megakaryocytopoiesis, and that a compensatory increase in proliferation was prevented by the cytotoxic agent. However, this interpretation is probably simplistic in view of the variable actions of the cytotoxic agents. For example, recovery of platelet counts occurs after radiation, even though rebound thrombocytosis does not develop, while the platelet counts stay below normal after vincristine or 5-fluorouracil in spite of discontinuance of antiplatelet serum. Thus, the different durations of action of the agents *in vivo* and their potential abilities to damage more than one compartment of cells or to induce cell damage that may have delayed manifestations make interpretation of results difficult. In contrast to these findings, administration of radiation or vincristine to animals with the more chronic thrombocytopoietic stimulation produced by 4 days of thrombocytopenia did not interfere with development of rebound thrombocytosis (Ebbe and Stohlman, 1970; Ebbe et al., 1975). These findings suggest that compensatory changes in a post-mitotic pool of megakaryocytes had accumulated during the more prolonged stimulus and were able to continue to overproduce platelets after administration of these cytotoxic agents. It will be of interest to see if additional experiments with other cytotoxic agents, with different mechanisms of action, will help to elucidate megakaryocytic responses to homeostatic mechanisms.

The inital bone marrow event that accounts for adjustments in megakaryocytopoiesis cannot be identified with the available data from experiments *in vivo*. Many of the findings would be explained if the peripheral homeostatic mechanism affected differentiation of cells from the non-proliferating into the proliferating compartment and also affected maturation through the compartment. In that case, local intramedullary mechanisms would account for maintenance of the non-proliferating committed compartment and pluripotential stem cell changes.

So far, consideration has been given only to those homeostatic mechanisms that may be dependent on the number or mass of circulating platelets. However, there are several interesting, and probably instructive, circumstances in which megakaryocytopoiesis appears to be stimulated even though the complement of circulating platelets is normal. One of these is the "non-specific" thrombocytosis associated with tissue injury (in contrast to the "specific" thrombocytosis associated with recovery from thrombocytopenia). The megakaryocytic findings that have been reported in non-specific thrombocytosis are somewhat variable (Table 3). Megakaryocytes have been found to be of increased number and decreased size in human diseases (Harker and Finch, 1969) or to have modest increases in ploidy (Lagerlöf, 1972; Penington et al., 1974). Megakaryocytes have also been found to be of normal number and increased size in association with non-specific thrombocytosis in parabiotic isogeneic mice (Ebbe et al., 1978b). With the less intense stimulation of sham surgery or exchange transfusion with platelet-rich fresh blood, rats developed non-specific thrombocytosis in association with megakaryocytes of normal size that matured more rapidly than normal (Ebbe et al., 1968a,b). These variable findings suggest that non-specific stimulation of megakaryocytes may produce somewhat different manifestations from the specific stimulus of thrombocytopenia but that the difference may be determined, in part, by the intensity of the stimulation.

Table 3. Megakaryocyte and Platelet Responses to "Non-Specific" Stimuli.

Megakaryocyte	
Number:	Normal – Increased
Ploidy:	Increased
Size:	Decreased – Normal – Increased
Maturation:	Accelerated
Platelet	
Number:	Increased
Size:	Normal
Production:	Increased

Several experimental conditions have been found in which the number of megakaryocytes are substantially below normal but in which the platelet counts are, paradoxically, normal (Table 4). An additional finding is that the megakaryocytes are macrocytic. Macromegakaryocytosis is known to occur when the system is stimulated by thrombocytopenia and also in some human myeloproliferative disorders such as polycythemia vera or primary thrombocythemia (Franzen et al., 1961; Branehog et al., 1975). The diversity of the experimental conditions under consideration is such that it seems unlikely that each one would be associated with a stem cell disorder analogous to the myeloproliferative disorders that might account for macromegakaryocytosis. Therefore, it has been proposed that the macrocytosis is due to stimulation of megakaryocytopoiesis. In two of the conditions, W/W^v and $S1/S1^d$ mice, platelet size and blood volumes have been found to be normal indicating that feed-back stimulation of megakaryocytes is not being produced by a reduced platelet mass (Ebbe and Phalen, 1978; Ebbe et al., 1978b). In the other two conditions (after administration of radiation or hydroxyurea), these measurements have not been made. In W/W^v mice, megakaryocytes are reduced to 50% of normal in the marrow and 20% in the spleen. In the marrow, they are disproportionately lower than the total cellularity. Megakaryocytes are about 30% larger than normal. $S1/S1^d$ mice have a comparable reduction in marrow megakaryocytes, but the cells are considerably larger (90% larger than normal). Because of their large size, the counts of megakaryocytes in splenic sections are difficult to interpret other than to say that they are not higher in $S1/S1^d$ mice than in controls. One of the hypotheses to explain the finding of macromegakaryocytosis in all these groups

Table 4. Megakaryocytopenia, Macromegakaryocytosis and Normal Platelet Production.

W/W^V mice
Sl/Sl^d mice
Sublethally irradiated mice
Hydroxyurea treated mice

of animals is that the megakaryocytopenia may serve as a stimulus which accounts for the macrocytosis. Comparison of the W/W^v and Sl/Sl^d mice suggests that megakaryocytopenia could not be the sole stimulus, however, because comparable degrees of megakaryocytopenia are associated with markedly different degrees of macrocytosis.

Our results in sublethally irradiated mice confirm those of Odell et al. (1971) in showing that recovery of the platelet count occurs even though there is persistent megakaryocytopenia. In this case, megakaryocyte numbers of about 50% of normal are associated with about a 50% increase in size. As a transient phenomenon after administration of hydroxyurea to normal mice, megakaryocyte numbers decreased about 50% and, simultaneously, sizes increased about 40% (Ebbe and Phalen, 1979). Clearly these four situations are of different etiologies and the pathogenetic mechanisms may be totally different. They serve to emphasize the point that a normal number of megakaryocytes cannot be presumed to be present from the finding of normal numbers of platelets in the blood. They also point to regulation of megakaryocytopoiesis by mechanisms that do not depend on the platelet count or mass. This notion is also supported by published observations on the effects of perturbations on the CFU-M. As discussed earlier, different investigators have found different effects of thrombocytopenic stimulation on CFU-M. However, there is uniformity in the finding that an increased proportion of the CFU-M are in cycle in marrow that is regenerating after irradiation (Williams and Jackson, 1978; Burstein et al., 1979b; Mizoguchi et al., 1979; Nakeff, 1979); the preliminary report of Burstein et al. (1979b) indicates that the stimulus for this proliferation is not over-ridden by the suppressive effect of peripheral thrombocytosis. These observations suggest that megakaryocytopoiesis may be stimulated in a hypoplastic marrow. In contrast, it has been found that growth of CFU-M *in vitro* appears to be stimulated by the presence of other cells, such that plating efficiency varies with the number of cells plated (Metcalf et al., 1975; Mizoguchi et al., 1979). While it may be interesting to speculate about the relationship between the *in vivo* and the *in vitro* observations, it would probably be premature to do so.

Some of the subjects that have been considered can be summarized with reference to an annotated schematic diagram of megakaryocytopoiesis *in vivo* (Fig. 1).

There appears to be an unrecognizable precursor compartment that is affected by three different cytotoxic agents: radiation, vincristine, and hydroxyurea. This compartment has, therefore, been identified as characterized by proliferation of cells. Each agent also has unique characteristics such as the tendency of radiation to damage earlier precursors, so that recovery from this agent takes longer than from the others. Vincristine, in the dose used, was toxic to recognizable megakaryocytes, but did not appear to damage the cells between the recognizable and the proliferating compartments. These cells are probably the ones which are becoming polyploid, and they have been tentatively identified as, in part, forming the group of small AChE-positive cells. Hydroxyurea appears to be toxic, for the most part, to dividing cells, but some degree of toxicity to the polyploid compartment cannot be ruled out with certainty. The cytotoxic agents have amply confirmed the existence of a proliferating megakaryocyte precursor, but the experiments in which they were used in combination with thrombocytopenia as a stimulus to megakaryocytopoiesis have not served to explain definitively the ways in which the stimulus affects the marrow.

It was tempting to add some highly speculative notations to this scheme, such as suggesting that (1) the platelet-dependent homeostatic feed-back mechanisms may act primarily at the interface between the committed non-proliferating stem cells and the cellular proliferation compartment, (2) cell-cell interactions may regulate traffic between the pluripotential and committed compartments, and (3) the committed but non-proliferating stem cell may be the CFU-M. However, these speculative notions served only to pose more questions than answers. A myriad of questions could already be asked about the cellular and chemical regulators of megakaryocytes and their stem cells and especially about the determinants of polyploidy. The list of titles for this meeting suggest that some of the questions will be answered by the new *in vitro* techniques.

The studies that have been summarized here have, for the most part, been done in animals, and it is important to ask how the results may apply to human beings. It is encouraging to note that the principles learned in human subjects seem to apply to laboratory animals, and the converse is also true. The only major discrepancy appears to be that in some animals the spleen may be a hemopoietic organ and that it may become even more important when blood cell production is stimulated. In contrast, expansion of human hemopoiesis is largely confined to the marrow except for disorders in which there are intrinsic proliferative abnormalities of hemopoietic cells. Otherwise, important discrepancies between human and animal megakaryocytopoiesis have not been identified.

Some of the human disorders of platelet production can be classified, with confidence, as disorders affecting one or another portion of the model of megakaryocytopoiesis. It should be emphasized that this type of classification may help in the diagnosis and treatment of diseases, but it may not be useful in determining etiology or basic cellular pathogenesis of a specific disorder. This classification has recently been reviewed so will be only briefly summarized here (Ebbe, 1979).

There is clearly a diverse group of disorders in which abnormal platelet production is a part of a panmyelopathy that occurs as a result of an acquired abnormality at the pluripotential stem cell level. These diseases include the myeloproliferative disorders, acute non-lymphocytic leukemias, aplastic states, and dyshemopoietic states such as paroxymal nocturnal hemoglobinuria and hemopoietic dysplasia. The major or initial clinical findings in these disorders may be related to under- or overproduction of platelets. The morphological abnormalities of megakaryocytes in some of these disorders may be sufficiently characteristic to be of diagnostic importance.

There is another somewhat heterogeneous group of hereditary disorders in which the hemopoietic abnormality appears to be restricted to the megakaryocytes. These diseases, which include hereditary thrombocytopenia and platelet dysfunctional states, may, therefore, be considered as disorders affecting a cell type committed to megakaryocytopoiesis. There are other genetically-determined abnormalities of platelet production in which cells other than megakaryocytes are involved, yet cannot be identified as pluripotential stem cell disorders. Classification of these disorders must await better definition of the hemopoietic abnormality.

It is to be anticipated that application of new techniques such as concentration of megakaryocytes and culture of megakaryocytes and their precursors will improve the understanding of this cell system and help to identify the ways in which it is regulated. With these insights we can then hope to better understand human diseases and prevent or treat them more effectively.

ACKNOWLEDGMENTS
Supported, in part, by research Grant R01-AM21355 from the National Institute of Arthritis, Metabolism, and Digestive Diseases and, in part, by the Office of Health and Environmental Research of the U.S. Department of Energy under Contract No. W-7405-ENG-48.

References

Aster, R. H. 1966. Pooling of platelets in the spleen: Role in the pathogenesis of ''hypersplenic'' thrombocytopenia. *J. Clin. Invest.* 45:645–657.

Branehog, I., B. Ridell, B. Swolin, and A. Weinfield. 1975. Quantifications in relation to thrombokinetics in primary thrombocythaemia and allied diseases. *Scand. J. Haematol.* 15:321–332.

Burstein, S. A., J. W. Adamson, D. Thorning, and L. A. Harker. 1979a. Characteristics of murine megakaryocytic colonies *in vitro*. *Blood* 54:169–179.

Burstein, S. A., J. W. Adamson, S. K. Erb, and L. A. Harker. , 1979b. Murine megakaryocytopoiesis: Evidence of early and late regulatory control. *Blood* 54 (Suppl. 1):165a.

Craddock, C. G., Jr., W. S. Adams, S. Perry, and J. S. Lawrence. 1955. The dynamics of platelet production as studied by a depletion technique in normal and irradiated dogs. *J. Lab. Clin. Med.* 45:906–919.

Cronkite, E. P. 1957. Regulation of platelet production. In *Homeostatic Mechanisms*. Brookhaven Symposia in Biology (No. 10). pp. 96–107.

Ebbe, S. 1979. Experimental and clinical megakaryocytopoiesis. *Clin. Haematol.* 8:371–394.

Ebbe, S. and E. Phalen. 1979. Does autoregulation of megakaryocytopoiesis occur? *Blood Cells* 5:123–138.

Ebbe, S. and E. Phalen. 1978. Regulation of megakaryocytes in W/W^v mice. *J. Cell. Physiol.* 96:73–80.

Ebbe, S. and F. Stohlman, Jr. 1970. Stimulation of thrombocytopoiesis in irradiated mice. *Blood* 35:783–792.

Ebbe, S., D. Howard, E. Phalen, and F. Stohlman, Jr. 1975. Effects of vincristine on normal and stimulated megakaryocytopoiesis in the rat. *Br. J. Haematol.* 29:593–603.

Ebbe, S., F. Stohlman, Jr., J. Donovan, and J. Overcash. 1968a. Megakaryocyte maturation rate in thrombocytopenic rats. *Blood* 32:787–795.

Ebbe, S., F. Stohlman, Jr., J. Overcash, J. Donovan, and D. Howard. 1968b. Megakaryocyte size in thrombocytopenic and normal rats. *Blood* 32:383–392.

Ebbe, S., E. Phalen, P. D'Amore, and D. Howard. 1978a. Megakaryocytic responses to thrombocytopenia and thrombocytosis in Sl/Sl^d mice. *Exp. Hematol.* 6:201–212.

Ebbe, S., E. Phalen, and D. Howard. 1978b. Parabiotic demonstration of a humoral factor affecting megakaryocyte size in Sl/Sl^d mice. *Proc. Soc. Exp. Biol. Med.* 158:637–642.

Ebbe, S., E. Phalen, J. Overcash, D. Howard, and F. Stohlman, Jr. 1971. Stem cell response to thrombocytopenia. *J. Lab. Clin. Med.* 78:872–881.

Evatt, B. L. and J. Levin. 1969. Measurement of thrombopoiesis in rabbits using ^{75}selenomethionine. *J. Clin. Invest.* 48:1615–1626.

Feinendegen, L. E., N. Odartchenko, H. Cottier, and V. P. Bond. 1962. Kinetics of megakaryocyte proliferation. *Proc. Soc. Exp. Biol. Med.* 111:177–182.

Franzen, S., G. Strenger, and J. Zajicek. 1961. Microplanimetric studies on megakaryocytes in chronic granulocytic leukaemia and polycythaemia vera. *Acta Haematol.* 26:182–193.

Goldberg, J., E. Phalen, D. Howard, S. Ebbe, and F. Stohlman, Jr. 1977. Thrombocytotic suppression of megakaryocyte production from stem cells. *Blood* 49:59–69.

Harker, L. A. 1968. Kinetics of thrombopoiesis. *J. Clin. Invest.* 47:458–465.

Harker, L. A. 1970. Regulation of thrombopoiesis. *Amer. J. Physiol.* 218:1376–1380.

Harker, L. A. and C. A. Finch. 1969. Thrombokinetics in man. *J. Clin. Invest.* 48:963–974.

Jackson, C. W. 1973. Cholinesterase as a possible marker for early cells of the megakaryocyte series. *Blood* 42:413–421.

Lagerlöf, B. 1972. Cytophotometric study of megakaryocyte ploidy in polycythemia vera and chronic granulocytic leukemia. *Acta Cytol.* 16:240–244.

Long, M. W. and R. L. Henry. 1979. Thrombocytosis-induced suppression of small acetylcholinesterase-positive cells in bone marrow of rats. *Blood* 54:1338–1346.

McDonald, T. P., T. T. Odell, Jr., and D. G. Gosslee. 1964. Platelet size in relation to platelet age. *Proc. Soc. Exp. Biol. Med.* 115:684–689.

Metcalf, D., H. R. MacDonald, N. Odartchenko, and B. Sordat. 1975. Growth of mouse megakaryocyte colonies *in vitro*. *Proc. Nat. Acad. Sci. USA* 72:1744–1748.

Mizoguchi, H., K. Kubota, Y. Miura, and F. Takaku. 1979. An improved plasma culture system for the production of megakaryocyte colonies *in vitro*. *Exp. Hematol.* 7:345–351.

Nakeff, A. 1979. Regulation of megakaryocyte proliferation: Comparison of the response of CFU-M to ionizing radiation and platelet depletion. *Exp. Hematol.* 7(Suppl. 6):63.

Nakeff, A. and J. E. Bryan. 1978. Megakaryocyte proliferation and its regulation as revealed by CFU-M analysis. In *Hematopoietic Cell Differentiation*. Golde, D. W., Cline, M. J., Metcalf, D., and Fox, C. F., eds. New York: Academic Press. pp.241–259.

Nakeff, A. and D. P. Floeh. 1976. Separation of megakaryocytes from mouse bone marrow by density gradient centrifugation. *Blood* 48:133–138.

Odell, T. T., Jr., C. W. Jackson, T. J. Friday, and K. Y. Du. 1971a. Assay of megakaryocytopoiesis in thrombocytopenic rats. *Br. J. Haematol.* 21:233–240.

Odell, T. T., Jr., C. W. Jackson, and R. S. Reiter. 1967. Depression of the megakaryocyte-platelet system in rats by transfusion of platelets. *Acta Haematol.* 38:34–42.

Odell, T. T., Jr., C. W. Jackson, and T. J. Friday. 1971b. Effects of radiation on the thrombocytopoietic system of mice. *Rad. Res.* 48:107–115.

Odell, T. T., Jr., C. W. Jackson, T. J. Friday, and D. E. Charsha. 1969. Effects of thrombocytopenia on megakaryocytopoiesis. *Br. J. Haematol.* 17:91–101.

Odell, T. T., Jr., C. W. Jackson, and D. G. Gosslee. 1965. Maturation of rat megakaryocytes studied by microspectrophotometric measurement of DNA. *Proc. Soc. Exp. Biol. Med.* 119:1194–1199.

Odell, T. T., Jr., C. W. Jackson, and T. J. Friday. 1970. Megakaryocytopoiesis in rats with special reference to polyploidy. *Blood* 35:775–782.

Odell, T. T., Jr., T. P. McDonald, and M. Asano. 1962. Response of rat megakaryocytes and platelets to bleeding. *Acta Haematol.* 27:171–179.

Paulus, J. M., J. Bury, and J. C. Grosdent. 1979a. Control of platelet territory development in megakaryocytes. *Blood Cells* 5:59–88.

Paulus, J. M., M. Prenant, and J. F. Deschamps. 1979b. Mitotic and endomitotic amplification in megakaryocyte (MKC) colonies. *Blood* 54 (Suppl. 1):166a.

Penington, D. G., and T. E. Olsen. 1970. Megakaryocytes in states of altered platelet production: Cell numbers, size and DNA content. *Br. J. Haematol.* 18:447–463.

Penington, D. G., K. Streatfield, and S. M. Weste. 1974. Megakaryocyte ploidy and ultrastructure in stimulated thrombopoiesis. In *Platelets: Production, Function, Transfusion, and Storage*. Baldini, M. G. and Ebbe, S., eds. New York: Grune and Stratton. pp. 115–130.

Radley, J. M., G. S. Hodgson, and J. Levin. 1980. Platelet production after administration of antiplatelet serum and 5-fluorouracil. *Blood* 55:164–166.

Shreiner, D. P. 1976. Thrombocytopoiesis in W/W^v mice. *J. Lab. Clin. Med.* 87:913–918.

Williams, N. and H. Jackson. 1978. Regulation of the proliferation of murine megakaryocyte progenitor cells by cell cycle. *Blood* 52:163–170.

Growth Factors

Published 1981 by Elsevier North Holland, Inc.
Evatt, Levine, and Williams, Editors
MEGAKARYOCYTE BIOLOGY AND PRECURSORS:
IN VITRO CLONING AND CELLULAR PROPERTIES

The Evolution of Techniques for the Study of Megakaryocyte Growth Factors

Bruce L. Evatt and Kathryn L. Kellar

Division of Hematology, Centers for Disease Control, Atlanta, Georgia

Investigation of the effect of growth factors on megakaryocytopoiesis began with observations of human disease. In 1960, Schulman et al. published a case report of an 8-year-old girl with chronic thrombocytopenia. When treated by infusions of normal plasma or whole blood, this patient responded differently than other patients with thrombocytopenia; she developed a rebound thrombocytosis. The elevated platelet count lasted for a period of several days and then returned to a baseline thrombocytopenia. Repeated intravenous doses of plasma preparations reproduced this phenomenon, but oral administration of the plasma had little or no effect on the circulating platelet count. These investigators postulated that this patient had a congenital deficiency of a growth factor affecting megakaryocytopoiesis, a thrombopoietin.

The lack of suitable human subjects and the limited nature of experiments which could be performed in humans caused other researchers interested in thrombopoietin to turn to animal studies in order to further study the megakaryocyte growth factors. In a classic study performed by Ted Odell (and repeated in a number of other laboratories, (Odell et al., 1965; de Gabriele and Penington, 1967; Harker, 1968; Evatt and Levin, 1969), animals were hypertransfused with platelets to a level that was 2 to 3 times baseline. The platelet count returned to baseline when the platelet transfusion ceased, but subsequently dropped below the baseline count when compared to control animals. These experiments suggested that suppression of platelet production had been induced by the platelet hypertransfusion. In other laboratories, investigators also observed that induction of acute thrombocytopenia in animals by platelet antisera or plasmapheresis was followed by a rebound thrombocytosis (Desforges et al., 1954; Krevans et al., 1955; Craddock et al., 1955; Odell et al., 1962; Matter et al., 1960; Ebbe et al., 1968). These authors suggested that a thrombopoietic factor had been induced by the thrombocytopenia and had stimulated the bone marrow to overcompensate in its production of platelets.

The first attempts to assay for growth factors in plasma and urine were performed in small animals (Spector, 1961; Odell et al., 1961; LeXuan et al., 1962; Linman and Pierre, 1963; Odell et al., 1971). These initial assays used the rate of increase in platelet numbers as a reflection of the rate of induction of thrombocytopoiesis following stimulation. It was soon apparent, however, that the platelet counts in small animals were very erratic and many of the experiments were not consistently reproducible when performed in other laboratories (Evatt and Levin, 1969; Odell et al., 1964). Therefore, other means had to be devised to study growth factors before reliable conclusions could be drawn.

Radioisotopic assays were the first reproducible assays for measuring the rate of platelet production and thus provided a substantial tool for the demonstration and study of circulating growth factors for the megakaryocyte cell line (Evatt and Levin, 1969; Penington, 1969; Cooper et al., 1970; Harker, 1970; McDonald, 1973). The basic principle of these assays was simple: an animal was injected with a radioactive amino acid. The labeled amino acid became incorporated into the proteins of developing megakaryocytes and as platelets were released, the increase in the number of radiolabeled platelets could be used to quantitate the rate of platelet production. Initially, these assays were used to examine the effect of circulating platelet counts on platelet production and the results supported the idea of a positive feedback control mechanism for platelet production. For example, when animals were hypertransfused with platelets prior to administration of isotope, platelet production was suppressed and the induced thrombocytopenia produced a subsequent increase in the rate of platelet production (Evatt and Levin, 1969).

The isotopic methods were further adapted to measure a circulating plasma factor which could stimulate megakaryocytopoiesis as follows: thrombocytopenia was induced in rabbits and the plasma was obtained for infusion into other animals in which platelet production was measured with isotope. The transferred plasma was seen to produce an increased rate of platelet production, suggesting that the plasma contained a growth factor for megakaryocytopoiesis. This plasma factor has been termed thrombopoietin (Evatt and Levin, 1969).

Attempts were made to improve the sensitivity of the assay for thrombopoietin by hypertransfusion of assay animals with platelets, but in practice such assays proved to be too laborious to be practical (Evatt and Levin, 1969; Shreiner and Levin, 1970; Harker, 1970; Cooper et al., 1973). The nature of the assays has made purification of the transferable thrombopoietic factor painfully slow. A fraction obtained with 60%–80% ammonium sulfate fractionation of the plasma from thrombocytopenic rabbits was found to contain the active thrombopoietin (Evatt et al., 1974). As shown in Table 1, further purification steps with DEAE Sephadex, Sephadex G-100, and CM-cellulose have increased the purity of this growth factor up to 1000-fold (Evatt and Levin, 1969; Evatt et al., 1974; Evatt et al., 1980; Levin and Evatt, 1979). Lectins have also been used in the purification (Levin et al., 1979).

Table 1. Purification of Thrombopoietin.

Step	Dose (mg/g body wt)[a]	Purification factor
Plasma	2.4-7.2	1
$(NH4)_2 SO_4$ (60%-80% sat.)	0.8-2.4	3
DEAE sephadex	0.05-0.06	87
Sephadex G-100	0.03-0.05	120
CM-cellulose	0.003-0.006	1067

[a]Dose of protein that produced an increased level of ^{75}Se-methionine in the platelets of normal mice (rabbits were utilized in the experiments in which whole plasma was administered).

The direct effects of growth factors on megakaryocytopoiesis have been studied using the measurement of platelet production with isotopes *in vivo*. Many of these studies have been hampered by the fact that platelet production is an endpoint, and many steps exist between stimulation and the resulting changes in the rate of platelet production. For example, when the effect of another hormone, erythropoietin, upon megakaryocytopoiesis was investigated, seemingly conflicting results were found (de Gabriele and Penington, 1967; Merino, 1950; Van Dyke, 1964; Shreiner and Levin, 1971; Siri et al., 1966). Very large doses of an exogenously produced erythropoietin preparation induced increased rates of platelet production (Evatt et al., 1976). However, endogenously produced erythropoietin induced by hypoxia in dimethyl silicone rubber membrane chambers did not increase platelet production (Evatt et al., 1976). These results suggested either that (1) the erythropoietin injected into the animals was contaminated with thrombopoietin or other substances such as endotoxin which could stimulate thrombopoiesis, or that (2) there was an indirect effect of the exogenously injected erythropoietin through unknown mechanisms. Similarly, when the effects of endotoxin upon platelet production were studied, it was observed that increasing amounts of endotoxin produced increased rates of platelet production (Alving et al., 1979). The endotoxin also induced thrombocytopenia proportional to the increase in the rate of platelet production, suggesting that endotoxin might induce megakaryocytopoiesis by its effect upon the levels of circulating platelets. However, in later experiments, when the purified active moiety of endotoxin, lipid A, was used, the same stimulation of platelet production was seen, but there was little or no effect upon circulating platelet counts (Ramsey et al., 1980). Likewise, when lipid A was incorporated into liposomes before injection, it produced no effect on circulating platelets although it still continued to stimulate platelet production. These latter studies suggested that the effect of lipid A and possibly endotoxin may be due to effects other than the effect upon circulating platelet counts; that is, an indirect mechanism that could not be studied with the *in vivo* system.

Fortunately, during the past decade a number of new techniques have been developed for studying megakaryocytes *in vitro*. There have also been a number of substances tested which will stimulate megakaryocytopoiesis under these *in vitro* conditions. The following papers examine a number of these different culture techniques and several of these growth factors, and discuss how these tools can be used to study the basic biology of megakaryocytopoiesis.

ACKNOWLEDGMENTS
Supported in part by a grant from Hemophilia of Georgia, Inc., Atlanta, Georgia. K. L. Kellar is supported by a National Research Service Award from the National Heart, Lung, and Blood Institute.

References

Alving, B. M., B. L. Evatt, J. Levin, W. R. Bell, R. B. Ramsey, and F. C. Levin. 1979. Platelet and fibrinogen production: Relative sensitivities to endotoxin. *J. Lab. Clin. Med.* 93:437–448.

Cooper, G. W., B. Cooper, and C. Y. Chang. 1970. Demonstration of a circulating factor regulating blood platelet production using ^{35}S-sulfate in rats and mice. *Proc. Soc. Exp. Biol. Med.* 134:1123–1127.

Cooper, G. W., B. Cooper, A. L. Ossias, and E. D. Zanjani. 1973. A hypertransfused mouse assay for thrombopoietin factors. *Blood* 42:423–428.

Craddock, C. G., W. S. Adams, S. Perry, and J. S. Laurence. 1955. The dynamics of platelet production as studied by a depletion technique in normal and irradiated dogs. *J. Lab. Clin. Med.* 45:906–919.

de Gabriele, G., and D. G. Penington. 1967. Physiology of the regulation of platelet production. *Br. J. Haematol.* 13:202–209.

Desforges, J. F., F. S. Bigelow, and T. C. Chalmers. 1954. The effects of massive gastrointestional hemorrhage on hemostasis. I. The blood platelets. *J. Lab. Clin. Med.* 43:501–510.

Ebbe, S., F. Stohlman, Jr., J. Overcash, J. Donovan, and J. Howard. 1968. Megakaryocyte size in thrombocytopenic and normal rats. *Blood* 32:383–392.

Evatt, B. L. and J. Levin. 1969. Measurement of thrombopoiesis in rabbits using ^{75}selenomethionine. *J. Clin. Invest.* 48:1615–1626.

Evatt, B. L., J. Levin, and K. M. Algazy. 1980. Partial purification of thrombopoietin from the plasma of thrombocytopenic rabbits. *Blood* 54:377–388.

Evatt, B. L., J. L. Spivak, and J. Levin. 1976. Relationship between thrombopoiesis and erythropoiesis: With studies of the effects of preparations of thrombopoietin and erythropoietin. *Blood* 48:547–558.

Evatt, B. L., D. P. Shreiner, and J. Levin. 1974. Thrombopoietic activity of fractions of rabbit plasma: Studies in rabbits and mice. *J. Lab. Clin. Med.* 83:364–371.

Harker, L. A. 1968. Kinetics of thrombopoiesis. *J. Clin. Invest.* 47:458–465.

Harker, L. A. 1970. Regulation of thrombopoiesis. *Am. J. Physiol.* 218:1376–1380.

Krevans, J. R. and D. P. Jackson. 1955. Hemorrhagic disorder following massive whole blood transfusions. *J. Amer. Med. Assoc.* 159:171–177.

Levin, J. and B. L. Evatt. 1979. Humoral control of thrombopoiesis. *Blood Cells* 5:101–121.

Levin, J., I. Tang, and J. L. Spivak. 1979. Thrombopoietin: Partial purification by affinity chromatography with immobilized lectins. *Blood* 54 (Suppl. (1):166a.

LeXuan, C. and S. Ebbe. 1962. Assay of plasma thrombopoietic activity in rats. *Med. Exp.* 7:317–323.

Linman, J. W. and R. V. Pierre. 1963. Studies on thrombopoiesis. III. Thrombocytosis-promoting effects of "thrombocythemic" and "polycythemic" plasmas. *J. Lab. Clin. Med.* 62:374–383.

McDonald, T. P. 1973. Bioassay for thrombopoietin utilizing mice in rebound thrombocytosis. *Proc. Soc. Exp. Biol. Med.* 144:1006–1011.

Matter, M., J. R. Hartmann, J. Kautz, Q. B. DeMarsh, and C. A. Finch. 1960. A study of thrombopoiesis in induced acute thrombocytopenia. *Blood* 15:174–185.

Merino, C. F. 1950. Studies on blood formation and destruction in the polycythemia of high altitude. *Blood* 5:1–31.

Odell, T. T., Jr., C. W. Jackson, T. J. Friday, and K. Y. Du. 1971. Assay of megakaryocytopoiesis in thrombocytopenic rats. *Br. J. Haematol.* 21:233–240.

Odell, T. T., Jr., C. W. Jackson, and R. S. Reiter. 1967. Depression of megakaryocyte platelet systems in rats by transfusion of platelets. *Acta Haematol.* 38:34–42.

Odell, T. T., Jr., T. P. McDonald, and F. L. Howsden. 1964. Native and foreign stimulators of platelet production. *J. Lab. Clin. Med.* 64:418–424.

Odell, T. T., Jr., T. P. McDonald, and M. Asano. 1961. Response of rat megakaryocytes and platelets to bleeding. *Acta Haematol.* 27:171–179.

Odell, T. T., Jr., T. P. McDonald, and T. C. Detwiler. 1961. Stimulation of platelet production by serum of platelet-depleted rats. *Proc. Soc. Exp. Biol. Med.* 108:428–431.

Penington, D. G. 1969. Assessment of platelet production with ^{75}Se-selenomethionine. *Br. Med. J.* 4:782–784.

Ramsey, R. B., M. B. Hamner, B. M. Alving, J. S. Finlayson, C. R. Alving, and B. L. Evatt. 1980. Effects of lipid A and liposomes containing lipid A on platelet and fibrinogen production in rabbits. *Blood* 56:307–310.

Schulman, I., M. Pierce, A. Lubens, and Z. Currimbhoy. 1960. Studies on thrombopoiesis. I. A factor in normal human plasma required for platelet production; chronic thrombocytopenia due to its deficiency. *Blood* 16:943–957.

Shreiner, D. P. and J. Levin. 1970. Detection of thrombopoietic activity in plasma by stimulation of suppressed thrombopoiesis. *J. Clin. Invest.* 49:1709–1713.

Shreiner, D. P. and J. Levin. 1971. The effects of hemorrhage and erythropoietin on thrombopoiesis. Scientific Program, American Society of Hematology, XIV Annual Meeting. p. 146.

Spector, B. 1961. *In vivo* transfer of thrombopoietic factor. *Proc. Soc. Exp. Biol. Med.* 108:146–149.

Van Dyke, D. 1964. Response of monkeys to erythropoietin of rabbit, sheep and human origin. *Proc. Soc. Exp. Biol. Med.* 116:171–174.

Sire, W. E., D. C. Van Dyke, H. S. Winchell, M. Pollycove, H. S. Parker, and A. S. Cleveland. 1966. Early erythropoietin, blood, and physiological responses to severe hypoxia in man. *J. Appl. Physiol.* 21:73–80.

Published 1981 by Elsevier North Holland, Inc.
Evatt, Levine, and Williams, Editors
MEGAKARYOCYTE BIOLOGY AND PRECURSORS:
IN VITRO CLONING AND CELLULAR PROPERTIES

Stimulation of DNA Synthesis in Megakaryocytes by Thrombopoietin *In Vitro*

Kathryn L. Kellar, Bruce L. Evatt, Cornelia R. McGrath, and Rosemary B. Ramsey

Division of Hematology, Centers for Disease Control, Atlanta, Georgia

Thrombopoietin is the term used to describe the humoral regulator of thrombocytopoiesis. The demonstration of a plasma substance that stimulated platelet production initiated investigations of the mechanism of action of thrombopoietin and prompted the development of assays to quantitate the activity. Elevated levels of thrombopoietin were seen in response to thrombocytopenia that was induced by transfusion of platelet-poor plasma or by injections of antiplatelet serum into experimental animals. Evatt and Levin (1969), Penington (1969), Harker (1968), and Cooper et al. (1970) developed assays for measuring the activity of thrombopoietin in plasma. These assays were based on the appearance of labeled platelets in the circulation following injection of ^{75}Se-selenomethionine or ^{35}S-sodium sulfate into animals previously administered with test or normal plasma.

Following the development of the bioassays, thrombopoietin was purified approximately 1000-fold (Evatt et al., 1979). The activity of thrombopoietin was associated with glycoproteins and chromatographically it eluted similarly to albumin and β-globulins. These properties were characteristic of the other hematopoietic regulators, erythropoietin and colony-simulating activity (CSA). It is generally acknowledged that one of the limiting factors in further purification and characterization of thrombopoietin is the assay. Although bioassays are the ultimate test of the activity of all biological substances, the bioassays currently used to measure thrombopoietin activity require a considerable volume of the purified product and a large number of animals. In addition, they are time-consuming and, at best, semiquantitative. It may be appreciated that the major contributions to the understanding of the role of erythropoietin and its purification were made only after the development of quantitative *in vitro* assays.

The functional characteristics of thrombopoietin have been studied almost exclusively by examining the response of the bone marrow in the thrombocytopenic animal. Within 48 hours after induction of thrombocytopenia or after injection of thrombopoietin preparations, the rate of platelet production is increased. The megakaryocytes in the bone marrow increase in size, ploidy and rate of maturation over 2–4 days and then return to normal (Ebbe et al., 1968; Penington and Olsen, 1970). Another observation, especially evident in sustained thrombocytopenia, has been that there is an increase both in the number of megakaryocytes and in the percentage of these cells that are immature (Harker, 1968; Odell et al., 1969; Penington and Olsen, 1970). These findings were proposed to indicate that one response to thrombopoietin *in vivo* was stimulation of endomitosis and a subsequent increase in the ploidy and the size of the megakaryocytes, which resulted in an increase in the potential platelet mass. Another response appeared to involve differentiation at the precursor level, resulting in an influx of immature megakaryocytes to replace the maturing cells (Harker, 1968; Ebbe et al., 1968; Penington and Olsen, 1970). The underlying assumption in these studies was that thrombopoietin interacted with the cells of the megakaryocyte compartment at some stage to produce the various changes, analogous to the interactions of erythropoietin and the erythropoietin responsive cell (Odell, 1974). There is in fact little evidence of a direct effect of thrombopoietin on megakaryocytopoiesis. The reports are limited to studies done by Freedman et al. (1977), Nakeff (1977), and preliminary reports of the data to be presented in this chapter (Kellar et al., 1979).

In the studies to be described, the effect of thrombopoietin on megakaryocytopoiesis was examined in cultures of megakaryocyte-enriched bone marrow cells. The response of megakaryocytes to thrombopoietin was measured in terms of the incorporation of ^{3}H-thymidine (^{3}H–TdR) into DNA and the results were analyzed to determine if this culture system could be used in the *in vitro* assay of thrombopoietin.

Methods and Materials

Megakaryocytes were purified from guinea pig bone marrow by using a modification of the method of Levine and Fedorko (1976). Bone marrow cells were separated by equilibrium density centrifugation followed by velocity sedimentation at 1 × g on bovine serum albumin (BSA, Fraction V, Sigma, St. Louis, Mo.) gradients. Purities of the megakaryocytes ranged from 50%–100%. These cell preparations were cultured in Modified Dulbecco's Minimal Essential Medium supplemented with 10% heat-inactivated guinea pig serum and 2.3% BSA for periods up to 3 days (Levine, 1977).

Thrombopoietin was prepared from the plasma of rabbits made thrombocytopenic by using the procedures described by Evatt and Levin (1969). Male New Zealand white rabbits were prebled by cardiac puncture to obtain normal plasma and to reduce the blood volume. After 18 hr, guinea pig antiserum to rabbit platelets was injected intravenously in a volume of 1–2 ml, and platelet counts were determined at 15 min. Rabbits with platelet counts less than 30 × 10^9/l were bled by cardiac puncture. Thrombopoietin was partially purified by precipitation of the plasma with saturated ammonium sulfate [$(Nh_4)_2SO_4$] at 4°C. Protein fractions of both normal and thrombocytopenic plasma were precipitated sequentially by adjusting to concentrations of 40%, 60%, and 80% $(NH_4)_2SO_4$. The 60%–80% $(NH_4)_2SO_4$ precipitate contained the thrombopoietin activity as assayed *in vivo*. These precipitates were reconstituted in water, dialyzed to remove $(NH_4)_2SO_4$, concentrated by ultrafiltration, and then dialyzed extensively with physiological saline before being added to the cell cultures.

The assay for thrombopoietin activity *in vivo* measured the incorporation of ^{75}Se-selenomethionine into the platelets of mice as described by Evatt et al. (1974). The assay was performed by injecting mice subcutaneously with four equal doses of the different fractions and then injecting with 1 μCi of ^{75}Se-selenomethionine. The percent dose uptake of isotope into platelets was measured 16 hr after the isotope was administered. There was a significant increase in the level of ^{75}Se-selenomethionine in the plasma fractions from thrombocytopenic rabbits that were precipitated at an $(NH_4)_2SO_4$ saturation of 60%–80%. These preparations were assayed for CSA by using the two-layer agar procedure (performed through the courtesy of Drs. E. F. Winton and J. M. Kinkade, Emory University School of Medicine, Atlanta, Ga.) and for erythropoietin by using an RIA (performed through the courtesy of Dr. E. Goldwasser, University of Chicago, Department of Biochemistry) (Winton et al., 1977; Sherwood and Goldwasser, 1979). Both of the assays indicated negligible activity.

Incorporation of ^{3}H–TdR into the DNA of the cultured cells was measured by labeling cells with either 1 μCi/ml ^{3}H–TdR (specific activity 5 Ci/mmol) for 3-hr or 0.25 μCi/ml ^{3}H–TdR (specific activity 0.5 Ci/mmol) for 19 hr before terminating the cultures. Cells were washed and then processed for scintillation counting or autoradiography. In some experiments, the ploidy of the cultured megakaryocytes was determined after staining the cells by the Feulgen procedure (Leuchtenberger, 1958).

Cultures were processed for scintillation counting by digesting the cells with 600 μg/ml Proteinase K (E-M Biochemicals, Cincinnati, O.), 10 μg/ml RNase A and 0.1% Triton X-100 in 0.1 M Tris-0.1 M NaCl, pH 7.5. For all experiments, four aliquots from each culture were counted. In most experiments only duplicate cultures were analyzed for any experimental point because the numbers of megakaryocytes available for each experiment were low. The average coefficient of variation for the duplicate cultures in the different experiments was 7 ± 2% (no. of experiments = 11). For comparisons of differences between means of samples, the Student's *t* test was used. A p value of 0.05 was considered a statistically significant difference.

Results and Discussion

The conditions for labeling the cultured cells with ^{3}H–TdR were determined by examining the effects of time and cell concentration on isotope incorporation. Megakaryocyte-enriched cell preparations were preincubated for 16 hr and then labeled with 1 μCi/ml ^{3}H–TdR for periods up to 24 hr (Fig. 1). Uptake of ^{3}H–TdR

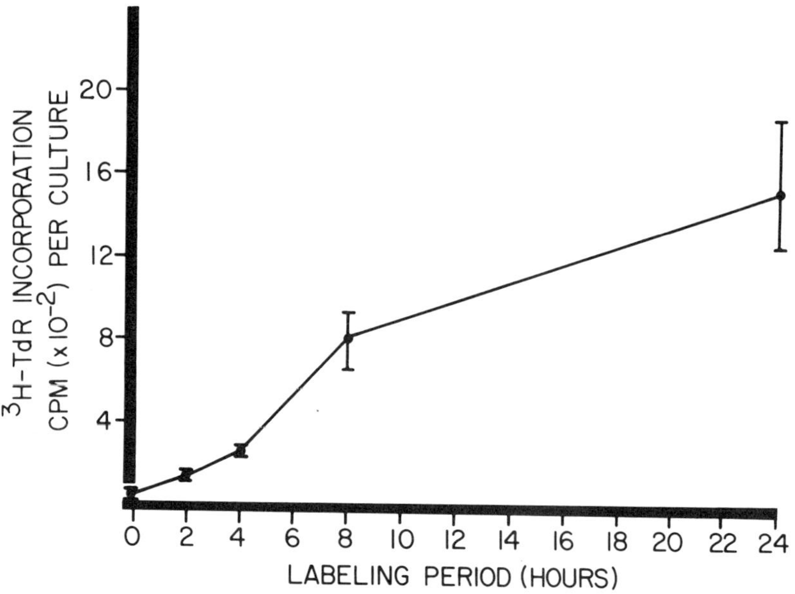

Figure 1. Effect of Time of Labeling with ^{3}H–TdR on Megakaryocyte-Enriched Cultures. Cultures contained 2 × 10^4 cells/ml in 2 ml and 68% megakaryocytes. After 16 hr of preincubation, 1 μCi/ml ^{3}H–TdR was added for the indicated time periods. Each point is the mean ± 1 S.E.M.

increased with the time of labeling. A labeling period of 3-hr was chosen to examine the effect of cell concentration. Isotope incorporation was found to be directly proportional to the cell concentration in the range tested, 2–6 × 10^4 cells/ml (Fig. 2).

The effects of thrombopoietin on the megakaryocyte-enriched cultures were examined by adding preparations in a concentration of 10% (v/v) at the initiation of incubation. Similar preparations of normal plasma or corresponding volumes of physiological saline were added to control cultures. After the cultures were labeled with ^{3}H–TdR for 3-hr, an increase in incorporation of ^{3}H–TdR in thrombopoietin-supplemented cultures compared to control cultures was observed (Fig. 3). A correlation between ^{3}H–TdR incorporation and the number of megakaryocytes in the culture was evident. Similar cultures were labeled for the entire period of incubation (19 hr) by using a reduced amount of ^{3}H–TdR, 0.25 μCi/ml. Thrombopoietin stimulated ^{3}H–TdR incorporation under these conditions as well. Stimulation was significant ($p < 0.05$) under both conditions except when comparing the effects of thrombopoietin to normal plasma in the cultures containing only 31% megakaryocytes. It was observed that some of the normal plasma fractions used in these cultures showed variability in the degree of stimulation or inhibition of ^{3}H–TdR incorporation exhibited compared to the saline controls. This observation suggested that the normal plasma fractions contained low levels of thrombopoietin and possibly

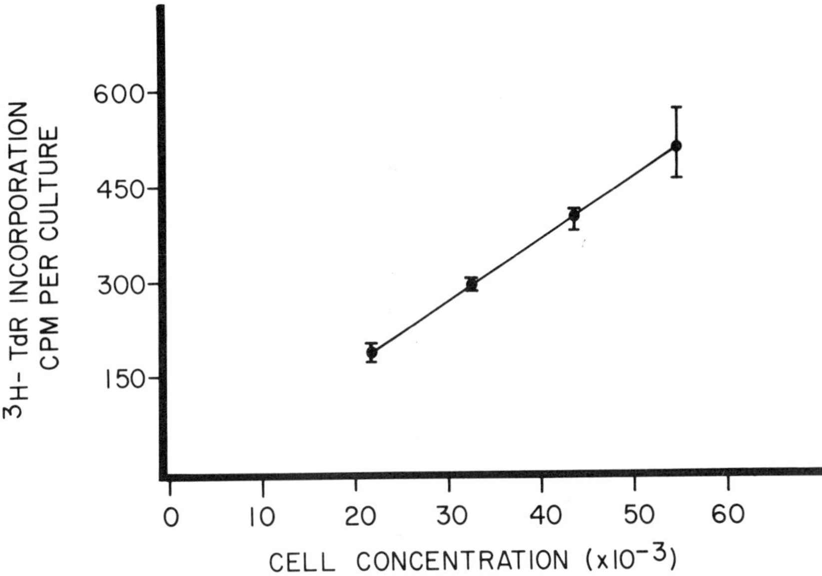

Figure 2. Effect of Cell Concentration on the Incorporation of ^{3}H–TdR into Megakaryocyte-Enriched Cultures. Cultures containing various concentrations of a 72% megakaryocyte preparation were preincubated for 16 hr and then labeled with 1 μCi/ml ^{3}H–TdR for 3 hr.

some inhibitory substances often reported to be present in plasma and serum fractions (Evatt and Levin, 1969).

Autoradiography was used to determine which cell types were incorporating ^{3}H–TdR in the cultures. Cells were classified as labeled or unlabeled and identified morphologically after they were stained with Wright stain. Labeling indices were calculated from counts of 600–1000 cells. The cells in the experiment described in Table 1 were labeled for 19 hr; similar results were obtained with a 3-hr labeling period. A higher labeling index was calculated for cells in the thrombopoietin-supplemented cultures compared to the controls. Identification of the labeled cells as megakaryocytes or nonmegakaryocytes indicated that the greatest proportion of the increase in the labeling indices in the cultures containing thrombopoietin was accounted for by the labeling of megakaryocytes (Table 2). The percentage of the megakaryocytes that labeled increased 400%–600% over the labeling in the saline control cultures while the labeling of the nonmegakaryocytes increased only 200%. The normal plasma fractions stimulated the labeling of nonmegakaryocytes similarly to thrombopoietin but there was very limited stimulation of the megakaryocytes. Since greater than 90% of the cultured cells were megakaryocytes, megakaryocytes accounted for almost all of the ^{3}H–TdR incorporated. A labeled megakaryocyte with the silver grains concentrated over the multilobed nucleus is shown in Figure 4A.

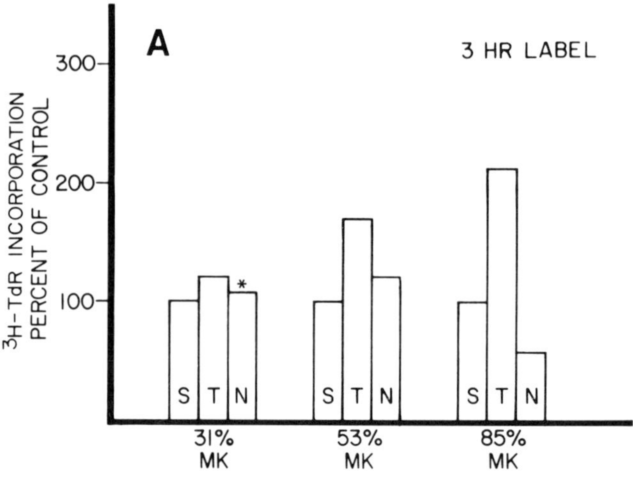

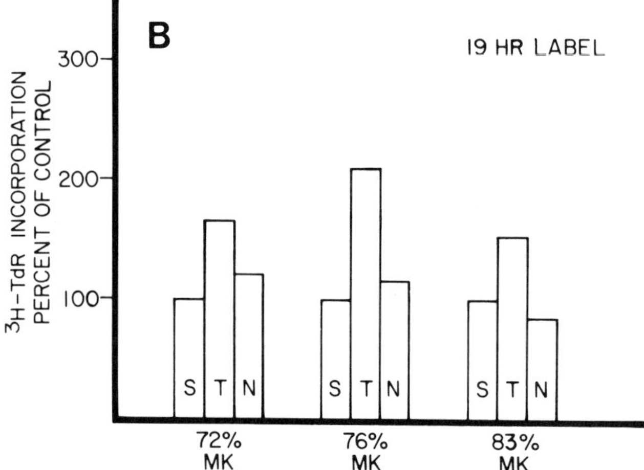

Figure 3. Stimulation of ^{3}H–TdR Incorporation by Thrombopoietin. Megakaryocyte cultures of the indicated purities were supplemented with thrombopoietin preparations (T), normal plasma fractions (N) or saline (S), incubated and labeled for 3-hr (A) or 19 hr (B) with ^{3}H–TdR. Each bar represents duplicate cultures. For comparisons of T to controls N or S, $p<0.05$ except where $p<0.1$ (*). The protein concentration of the thrombopoietin preparations was approximately 20 mg/ml.

A differential count of cells in a similar experiment indicated that the blasts as well as the megakaryocytes accounted for the greatest proportion of the total labeled cells in the thrombopoietin-supplemented cultures compared to the controls (Table 3). Both of these cell types increased in the percentage that labeled and consequently accounted for 81% of the labeled cells in the thrombopoietin cultures. The blasts

Table 1. Labeling Indices at 19 Hours.[a]

Supplement added to cultures	% Cells labeled
Saline	1.8
Normal plasma	4.8
Thrombopoietin I	8.6
Thrombopoietin II	11.0

[a]Cultures were labeled for 19 hr with 0.5 μCi ^{3}H-TdR (0.5 Ci/mmol). The percentages of the cells labeled were determined by counting 600-1000 cells in the autoradiographs. Thrombopoietin I and II were different preparations.

were identified on the basis of a high nuclear-to-cytoplasmic volume ratio and an agranular, basophilic cytoplasm. Some of these cells were larger and had a higher grain density over the nucleus than other blasts. One of these cells is seen in Figure 4B. When visual grain counts ($\leq$ 400 grains per cell) were performed, it was evident that there was a heavy concentration of label in the megakaryocyte compartment (Fig. 5). In addition, it was observed that many of the blasts had grain counts in the range observed for identifiable megakaryocytes. On this basis these cells were tentatively identified as polyploid megakaryoblasts and other means of identifying these blasts are being investigated.

Since the incorporation of ^{3}H–TdR was found to be proportional to the cell concentration, incorporation increased with the time of labeling, incorporation was stimulated by thrombopoietin compared to controls, and the number of megakaryocytes and blasts that labeled increased with thrombopoietin, the incorporation of ^{3}H–TdR was examined further as a possible *in vitro* assay for thrombopoietin. In an additional experiment, the relationship of ^{3}H–TdR incorporation to the concentration of thrombopoietin was examined (Fig. 6). For both the 3-hr and 19-hr

Table 2. Effect of Thrombopoietin on ^{3}H-TdR Labeling in Megakaryocyte Cultures.

Supplement added to cultures	% Increase of labeled megakaryocytes[a]	% Increase of labeled nonmegakaryocytes[a] (including blasts)
Normal plasma	133%	204%
Thrombopoietin I	483%	220%
Thrombopoietin II	608%	229%

[a]The labeling indices of the megakaryocytes or nonmegakaryocytes in the saline-supplemented cultures were compared to the labeling indices in the normal plasma or thrombopoietin-supplemented cultures to calculate the percent increases.

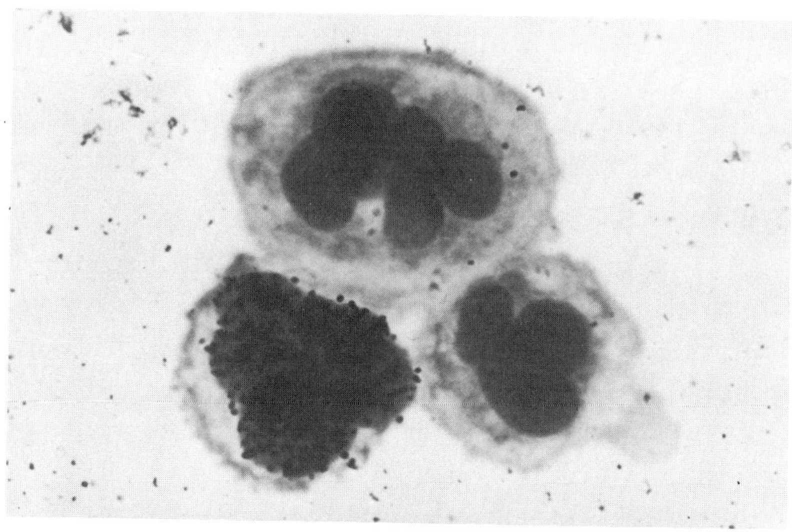

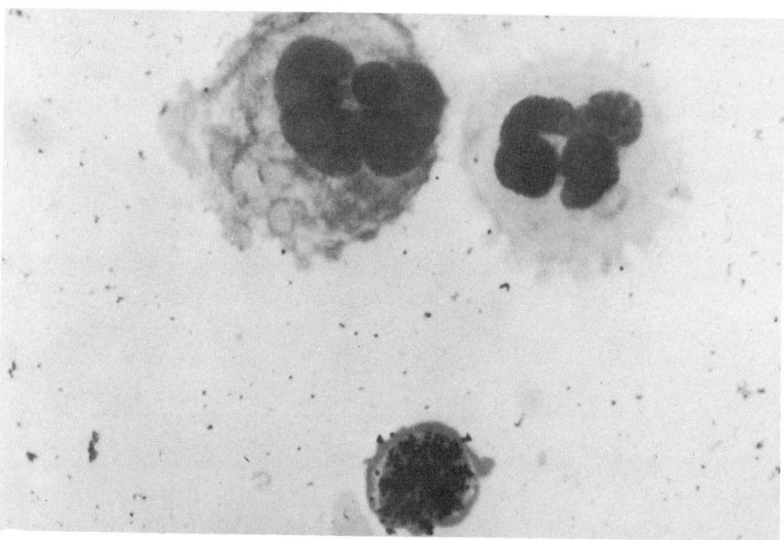

Figure 4. Autoradiography of ^{3}H–TdR Incorporation into Cultured Cells. (A) Labeled and unlabeled megakaryocytes from a culture containing 75% megakaryocytes after labeling with 0.25 μCi/ml ^{3}H–TdR for 19 hr (×1000). (B) A labeled blast that appears to be a megakaryoblast from megakaryocyte-enriched cultures.

Table 3. ^{3}H-TdR Labeling of Different Cell Types.

Supplement added to cultures	% Total labeled cells MK	BL	MO	GR	ER
Saline	46.2	11.8	11.8	30.2	0
Normal plasma	31.2	15.8	40.6	12.4	0
Thrombopoietin	53.8	27.2	9.5	9.5	0

Cultured cells were classified as megakaryocytes (MK), blasts (BL), monocytes (MO), granulocytes (GR), or erythroid cells (ER). These cells were further categorized as to the percent of the total labeled cells that cell type represented.

labeling periods, isotope incorporation was linearly related to the concentration of thrombopoietin. The slopes for the two curves were similar, but the 3-hr labeling period did not appear to be sensitive at the lower concentration of thrombopoietin.

The ploidy of the megakaryocytes from the cultures was measured to determine if the increase in DNA synthesis stimulated by thrombopoietin was detectable as an increase in ploidy. The cultured cells were stained by the Feulgen procedure and densitometric measurements of the nuclei were made on a Vickers M85 microdensitometer. A representative plot of the relative density values is shown in Figure 7.

Figure 5. Grain Counts on Labeled Cultured Cells. Up to 400 grains per cell were counted from autoradiographs of the cultures.

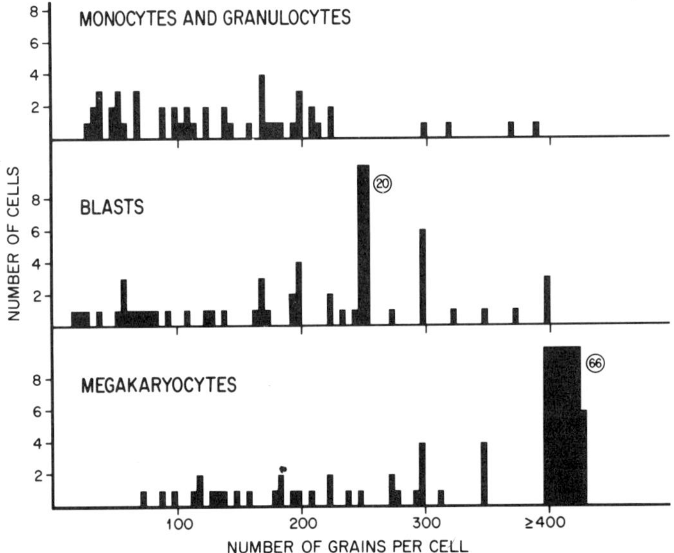

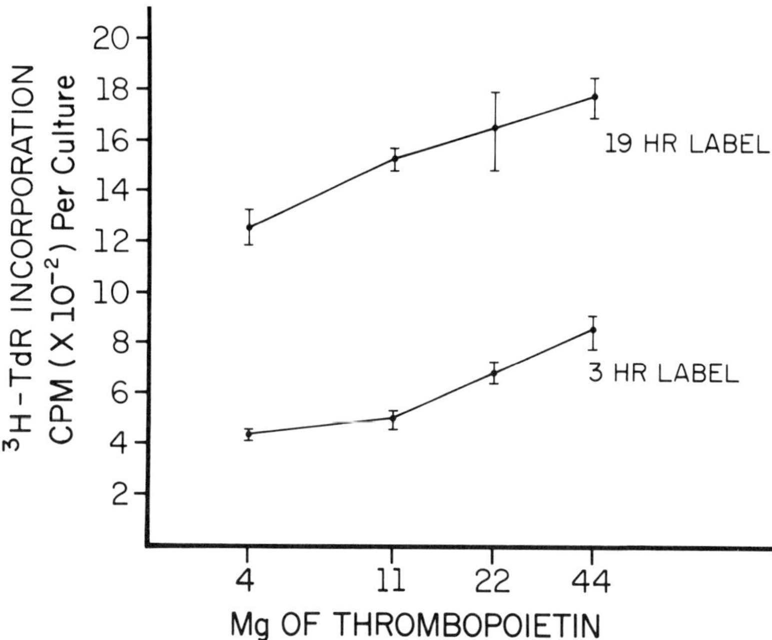

Figure 6. Effect of Different Concentrations of Thrombopoietin on ^{3}H–TdR Incorporation. Cultures containing 83% megakaryocytes were supplemented with the indicated amounts of thrombopoietin and labeled with ^{3}H–TdR for the indicated time periods.

Studies of megakaryocytes cultured for 3 days demonstrated that there were daily shifts in the ploidy distributions (Fig. 8). In a representative experiment, it was observed that the relative number of megakaryocytes at the 8N ploidy level increased and the 16N and 32N megakaryocytes decreased over 3 days in the saline-control cultures. In contrast, in the thrombopoietin-supplemented cultures, the 8N and 16N populations both decreased after day 1, followed by a subsequent rise in the 32N population that continued to day 3. The megakaryocytes in the normal plasma-supplemented cultures followed the same pattern as in the saline-controls, with the exception of the 32N cells. On day 3 these cells increased in number, again suggesting the presence of low levels of thrombopoietin in the normal plasma fractions.

The results of these studies indicated that polyploidization proceeded *in vitro* and that an increase in the 32N population by days 2 and 3 was stimulated by thrombopoietin. The megakaryocytes from saline and normal plasma-supplemented cultures showed little maturation to higher ploidy levels, while thrombopoietin induced additional endoreduplication. These observations confirmed an effect of thrombopoietin on the stimulation of megakaryocytopoiesis *in vitro*.

The need for *in vitro* systems to study megakaryocytopoiesis has been recognized, and several different culture systems are available. These include the plasma clot and the semisolid agar systems. These systems were designed to measure the fraction

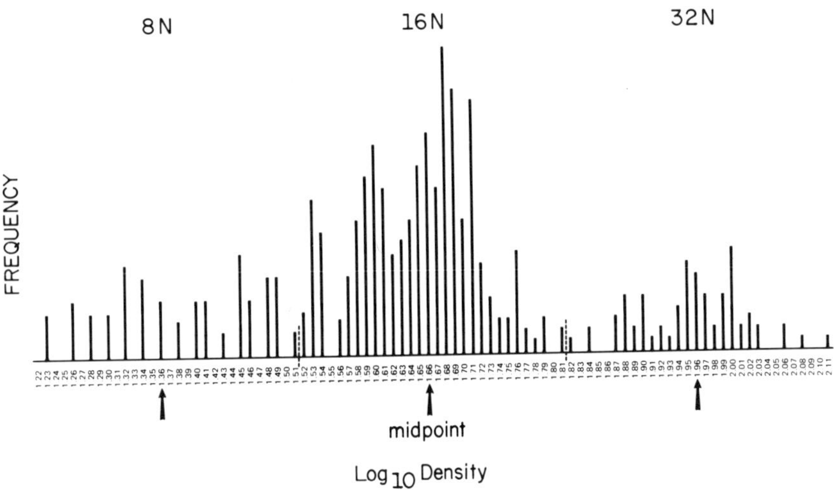

Figure 7. Density Distribution of Feulgen-Stained Nuclei. A computer-generated frequency distribution of the $\log_{10}$ of the density readings for the Feulgen-stained nuclei in a megakaryocyte culture; 300–500 megakaryocytes were counted per culture. A Gaussian curve fitting program was used to determine the midpoint of the curve for the 16N population and then the midpoints and the limits of the 8N and 32N populations were assigned. Several 4N megakaryocytes present in this culture are not shown.

of CFU-M in bone marrow stimulated by thrombopoietin or thrombocytopoiesis stimulating factor (TSF), or various other megakaryocyte colony-stimulating factors. Both systems support colony formation, but they also support limited maturation of identifiable megakaryocytes. Studies with these systems and observations made *in vivo* have indicated that there may be several factors involved in the stimulation and control of stem cell commitment and subsequent megakaryocytopoiesis. Williams et al. (1979) have suggested that thrombopoietin may be a circulating regulator controlling endomitosis and platelet production while megakaryocyte colony-stimulating factors control proliferation. The colony assay systems are suitable for analyses of the early proliferative events that involve the progenitor cells of the megakaryocyte compartment. However, once endoreduplication begins, an *in vitro* system of isolated megakaryocytes relatively free of other cell types is useful for investigating the mechanisms and factors regulating the later processes. The stimulation of DNA synthesis by thrombopoietin in the culture system just described appears to be a physiological response that correlates with some of the early effects of thrombocytopenia on megakaryocytes *in vivo*. Furthermore, the response can be measured within a period of 24 hours of culture and can be quantitated by a sensitive assay using ^{3}H–TdR. We propose that this system may be useful in the assay of thrombopoietin *in vitro* and may facilitate further purification and characterization of this proposed hormone. In addition, this system should serve as an excellent model for investigating many aspects of megakaryocyte maturation.

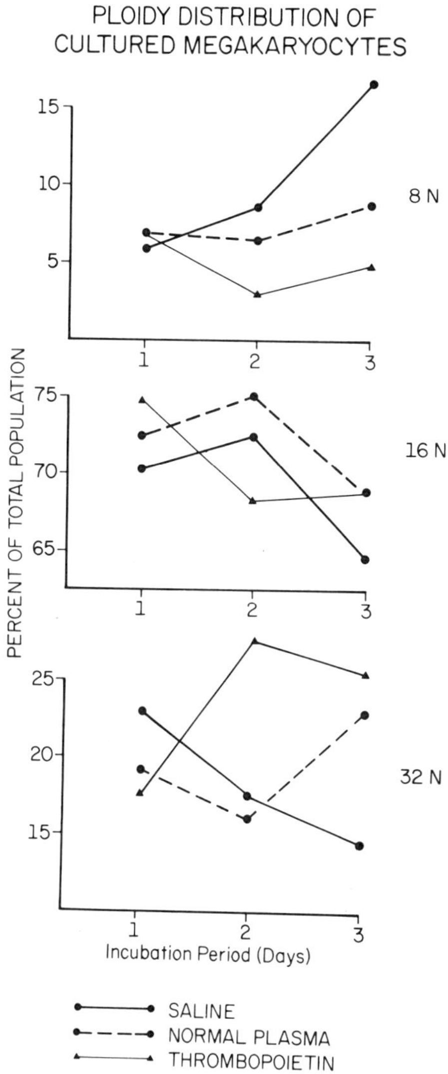

Figure 8. Ploidy Distribution of Megakaryocytes Cultured for Three Days. Cells were incubated for 1, 2, or 3 days and labeled with ^{3}H–TdR for 19 hr prior to terminating the cultures. Slide preparations were stained by the Feulgen procedure and megakaryocyte nuclei were measured by microdensitometry. The 4N and 64N cells represented less than 4% of the total and are not shown.

Summary

The response of megakaryocytes to thrombopoietin *in vitro* was studied by measuring the incorporation of ^{3}H–TdR into DNA. Guinea pig megakaryocytes were isolated in purities ranging from 50%–100% by using sequential BSA gradients.

Liquid cultures of the megakaryocyte preparations were then incubated for up to 3 days and labeled for either 3 hr or 19 hr with ^{3}H–TdR. Isotope incorporation increased with the time of labeling and was proportional to the cell concentration. Thrombopoietin prepared by ammonium sulfate precipitation of thrombocytopenic rabbit plasma increased the incorporation of ^{3}H–TdR in megakaryocyte cultures compared to similar fractions of normal plasma or 0.9% NaCl. A linear relationship between isotope incorporation and the concentration of thrombopoietin was demonstrated. Autoradiography revealed that the increase in ^{3}H–TdR uptake resulted specifically from an increase in the percentage of the megakaryocytes and blasts labeled. The megakaryocytes accounted for the highest percentage of the total cells labeled in the presence of thrombopoietin. Many of the labeled blasts were thought to be megakaryoblasts because the grain density over the nuclei suggested that they were polyploid cells. The ploidy levels of megakaryocytes in culture were examined over a 3-day period. Day-to-day shifts in the different ploidy classes indicated that polyploidization proceeded *in vitro*. By days 2 and 3, the cultures stimulated by thrombopoietin had increased numbers of 32N mepakaryocytes compared to day 1 levels and also compared to control cultures. These studies demonstrate that thrombopoietin stimulates DNA synthesis in megakaryocytes *in vitro*, and this stimulation is reflected by increased numbers of megakaryocytes with higher ploidy.

ACKNOWLEDGMENTS

We gratefully acknowledge the technical assistance of Peggy Fett and Evelyn Duval. K. L. Kellar is supported by National Research Service Award F32 HLO5940 from the National Heart, Lung, and Blood Institute and by Hemophilia of Georgia, Inc., Atlanta, Georgia.

References

Cooper, G. W., B. Cooper, and C. Chang. 1970. Demonstration of a circulating factor regulating blood platelet production using ^{35}S-sulfate in rats and mice. *Proc. Soc. Exp. Biol. Med.* 134:1123–1127.

Ebbe, S., F. Stohlman, Jr., J. Overcash, J. Donovan, and D. Howard. 1968. Megakaryocyte size in thrombocytopenic and normal rats. *Blood* 32:383–392.

Evatt, B. L. and J. Levin. 1969. Measurement of thrombopoiesis in rabbits using ^{75}selenomethionine. *J. Clin. Invest.* 48:1615–1626.

Evatt, B. L., J. Levin, and K. M. Algazy. 1979. Partial purification of thrombopoietin from the plasma of thrombocytopenic rabbits. *Blood* 54:377–388.

Evatt, B. L., D. P. Shreiner, and J. Levin. 1974. Thrombopoietic activity of fractions of rabbit plasma: Studies in rabbits and mice. *J. Lab. Clin. Med.* 83:361–371.

Freedman, M. H., T. P. McDonald, and E. F. Saunders. 1977. The megakaryocyte progenitor (CFU-M): *In vitro* proliferative characteristics. *Blood* 50 (Suppl. 1):146.

Harker, L. A. 1968. Kinetics of thrombopoiesis. *J. Clin. Invest.* 47:458–465.

Kellar, K. L., C. R. McGrath, B. L. Evatt, and R. B. Ramsey. 1979. Stimulation of DNA synthesis in megakaryocytes by thrombopoietin *in vitro*. *Blood* (Suppl. 1) 54:165a.

Leuchtenberger, C. 1958. Quantitative determination of DNA in cells by Feulgen microspectrophotometry. In *General Cytochemical Methods* (Vol. 1). Danielli, J. F., ed. New York: Academic Press. pp. 219–278.

Levin, J., I. Tang, and J. L. Spivak. 1979. Thrombopoietin: Partial purification by affinity chromatography with immobilized lectins. *Blood* (Suppl. 1) 54:166a.

Levine, R. F. 1977. Culture *in vitro* of isolated guinea pig megakaryocytes: Recovery, survival, morphologic changes, and maturation. *Blood* 50:713–725.

Levine, R. F. and M. E. Fedorko. 1976. Isolation of intact megakaryocytes from guinea pig femoral marrow. *J. Cell Biol.* 69:159–172.

McDonald, T. P. and C. Nolan. 1979. Partial purification of a thrombocytopoietic-stimulating factor from kidney cell culture medium. *Biochem. Med.* 21:146–155.

Nakeff, A. 1977. Colony-forming unit, megakaryocyte (CFU-M): Its use in elucidating the kinetics and humoral control of the megakaryocytic committed progenitor cell compartment. In *Experimental Hematology Today*. Baum, S. G., and Ledney, D. C., eds. New York: Springer-Verlag. pp. 111–123.

Odell, T. T., Jr. 1974. Megakaryocytopoiesis and its response to stimulation. In *Platelets: Production, Function, Transfusion, and Storage*. Baldini, M. G., and Ebbe, S., eds. New York: Grune and Stratton. pp. 11–20.

Odell, T. T., Jr., C. W. Jackson, T. J. Friday, and D. E. Charsha. 1969. Effects of thrombocytopenia on megakaryocytopoiesis. *Br. J. Haematol.* 17:91–101.

Penington, D. G. 1969. Assessment of platelet production with ^{75}Se selenomethionine. *Br. Med. J.* 4:782–784.

Penington, D. G. and T. E. Olsen. 1970. Megakaryocytes in states of altered platelet production: Cell numbers, size and DNA content. *Br. J. Haematol.* 18:447–463.

Sherwood, J. B. and E. Goldwasser. 1979. A radio-immunoassay for erythropoietin. *Blood* 54:885–893.

Williams, N., T. P. McDonald, and E. M. Rabellino. 1979. Maturation and regulation of megakaryocytopoiesis. *Blood Cells* 5:43–53.

Winton, E. F., W. R. Vogler, K. L. Kellar, M. B. Parker, and J. F. Kinkade, Jr. 1977. Slide chamber culture system for the *in vitro* study of humoral regulation of granulocyles and monocyte-macrophage proliferation and differentiation. *Blood* 50:289–302.

Discussion

Jackson: Did you see an increase in the ploidy levels of labeled cells?

Kellar: We have not looked at both simultaneously yet.

Jackson: The use of labeling indices may provide a more sensitive procedure to determine if you do get an increase in ploidy.

Kellar: Those are the experiments we are performing. We are interested in whether these cultures reflect what is seen *in vivo*. For example, at initial times in culture (about 1–4 hr), thymidine incorporation is seen in about 40% of all Stage I megakaryocytes. Unfortunately, these Stage I cells represent only about 1–2% of the total purified cells.

Jackson: Have you looked at grain counts in your stimulated versus your normal plasma control cultures?

Kellar: We haven't put that data together but we are now employing a microdensitometer to do these grain counts quantitatively.

Levine: Since changes in cAMP levels affect cell proliferation in many sorts of cells and since adenosine and theophylline, as I had used, can alter those cell levels, did you just use medium containing these substances in the disaggregation phase? How long were your cells exposed to those agents?

Kellar: We used it as you do, in the early stages. For the latter stages of the purification procedure, we used only Ca++ and Mg++ free Hank's. It would be interesting to examine the differences.

McDonald: You showed an experiment comparing normal plasma with thrombopoietin batches I and II. Did you use the same fractionation procedure for the normal plasma controls?

Kellar: Yes.

Burstein: Have you added erythropoietin to these cultures?

Kellar: Yes, we have. We observed an increase in the stimulation of thymidine incorporation that was comparable to that seen with thrombopoietin. Autoradiography has not been done, so we don't know if the same cells are responding. They may be different, but it has been shown that erythropoietin can directly stimulate megakaryocyte colony formation (*Blood Cells* 5:25, 1979). Autoradiography combined with cell morphology should be performed.

Evatt: It should be noted that we have not used purified erythropoietin, and our preparations may contain thrombopoietin.

Kellar: We used Connaught Step III preparations of erythropoietin. Samples of our thrombopoietin preparations have been sent to Dr. Goldwasser to determine if there is cross-reactivity with erythropoietin. We would like to obtain purified erythropoietin to test for stimulation of megakaryocytes.

B. Schick: Did you add the tritiated thymidine to your cultures immediately after they were isolated?

Kellar: Yes. We labeled the cells overnight with thymidine or in earlier experiments we incubated the cells overnight and then used a 3-hr exposure to thymidine. We add the thymidine for longer periods to detect all responsive cells.

B. Schick: Did you see a significant change in the incorporation of counts between the cells that were labeled as fresh preparations and those incubated overnight?

Kellar: No.

Bunn: I will show tomorrow that one of the results of the purification procedure is the large bias in selection for higher ploidy cells. Your data also showed relatively few 4N and 8N cells. You may have seen a much greater effect with a less-enriched cell population, which was not so depleted of younger megakaryocytes.

Kellar: Yes, that is correct. We are approaching this by making modifications to the second gradient so that we can obtain younger megakaryocytes.

Levin: You concluded that thrombopoietin increased the ploidy of the megakaryocytes. I wonder if you aren't seeing some maintenance effect, rather than a true increase in ploidy?

Kellar: That is possible. However, there is a decrease in the 8N and 16N cells at the same time that there is an increase in the 32N cells. We may be seeing an increase in the rate of maturation and maintenance of endoreduplication *in vitro* in the presence of thrombopoietin.

Evatt: It appears that thrombopoietin maintains the usual progression of ploidy compared to a slowdown in its absence.

Vignuelle: Would you comment on the biological half-life of the thrombopoietin preparations?

Evatt: I don't think anyone can tell what the biological half-life of thrombopoietin is.

Nakeff: In your saline controls there was a substantial increase in the proportion of 8N cells, but 16N and 32N dropped off at days 2 and 3. Do you know what is happening there?

Kellar: I think that this is a positive indication that we are able to maintain a slight increase in ploidy without any stimulation at all. Also, there are probably some replicating cells in the cell preparation. We would like to believe that in the presence of saline, replication is continuing to feed in and show up in the 4N and then the 8N cell compartments by endoreduplication. We believe that these cells are being shifted into the highly ploidy groups in the presence of thrombopoietin.

Evatt: These cultures do contain megakaryoblasts. I think that there are two compartments. The blast cells go through two divisions to get into the first compartment (4N and 8N cell classes); without thrombopoietin the cells go ahead and mature in the 4N and 8N levels. Further endoreduplication requires thrombopoietin. There may be some progression with fractions from normal plasma, since they contain some thrombopoietin. With the strong stimulation of thrombopoietin the cells appear to progress through the ploidy classes.

Published 1981 by Elsevier North Holland, Inc.
Evatt, Levine, and Williams, Editors
MEGAKARYOCYTE BIOLOGY AND PRECURSORS:
IN VITRO CLONING AND CELLULAR PROPERTIES

Thrombopoetin and Its Control of Thrombocytopoiesis and Megakaryocytopoiesis*

T. P. McDonald

University of Tennessee College of Medicine, Knoxville Unit, Department of Medical Biology, Memorial Research Center Knoxville, Tennessee

The clarification and study of a thrombocytopoiesis-stimulating factor (TSF or thrombopoietin) that controls not only thrombocytopoiesis but also megakaryocytopoiesis has aroused considerable interest over the past decade (Abildgaard and Simone, 1967; Cooper, 1970; McDonald, 1977a; Levin and Evatt, 1979). For these studies thrombopoietin has been used from various sources, i.e., plasma and urine from thrombocytopenic and thrombocytotic patients (McDonald, 1975), thrombocytopenic animals (Evatt et al., 1974; McDonald, 1973a,b; McDonald et al., 1977a), animals made thrombocytopenic by X-rays (McDonald et al., 1977a), and thrombopoietin from kidney cell culture medium (McDonald et al., 1975, 1976, 1979; McDonald, 1976, 1977b; Clift and McDonald, 1979). Thrombopoietin from kidney cell culture medium has many similarities to the hormone found in the plasma of thrombocytopenic animals; it is the purpose of this presentation to demonstrate similarities in the ability of hormones to stimulate both megakaryocytopoiesis and thrombocytopoiesis and also to delineate similarities in thrombopoietin from different sources.

Summarized below are the effects of thrombopoietin from different sources on platelet production in mice, i.e., % ^{35}S incorporation into platelets, platelet counts, and platelet sizes. Also presented in this report are the effects of thrombopoietin on megakaryocytopoiesis *in vivo* [acetylcholinesterase (AChE)-positive cells in mouse marrow, megakaryocytic spleen colonies in irradiated-bone marrow transplanted mice, and megakaryocyte endomitosis], and the effects of a thrombopoietin-rich preparation on megakaryocytopoiesis *in vitro*.

*This work was supported by the National Heart, Lung, and Blood Institute, Grant No. HL 14637.

I. Effects of Thrombopoietin on Thrombocytopoiesis

Rabbit anti-mouse platelet serum (RAMPS), produced as previously described (McDonald, 1973; McDonald and Clift, 1976), was used to produce immunothrombocythemic mice for the assay of TSF (McDonald, 1973a,c, 1974).

Using this assay technique, Figure 1 shows an experiment in which normal mouse plasma and saline are compared to three sources of thrombopoietin, i.e., plasma from thrombocytopenic mice whose thrombocytopenia was produced by RAMPS (McDonald et al., 1977a), plasma from thrombocytopenic mice whose thrombocytopenia was induced by 750 R X-rays (McDonald et al., 1977a), and TSF from kidney cell culture medium (Clift and McDonald, 1979). Normal mouse plasma and saline injected into rebound-thrombocytotic mice resulted in similar responses. However, mice injected with the three sources of thrombopoietin showed signifi-

Fig. 1. The effects of three different thrombopoietin preparations on % ^{35}S incorporation into platelets of mice in rebound-thrombocytosis. TSF-1, 0.75 ml of serum per mouse obtained from thrombocytopenic RAMPS-injected donor mice, sacrificed at 4 hours; TSF-2, 0.75 ml of serum per mouse from irradiated thrombocytopenic mice that were killed 7 days after treatment with X-rays; TSF-3, TSF from kidney cell culture medium (injected at the dose of 30 mg protein per mouse). In all cases, the material was injected in 4 equal doses and the 24-hour % ^{35}S incorporated into platelets was measured on day 8 after RAMPS injection. The numbers on the bars indicate the number of mice per treatment and the vertical lines indicate the standard errors. Values were significantly elevated over suitable control values: $^{*}P \leq 0.025$, $^{**}P \leq 0.005$, $^{***}P \leq 0.0005$. Some of the data were taken from McDonald et al. (1977a) and reproduced by permission of the publisher.

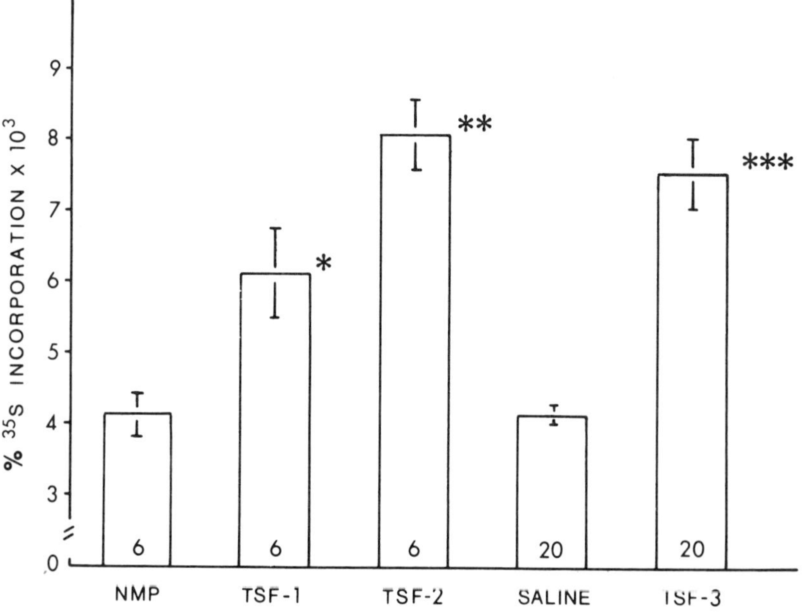

cantly ($P \leq 0.025 - P \leq 0.0005$) elevated % ^{35}S incorporation into their platelets, thus indicating that all three preparations contained a substance that stimulates platelet production in assay mice.

As shown in Table 1, not only did TSF increase the % ^{35}S incorporation into platelets of mice, but an increase in platelet counts also occurred. Plasma from thrombocytopenic mice (McDonald et al., 1977a) caused increased ($P \leq 0.025$) platelet counts of recipient mice three days after injection when compared to counts of other mice injected with normal mouse plasma. Moreover, mice injected with TSF from kidney cell culture medium had higher ($P \leq 0.025$) platelet counts than mice injected with production medium, control (PMC).

II. Effects of Thrombopoietin on Platelet Sizes of Mice

It is now clear that TSF from kidney cell culture medium will stimulate platelet production in recipient mice (McDonald et al., 1975, 1976, 1979; McDonald, 1976, 1977b; Clift and McDonald, 1979). In addition, several experiments have shown that large-heavy platelets are "younger" than small-light platelets (Karpatkin and Garg, 1974). Therefore, "true" thrombocytopoiesis-stimulating factor should increase platelet size as well as platelet count and ^{35}S incorporation into platelets. Therefore, platelet sizing was investigated in mice after the animals had received an injection of highly potent TSF preparation.

C_3H male mice were injected with RAMPS or TSF-rich production medium. For controls, other mice were injected with normal rabbit serum or saline. TSF and saline were injected subcutaneously (s.c.) into mice two times (once in the morning and once in the afternoon) on the first day of the experiment (day 0). The mice

Table 1. Effects of Plasma From Thrombocytopenic Mice and Kidney Cell Culture Medium On Platelet Counts of Rebound-Thrombocytotic Mice.

Treatment	No. of mice	Platelet count $\times 10^{-5} \pm$ SE
1. Normal mouse plasma, 0.75 ml/mouse; platelet count of donor mice 999,000/mm^3 [b]	5	8.1 ± 0.2
2. Plasma from thrombocytopenic mice, 0.75 ml/mouse; platelet count of donor mice 80,000/mm^3, collected at 4 hr after RAMPS injection[a]	5	9.7 ± 0.5[a]
3. Production medium, control, 43.9 mg protein/mouse in 2.0 ml saline	4	8.5 ± 0.4
4. TSF-rich kidney cell culture medium, 48 mg protein/mouse in 2.0 ml saline	5	10.0 ± 0.8[a]

[a]Values were significantly higher than control counts: $P < 0.025$.

[b]Data taken from McDonald et al., 1977a.

were killed 1, 2, 3, or 4 days later, platelet sizes and platelet counts were determined as previously outlined (McDonald, 1976, 1980).

Typical histograms of platelet size distributions obtained from platelets of mice injected with normal rabbit serum, RAMPS, saline, or TSF-rich culture medium are shown in Figure 2. The distribution of platelets sized by the Particle Data

Fig. 2. Typical histograms of platelet sizes from normal mice, a mouse injected with RAMPS, and a mouse injected with TSF-rich kidney cell culture medium. In Panel A, (o) indicates the size distribution of a mouse injected with normal rabbit serum. Mathematical characteristics are: mode, 3.56 μm^3; median, 3.68 μm^3; geometric mean, 3.73 ± 2.46 (SD) μm^3, skewness, 0.27 μm^3; skewness, 0.07 μm^3; arithmetic mean, 4.27 ± 2.65 μm^3, skewness, 0.27 μm^3. (•) represents a mouse that was injected with RAMPS 2 days prior to platelet sizing; mode, 4.54 μm^3; median, 4.7 μm^3; geometric mean, 4.56 ± 3.81 μm^3; skewness, 0.01 μm^3; arithmetic mean, 5.46 ± 3.70 μm^3; skewness, 0.25 μm^3. In Panel B, (o) shows the platelet size distribution of a mouse that was injected with saline; mode, 3.56 μm^3; median, 3.56 μm^3; geometric mean, 3.58 ± 2.3 μm^3; skewness, 0.01 μm^3; arithmetic mean, 4.10 ± 2.6 μm^3; skewness, 0.21 μm^3. Platelet size distribution of a mouse that was injected with TSF-rich kidney cell culture medium 2 days prior to platelet sizing is shown by (•): mode, 3.81 μm^3; median, 3.81 μm^3; geometric mean, 3.95 ± 2.57 μm^3; skewness, 0.05 μm^3; arithmetic mean, 4.57 ± 3.06 μm^3; skewness, 0.25 μm^3.

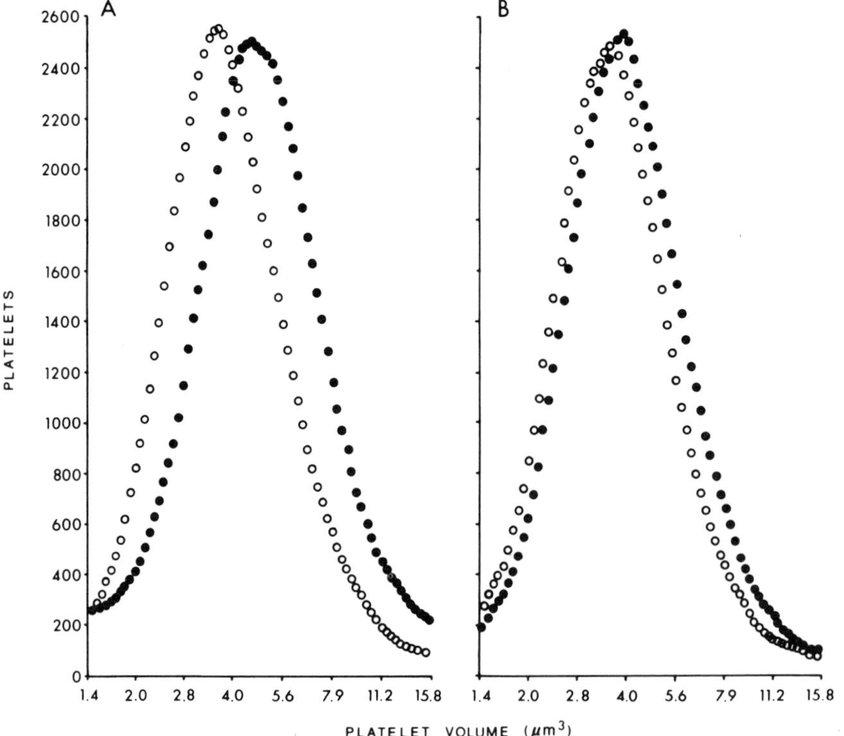

instrument showed essentially a log-normal symmetry as indicated by the similarity in the values for the mean, mode, and median. The average platelet sizes were increased in mice injected with RAMPS or TSF, as evidenced by the displacement of the size distribution histograms to the right.

In Figure 3 is shown a typical experiment in which normal rabbit serum and RAMPS were injected into normal mice and platelet counts and platelet sizes were determined at daily intervals thereafter. When RAMPS was given at doses sufficient to decrease the platelet count to approximately1/5 of the normal count at day1, the average platelet size was significantly increased; platelet sizes remained significantly elevated through day 4. By day 5 the platelet size distributions had decreased to slightly below normal values.

The results of four experiments in which normal mice were injected with saline or TSF from kidney cell culture medium and the platelet sizes measured are summarized in Figure 4. Platelets from control mice were sized at 1–4 days after saline injection. Platelet sizes, as measured by the arithmetic mean, were significantly elevated after TSF injection. The average size of platelets was increased at day 1 ($P \leqslant 0.0005$), reached maximum on day 2 ($P \leqslant 0.0005$), and had returned to normal by day 4.

III. Effects of Thrombopoietin on Megakaryocytopoiesis *In Vivo*

Acetylcholinesterase-Positive Cells in Mouse Bone Marrow

In an attempt to demonstrate similarities between thrombopoietin preparations, the effects of exogenous sources of TSF (kidney cell culture medium) and RAMPS on the number of small acetylcholinesterase (AChE)-positive cells in the bone marrow of mice were determined (Kalmaz and McDonald, unpublished results). Male C_3H mice were injected intraperitoneally (i.p.) with normal rabbit serum, RAMPS, saline, or TSF-rich kidney cell culture medium and the percentages of small AChE + cells were determined from marrow smears at 2–48 hr later. The smears were stained by use of the "direct coloring" thiocholine-method for AChE (Karnovsky and Roots, 1964).

The effects of various treatments on the percentage of small AChE + megakaryocytes are shown in Table 2. RAMPS, which causes severe thrombocytopenia leading to the elaboration of a thrombocytopoietic-stimulating factor, results in increased megakaryocytopoiesis by action on a precursor cell to produce more megakaryocytes. Significant increases were found in the number of small AChE + cells at 6–14 hr; the number of these cells peaked at 8–10 hr. Similarly, when TSF was injected into other mice, an increase in the percentage of small AChE + cells was also seen with the peak occurring at 8–10 hr. Furthermore, the same time period was observed for mice that had high titers of endogenous thrombopoietin caused by RAMPS. Similarities in the numbers of AChE + cells in response to the two preparations are obvious.

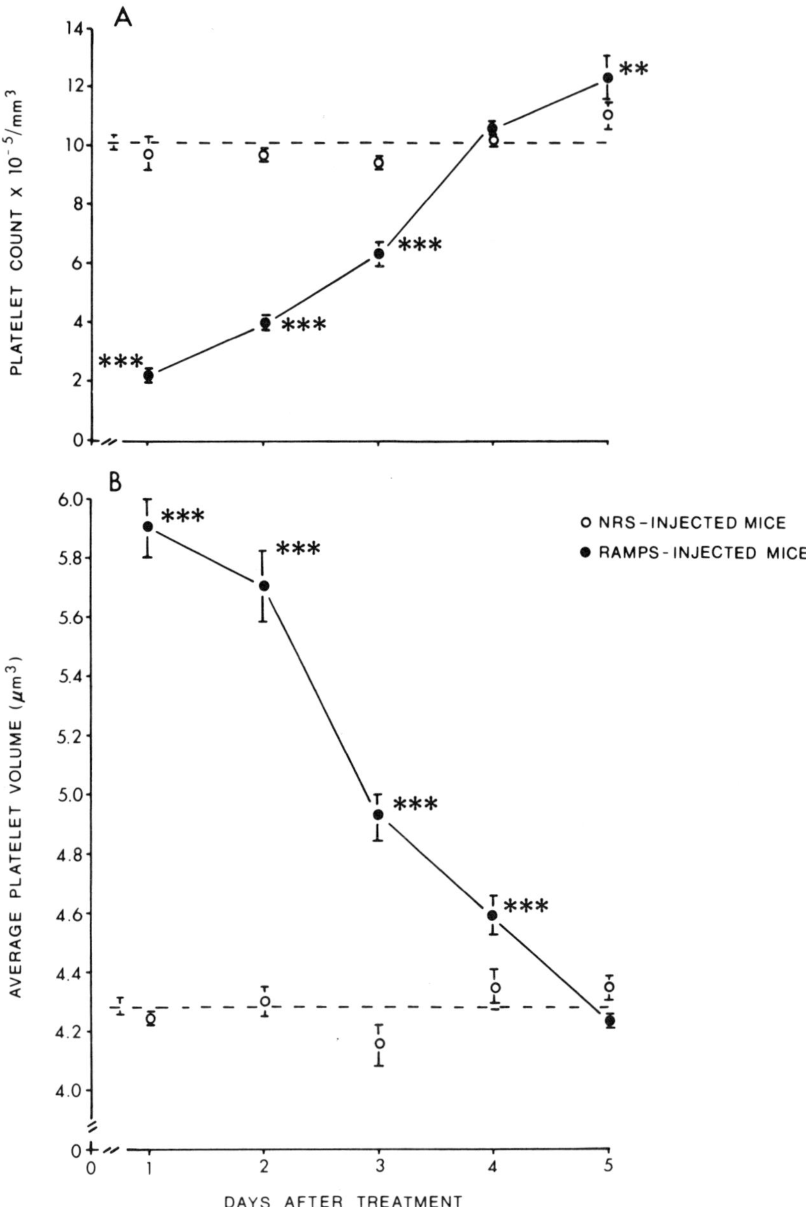

Fig. 3. Platelet counts and average platelet sizes (arithmetic means) of mice following a single intraperitoneal injection of normal rabbit serum or RAMPS. Each point represents the mean of 6 mice; the horizontal dashed lines (– –) represent the average values of mice injected with normal rabbit serum; vertical lines indicate the standard errors. The values were significantly different from control values: $^{**}P \leqslant 0.005$, $^{***}P \leqslant 0.0005$.

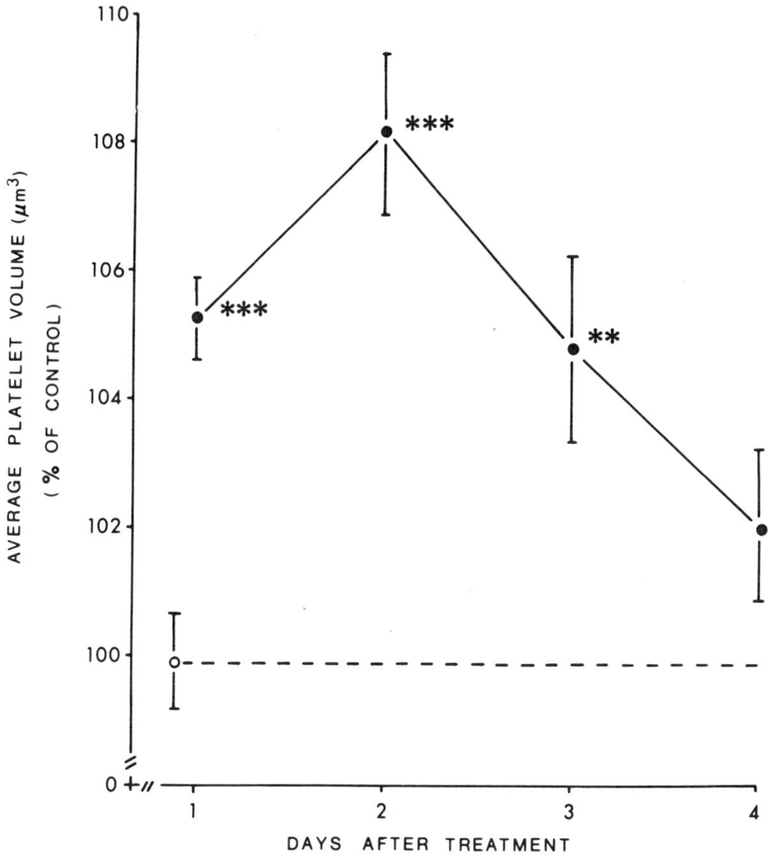

Fig. 4. Average mean sizes of platelets from mice after treatment with saline (o) or TSF-rich kidney cell culture medium (•). The data are the pool of 4 separate experiments in which 3–5 mice were used in each experimental treatment group. The vertical lines indicate the standard errors; values were significantly different from 31 saline-injected control mice: $^{**}P \leqslant 0.005$, $^{***}P \leqslant 0.0005$.

Megakaryocytic Spleen Colonies

Another set of experiments tested the possibility that kidney cell culture medium, shown to stimulate thrombocytopoiesis in mice, might also stimulate the numbers of megakaryocytic colonies in lethally-irradiated bone marrow transplanted mice (McDonald et al., 1978). Approximately 1–2 hr after whole-body irradiation, mice were injected intravenously (i.v.) with 1×10^5 nucleated bone marrow cells obtained from normal C_3H donor mice by the use of sterile techniques. Four days after irradiation and bone marrow transplantation, the mice were given two s.c. injections of PMC (production medium, control) or TSF-rich kidney cell culture medium. On day 7 the mice were sacrificed, and their spleens removed, fixed in

Table 2. Percentage of Small Acetylcholinesterase (AChE) Positive Megakaryocytes In Marrow of Mice With Time After Various Treatments.

Time (hr)	NRS[a]	RAMPS	Saline	TSF
2	10.7 ± 1.9 (3)	12.8 ± 0.3 (6)	11.1 ± 1.9 (3)	10.4 ± 0.9 (6)
4	9.7 ± 2.3 (3)	14.1 ± 2.0 (6)	12.0 ± 2.3 (3)	11.4 ± 1.3 (6)
6	19.3 ± 1.5 (3)	17.5 ± 1.1 (6)*	10.0 ± 0.9 (3)	18.6 ± 1.4 (6)**
8	10.9 ± 1.0 (3)	27.7 ± 2.5 (6)**	9.3 ± 2.5 (5)	25.9 ± 1.6 (9)***
10	9.0 ± 0.0 (3)	26.7 ± 2.4 (3)***	7.7 ± 0.3 (3)	26.0 ± 0.6 (3)***
12	8.0 ± 0.5 (3)	24.3 ± 1.8 (3)**	7.3 ± 0.3 (3)	16.3 ± 2.0 (3)*
14	6.2 ± 1.3 (3)	24.3 ± 2.8 (3)**	8.3 ± 2.4 (3)	13.0 ± 1.7 (3)
16	7.8 ± 0.5 (3)	7.6 ± 0.9 (3)	7.3 ± 1.3 (3)	19.0 ± 2.3 (6)*
48	–	7.7 ± 1.1 (3)	6.7 ± 0.3 (3)	8.0 ± 0.6 (3)
$\bar{x}$	9.1 ± 0.6 (8)		8.9 ± 0.6 (9)	

[a]NRS, normal rabbit serum; RAMPS, rabbit anti-mouse platelet serum; TSF, thrombopoiein-rich kidney cell culture medium.

The number of mice at each time and treatment is shown in parenthesis after the mean and SE. For the overall average of the control groups each time period was treated as a single sample. The values were significantly higher than controls: $^{*}P < 0.05$, $^{**}P < 0.005$, $^{***}P < 0.0005$.

Telley's solution, cut into longitudinal sections, and stained with hematoxylin-eosin (McDonald et al., 1978). The central 5 μm section of each spleen was examined for the number of megakaryocytes and megakaryocytic colonies. In order to be scored as megakaryocytic, a colony had to contain at least three megakaryocytes.

The average number of megakaryocytes per megakaryocytic colony and the percentage of megakaryocytic colonies in spleens after no further treatment or after the injection of PMC and TSF-rich kidney cell culture medium into lethally irradiated-bone marrow transplanted mice are shown in Figure 5. Thrombopoietin increased significantly ($P \leqslant 0.05$) the number of megakaryocytic colonies in the spleens of mice when compared to mice given X-rays and bone marrow alone. However, the values were not statistically increased when compared to PMC-treated mice (probably because of the large amount of variation associated with measuring megakaryocytic colonies). In addition, the average number of megakaryocytes per megakaryocytic colony was higher in the TSF-treated group even though the values were not statistically different. Although not shown, TSF did not greatly effect the number of surface colonies or the total number of colonies per spleen section.

Megakaryocyte Endomitosis

It has been proposed that TSF causes the stimulation of megakaryocyte endomitosis. Consistent with this hypothesis was the finding that stimulation of megakaryocytopoiesis by induced thrombocytopenia initiates an increase in the endomitotic index of rats (Odell et al., 1969, 1976) and mice (Odell and Boran, 1977). Therefore, the endomitotic index of megakaryocytes was determined in mice injected with TSF-rich culture medium and plasma obtained from thrombocytopenic donor rats (Odell et al.,1979). The finding that endomitosis was stimulated by these TSF-rich preparations provides additional evidence for the presence of a humoral thrombocytopoietic agent in such materials and for the ability of the megakaryocyte cell line to respond to it.

In the first study, TSF-rich plasma was obtained by making CD rats thrombocytopenic by injecting them with rabbit anti-rat platelet serum. Plasma was collected approximately 4 hr after the injection of the platelet antiserum. Both normal and thrombocytopenic rats were bled and their plasma harvested. The plasma was then injected s.c. into C_3H mice with the total dose divided into two equal parts at 0 and 6 hr. The mice were sacrificed 32 hr after the first injection, and their femurs collected, fixed in Zenker-Formol solution, cut into longitudinal sections of about 5 μm thick, and stained with hematoxylin-eosin. The total number of megakaryocytes and number of megakaryocytes in endomitosis were scored for each mouse and the percentage of megakaryocytes in endomitosis was calculated.

The results of this study are shown in Figure 6A. Plasma from normal rats, in the doses used, had no effect on the mitotic index. However, in mice injected with plasma from platelet-poor rats the mitotic index was significantly increased ($P \leqslant 0.001$) at both dose levels over that of any control group. In addition, a dose response relationship in the groups treated with plasma from thrombocytopenic rats was indicated by a significantly higher ($P \leqslant 0.05$) value in the group receiving a dose of 2 ml per mouse as compared to the group injected with 1 ml per mouse.

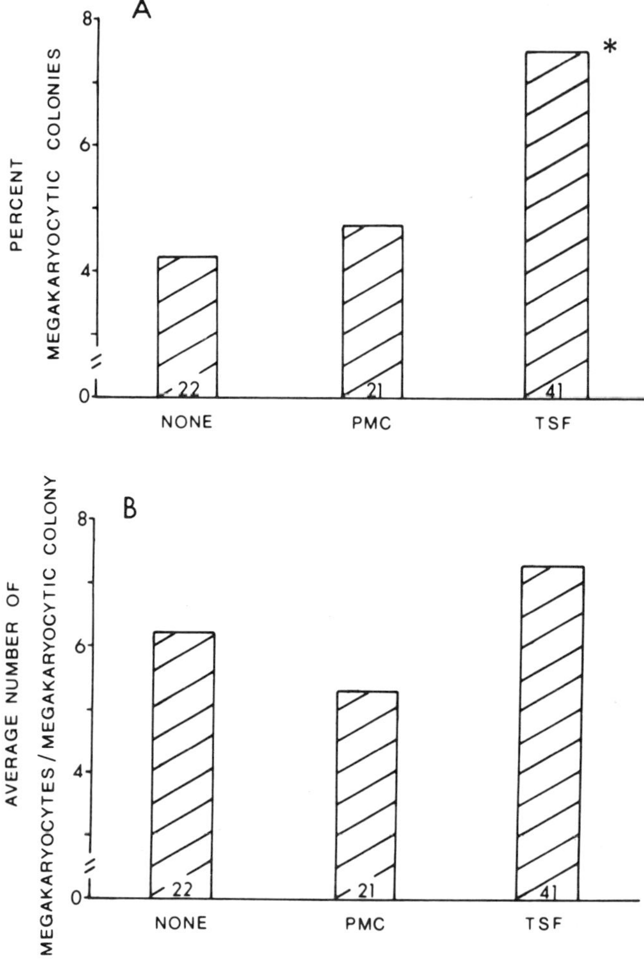

Fig. 5. The effects of injection of control medium and TSF-rich kidney cell culture medium on megakaryocytic spleen colonies of mice. Values are the average of 4 separate experiments: 20 mg protein per mouse of kidney cell culture medium [(either TSF-rich or production medium, control (PMC)] were injected s.c. into each mouse on day 4 after irradiation and bone marrow transplantation: mice were sacrificed on day 7. Values were significantly higher than those from uninjected control mice: $^*P \leqslant 0.05$. Some of the data were taken from McDonald et al. (1978).

The second study was aimed at determining whether TSF-rich kidney cell culture medium would also induce an increase in the endomitotic index of mouse megakaryocytes (Odell and McDonald, unpublished observations). As before, C_3H mice were injected s.c. two times on the first day with concentrated medium from human embryonic kidney cell cultures. The techniques were the same as described above;

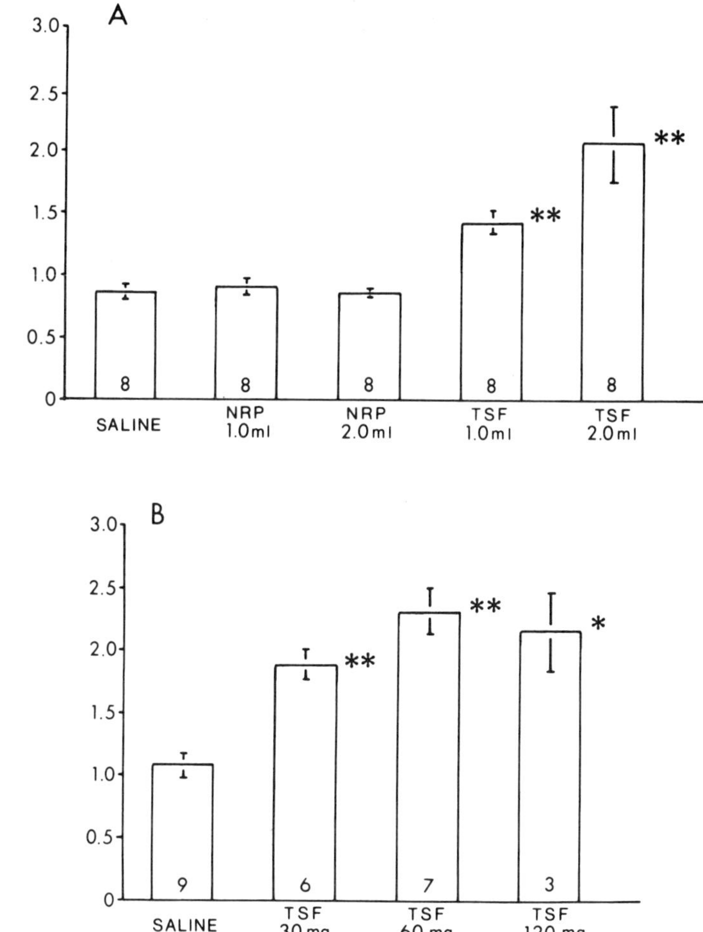

Fig. 6. Endomitotic indices of mouse megakaryocytes 32 hours after various treatments. Panel A, NRP-plasma from normal rats; TSF-plasma from thrombocytopenic rats. The dose of plasma for each mouse is indicated on the horizontal axis. Panel B, the TSF used was various doses of kidney cell culture medium. The numbers on the bars represent the number of mice per treatment and vertical lines indicate the standard errors. Values were significantly higher than control: $*P \leq 0.005$, $**P \leq 0.001$. Some of the data were taken from Odell et al. (1979).

32 hr later the mice were killed and sections of the femoral marrow were made. The endomitotic index of the megakaryocyte population for each mouse was determined by scoring the percentage of megakaryocytes in endomitosis. As shown in Figure 6B, the TSF from kidney cell culture medium caused a significant increase ($P \leq 0.005 - P \leq 0.001$) in mitotic indices of megakaryocytes of mice when

compared to those of saline-injected mice. In addition, the endomitotic indices of the 60 mg and 120 mg protein groups were greater than that of the 30 mg group, although the differences were not statistically significant at the 0.05 level.

IV. Effects of Thrombopoietin on Megakaryocytopoiesis *In Vitro*

Tissue culture studies of the growth of bone marrow cells have greatly enhanced our knowledge of hemopoietic differentiation (Bradley and Metcalf, 1966). Clonal assays have greatly improved the clarification of the erythropoietin-response of erythroid precursors (Axelrad et al.,1973). Therefore, we sought to determine if TSF from kidney cell culture medium would increase the number of committed murine megakaryocyte precursor cells (CFU-M). In these studies, megakaryocyte colonies were grown using a modification of the plasma clot tissue culture system used by McLeod et al. (1976). Mouse bone marrow cells were suspended in NCTC 109 and mixed with bovine serum albumin, beef embryo extract, asparagine, TSF in varying concentrations, tissue culture medium, and untreated fetal calf serum (Freedman et al., in press). Finally, bovine plasma was added and mixed with the ingredients. The cultures were incubated at 37°C with 5% CO_2 in air and high humidity. At harvest the clots were reamed free of the walls, fixed with gluteraldehyde on slides and stained by the "thiocholine" method (Karnovsky and Roots, 1964) to detect the megakaryocyte enzyme marker AChE. A colony was defined as three or more contiguous morphologically recognizable megakaryocytes showing the brownish stain for AChE.

The results, depicted in Table 3, show the effect of adding increasing doses of TSF to cultures. A dose-response was found between the number of colonies and

Table 3. Response of Mouse Marrow Megakaryocyte Colony (CFU-M) Growth to TSF From Kidney Cell Culture Medium.

Dose of TSF μg protein/culture	Number of megakaryocytic colonies ± SE (CFU-M/10^5 cells plated)
0.5	0 ± 0
1.0	1.0 ± 0.6
2.0	2.0 ± 0.7
4.0	2.2 ± 0.9
8.0	4.0 ± 0.4
16.0	5.8 ± 0.9
32.0	7.2 ± 1.0
65.0	4.0 ± 0.5
130.0	0.7 ± 0.3
250.0	0 ± 0

Each culture contained 1 × 10^5 nucleated mouse bone marrow cells; megakaryocyte colonies were determined after 7 days of culture. Each determination is the mean and SE of quadruplicate values from four studies.

Source: Data taken from Freedman et al., in press.

the doses of TSF between concentrations of 1 and 32 μg per 1 $\times 10^5$ cells; above this amount the numbers of CFU-M declined. These data indicate that TSF from kidney cell culture medium will stimulate megakaryocytopoiesis *in vitro*.

V. Discussion

TSF Assay Techniques

Several attempts have been made to find the optimum conditions for the assay of TSF (McDonald, 1973a, 1976, 1977b; McDonald et al., 1976, 1979). As shown previously (McDonald,1973a), mice in rebound-thrombocytosis are more sensitive to exogenous TSF preparations than are normal mice. In additional experiments, we have shown that if the first of four TSF injections is given on day 5 after RAMPS treatment and the 24-hr % ^{35}S incorporation determined on day 8, greater ^{35}S values can be found than at the other times tested (McDonald et al., 1976). The route of administering TSF does not seem to be important since TSF injected either s.c. or i.p. gave essentially the same result (McDonald, 1977b). Multiple injections were more effective than single injections (McDonald, 1977b). Isotope incorporation into platelets gave a more sensitive measurement of thrombocytopoiesis than platelet counting or platelet sizing (McDonald, 1976), although both platelet counting (McDonald et al., 1977a) and platelet sizing (McDonald, 1980) have shown modest responses to highly potent exogenous preparations of thrombopoietin. Recent work (McDonald et al., 1979) revealed that the time of measurement of ^{35}S incorporation into platelets and selection of the mouse strain determines to a large degree the sensitivity of the TSF assay. Sex of the mice does not appear to be important; similar responses were found in both male and female mice (McDonald et al., 1979). In addition, recent results (Clift and McDonald, 1979) have shown that, although ^{35}S gave significantly greater values at the higher TSF doses, either ^{35}S-sodium sulfate or ^{75}Se-selenomethionine can be used to measure platelet production rates in mice stimulated with thrombopoietin.

Platelet Sizing

In the studies presented herein and elsewhere (McDonald, 1976; McDonald et al., 1979), we have shown an increase in the average size of platelets from mice recovering from severe thrombocytopenia induced by RAMPS and after injections of exogenous sources of TSF (McDonald, 1980). Previously, Weiner and Karpatkin (1972) injected plasma from thrombocytopenic guinea pigs into recipient animals in a study to test the hypothesis that changes in platelet size can be used to measure TSF. Their results showed an increase in the number of large platelets compared with an increase in platelet counts 4–5 days after injection. The study was confirmed in rabbits (Weintraub and Karpatkin, 1974).

Similarities in TSF from Different Sources

The data presented in this report extend and confirm our previous findings that TSF from kidney cell culture medium has many similarities to TSF from thrombocytopenic animals and patients. We have demonstrated herein and previously that kidney cell culture medium stimulates platelet production in mice as evidenced by increased ^{35}S (McDonald et al., 1975, 1976, 1979; McDonald, 1976, 1977b; Clift and McDonald, 1979) and ^{75}Se (Clift and McDonald, 1979) incorporation into platelets, elevated platelet counts (McDonald,1976), and an increase in the size of platelets (McDonald, 1980). As shown in the present report and elsewhere, TSF-rich kidney cell culture medium stimulates megakaryocytopoiesis both *in vivo* [stimulation of endomitiosis in megakaryocytes (Odell and McDonald, unpublished observations), an increase in the numbers of small AChE + cells in the marrow of mice (Kalmaz and McDonald, unpublished results), and an increase in the numbers of megakaryocytic spleen colonies in lethally irradiated-bone marrow transplanted mice (McDonald et al., 1978)] and *in vitro* [stimulation of CFU-M *in vitro* (Freedman et al., in press)]. In addition, evidence of immunologic similarities between TSF from kidney cell culture medium and human urine (McDonald et al., 1977) and plasma (McDonald, 1978) have been found. Recently, we have shown similarities in the chemical and physical characteristics of TSF from kidney cell culture medium and TSF extracted from plasma of thrombocytopenic animals (McDonald and Nolan, 1979).

Megakaryocytopoiesis In Vivo

Several studies (Odell et al., 1969, 1976; Odell and Boran, 1977; Jackson, 1973) have shown that the number of all stages of megakaryocytes increases after acute thrombocytopenia that had been induced by an injection of platelet specific antisera. Jackson (1973,1974) found a two-fold increase in the number of small AChE + cells that occurred by 6 hr after the injection of a single dose of platelet specific antisera. Similarly, Nakeff and Bryan (1978) found a 1.5-fold increase at 6 hr in AChE + cells after platelet specific antisera injection. The results of the present work agree with these previous studies and show that, following RAMPS injection, the number of small AChE + cells is increased at 2–14 hr with peak values occurring at 8–10 hr. In addition, TSF from kidney cell culture medium increased the percentage of small AChE + cells. The increase was seen at the same time as was observed in mice whose megakaryocytic precursors cells were stimulated by RAMPS injection. Therefore, exogenous sources of TSF from kidney cell culture medium apparently stimulate megakaryocytopoiesis in a similar manner as does TSF produced in the animal in response to severe thrombocytopenia.

In addition to an increase in the numbers of small AChE + cells in the marrow of mice injected with TSF, other studies show that TSF from kidney cell culture medium increases megakaryocyte production from transplanted marrow stem cells in the spleens of lethally-irradiated mice. These data, therefore, support the conclusion of other studies that kidney cell culture medium contains the TSF. In support of this work, Goldberg et al. (1977) have shown that megakaryocyte differentiation in the spleens of lethally-irradiated mice transplanted with bone marrow cells can be suppressed by platelet transfusions. Inhibition of cell proliferation or differentiation of megakaryocyte precursors by increased levels of platelets to decrease the TSF titer probably led to the alterations observed in megakaryocytopoiesis. Ebbe et al. (1968) showed that changes occur in megakaryocyte size in response to alterations in platelet number and suggested that the number of megakaryocyte precursors may be influenced by the level of circulating thrombopoietin. The work presented herein and previously (McDonald et al., 1978) agrees with these studies and shows that injections of TSF from kidney cell culture medium can influence the number of megakaryocytes in the spleens of bone marrow-transplanted mice.

Thrombopoietin, produced in response to thrombocytopenia or administered from exogenous sources, will stimulate megakaryocytopoiesis in mice. Previous studies have shown that mice respond to severe thrombocytopenia with an increase in the endomitotic index of megakaryocytes (Odell and Boran, 1977). In the present work, recipient mice responded with increased endomitosis of megakaryocytes after receiving an injection of plasma from thrombocytopenic rats or after injections of TSF-rich kidney cell culture medium. The findings of the present report provide a hemopoietic basis for previous reports that have shown an increase in the number of circulating platelets and an increased uptake of platelet radioactive labels in mice, rats, and rabbits treated with plasma from thrombocytopenic donors (McDonald, 1977a). These results are also consistent with *in vitro* studies in which putative TSF promoted the appearance and maturation of megakaryocytes in cultures of marrow (Nakeff et al., 1975). The fact that exogenous sources of TSF from kidney cell culture medium increase the endomitotic index of megakaryocytes in marrow of mice supports the thesis that the kidney elaborates a thrombocytopoietic factor. In addition to the effects of thrombocytopoiesis already reported (McDonald et al., 1975, 1976, 1979; McDonald, 1976, 1977b; Clift and McDonald, 1979), kidney cell culture medium stimulates an increase in the endomitotic index of megakaryocytes (Odell and McDonald, unpublished observations). Moreover, the increase was of the same magnitude as previously observed in C_3H mice made severely thrombocytopenic by an injection of anti-platelet serum (Odell and Boran, 1977) or of plasma from thrombocytopenic rats (Odell et al., 1979); the increase occurred 32 hr after injection of the test material, just as it did in mice injected with antiserum (Odell and Boran, 1977). It is concluded, therefore, that kidney cells produce a humoral agent that contributes to the regulation of both megakaryocytopoiesis and platelet production.

Megakaryocytopoiesis In Vitro

As described herein, differentiation of mouse marrow megakaryocyte progenitors (CFU-M) was studied *in vitro* by means of a colony assay technique using a plasma clot system. As shown, TSF had a stimulatory function in megakaryocyte differentiation at a precursor cell level. These data, obtained by using TSF from kidney cell culture medium, appear to be in agreement with other results (Metcalf et al., 1975; Nakeff and Daniels-McQueen, 1976). The low colony yield found could be explained by the fact that the TSF used in the present and other studies was relatively crude and impure. The material used by Williams et al. (1978), which also can stimulate CFU-M, cannot be compared to other preparations because none have been fully characterized, thereby making it difficult to assess their similarities and differences.

The findings of the present report are summarized in Table 4. The nature, precise site of production, and mechanism of action of thrombopoietin remain unknown. During the last decade, new evidence partially clarified some of these problems. The simplest mechanism of action indicates that the response of megakaryocytes to thrombopoietin, like thrombocytopenia, is characterized by an increase in the rate of maturation, number, ploidy, and size of megakaryocytes. Thrombocytopoiesis is, therefore, elevated by an increased rate of megakaryocytopoiesis.

Summary

A thrombocytopoiesis-stimulating factor (TSF or thrombopoietin) controls platelet production by acting on precursor or immature cells to cause an increase in megakaryocytopoiesis, leading to elevated thrombocytopoiesis. TSF has been detected both in plasma from thrombocytopoietic donors after they have been made platelet-deficient by the injection of platelet-specific antisera or by whole-body irradiation, and in culture medium following growth of human embryonic kidney cells. Although TSF has been measured in several assay systems, the most sensitive method appears to be the one using immunothrombocythemic mice. After the animals have been injected with putative TSF-rich materials, alterations in the rate of platelet production are measured by the amount of ^{35}S-sodium sulfate incorporation into the platelets of mice in rebound-thrombocytosis. Our results show that TSF from kidney cell culture medium and from the plasma of thrombocytopoietic animals increases % ^{35}S incorporation into platelets; potent TSF preparations elevate platelet counts of recipient mice and cause the production of "larger-than-normal" platelets. These TSF-rich preparations also result in increases in megakaryocytopoiesis both *in vivo*

Table 4. Summary of Stimulatory Characteristics of TSF From Kidney Cell Culture Medium, TSF Generated Endogenously, and TSF-Rich Plasma From Thrombocytopenic Animals on Mouse Thrombocytopoiesis and Megakaryocytopoiesis.

		RSF-rich kidney cell culture medium	RAMPS-injected mice	Plasma from thrombocytopenic animals
I.	Thrombocytopoiesis			
A.	% ^{35}S incorporation into platelets	+	+	+
B.	Platelet count	+	+	+
C.	Platelet size	+	+	NT
II.	Megakaryocytopoiesis			
A.	Acetylcholinesterase + cells	+	+	NT
B.	Megakaryocytic spleen colonies	+	NT	NT
C.	Megakaryocyte endomitosis	+	NT	+
D.	CFU-M *in vitro*	+	NT	NT

+ — positive results. NT — not tested in this study.

and *in vitro*. Specifically, TSF causes an increase in the acetylcholinesterase-positive cells in mouse bone marrow, an elevation in the numbers of megakaryocytic spleen colonies in lethally-irradiated bone marrow reconstituted mice, and an increase in the endomitosis of megakaryocytes in the marrow of mice. An increase in the numbers of CFU-M by TSF after the culture of mouse bone marrow was also shown.

ACKNOWLEDGMENTS
My thanks are due to Rose Clift and Marilyn Cottrell for technical assistance, to Pat Taylor for stenographic aid, and to collaborators both past and present, Drs. G. D. Kalmaz, T. T. Odell, M. H. Freedman, E. F. Saunders, C. C. Congdon, Chris Nolan, R. D. Lange, and Otto Walasek.

References

Abildgaard, C. F., and J. V. Simone. 1967. Thrombopoiesis. *Semin. Hematol.* 4:424–452.

Axelrad, A. A., D. L. McLeod, M. M. Shreeve, and D. S. Heath. 1973. Properties of cells that produced erythrocytic colonies *in vitro*. In *Proceedings of the Second International Workshop on Hemopoiesis in Culture*. Robinson, W. A. (ed.). New York: Grune and Stratton, pp.226–237.

Bradley, T. R., and D. Metcalf. 1966. The growth of mouse bone marrow cells *in vitro*. *Aust. J. Exp. Biol. Med. Sci.* 44:287–299.

Clift, R. and T. P. McDonald. 1979. A comparison of ^{35}S sodium sulfate and ^{75}Se selenomethionine as platelet labels for the assay of thrombopoietin. *Proc. Soc. Exp. Biol. Med.*162:380–382.

Cooper, G. W. 1970. The regulation of thrombopoiesis. In *Regulation of Hematopoiesis* (Vol. II). Gordon, A. S. (ed.). New York: Appleton-Century-Crofts, pp.1611–1629.

Ebbe, S., F. Stohlman, Jr., J. Overcash, J. Donovan, and D. Howard. 1968. Megakaryocyte size in thrombocytopenic and normal rats. *Blood* 32:383–392.

Evatt, B. L., D. P. Shreiner, and J. Levin. 1974. Thrombocytopoietic activity of fractions of rabbit plasma: Studies in rabbits and mice. *J. Lab. Clin. Med.* 83:364–371.

Freedman, M. H., T. P. McDonald, and E. F. Saunders. 1981. Differentiation of murine marrow megakaryocyt e progenitors (CFU-M): Humoral control *in vitro. Cell Tissue Kinet,* 14:53-58.

Goldberg, J., E. Phalen, D. Howard, S. Ebbe, and F. Stohlman, Jr. 1977. Thrombocytotic suppression of megakaryocyte production from stem cells. *Blood* 49:59–69.

Jackson, C. W. 1973. Cholinesterase as a possible marker for early cells of the megakaryocytic series. *Blood* 42:413–421.

Jackson, C. W. 1974. Some characteristics of rat megakaryocyte precursors identified using cholinesterase as a marker. In *Platelets: Production, Function, Transfusion, and Storage*. Baldini, M. G. and Ebbe, S. (eds.). New York: Grune and Stratton, pp.33–40.

Kalmaz, G. D., and T. P. McDonald. Unpublished results.

Karnovsky, M. J. and L. Roots. 1964. A "direct-coloring"thiocholine method for cholinesterases. *J. Histochem. Cytochem.*12:219–221.

Karpatkin, S. and S. K. Garg. 1974. The megathrombocyte as an index of platelet production. *Br. J. Haematol.* 26:307–311.

Levin, J. and B. L. Evatt. 1979. Humoral control of thrombopoiesis. *Blood Cells* 5:105–121.

McDonald, T. P. 1975. Assay of thrombopoietin utilizing human sera and urine fractions. *Biochem. Med.*13:101–110.

McDonald, T. P. 1977a. Assays for thrombopoietin. *Scand. J. Haematol.* 18:5–12.

McDonald, T. P. 1973a. Bioassay for thrombopoietin utilizing mice in rebound thrombocytosis. *Proc. Soc. Exp. Biol. Med.*144:1006–1012.

McDonald, T. P. 1976. A comparison of platelet size, platelet count, and platelet ^{35}S incorporation as assays for thrombopoietin. *Br. J. Haematol.* 34:257–267.

McDonald, T. P. 1980. Effect of thrombopoietin on platelet size of mice. *Exp. Hematol.* 8:527–532.

McDonald, T. P. 1977b. Effects of different routes of administration and injection schedules of thrombopoietin on ^{35}S incorporation into platelets of assay mice. *Proc. Soc. Exp. Biol. Med.*155:4–7.

McDonald, T. P. 1973b. The hemagglutination-inhibition assay for thrombopoietin. *Blood* 41:219–233.

McDonald, T. P. 1974. Immunoassay and bioassay for thrombopoietin. In *Platelets: Production, Function, Transfusion and Storage*. Baldini, M. G. and Ebbe, S. (eds.). New York: Grune and Stratton, pp. 81–92.

McDonald, T. P. 1978. Neutralizing antiserum to thrombopoietin. *Proc. Soc. Exp. Biol. Med.* 158:557–560.

McDonald, T. P. 1973c. Regulation of thrombopoiesis. *Medicina* 33:459–466.

McDonald, T. P. and R. Clift. 1976. Mechanism of thrombocytopenia induced in mice by anti-platelet serum. *Haemostasis* 5:38–50.

McDonald, T. P. and C. Nolan. 1979. Partial purification of a thrombocytopoietic-stimulating factor from kidney cell culture medium. *Biochem. Med.* 21:146–155.

McDonald, T. P., R. Clift, and M. Cottrell. 1979. Assay for thrombopoietin: A comparison of time and isotope incorporation into platelets and the effects of different strains and sexes of mice. *Exp. Hematol.* 7:289–296.

McDonald, T. P., R. Clift, C. Nolan, and I. I. E. Tribby. 1976. A comparison of mice in rebound-thrombocytosis with platelet-hypertransfused mice for the assay of thrombopoietin. *Scand. J. Haematol.*16:326–334.

McDonald, T. P., M. Cottrell, and R. Clift. 1977a. Hematologic changes and thrombopoietin production in mice after X-irradiation and platelet-specific antisera. *Exp. Hematol.* 5:291–298.

McDonald, T. P., M. Cottrell, C. Nolan, and O. Walasek. 1977b. Immunologic similarities of thrombopoietin from different sources. *Scand. J. Haematol.* 18:91–97.

McDonald, T. P., M. Cottrell, C. C. Congdon, O. Walasek, and G. H. Barlow. 1978. Stimulation of megakaryocytic spleen colonies in mice by thrombopoietin. *Life Sci.* 22:1853–1858.

McDonald, T. P., R. Clift, R. D. Lange, C. Nolan, I. I. E. Tribby, and G. H. Barlow. 1975. Thrombopoietin production by human embryonic kidney cells in culture. *J. Lab. Clin. Med.* 85:59-66.

McLeod, D. L., M. M. Shreeve, and A. A. Axelrad. 1976. Induction of megakaryocyte colonies with platelet formation *in vitro*. *Nature* 261:492–494.

Metcalf, D., H. R. MacDonald, N. Odartchenko, and B. Sordat. 1975. Growth of mouse megakaryocyte colonies *in vitro*. *Proc. Nat. Acad. Sci. USA* 72:1744–1748.

Nakeff, A. and J. E. Bryan. 1978. Megakaryocyte proliferation and its regulation as revealed by CFU-M analysis. In *Hematopoietic Cell Differentiation*. Golde, D. W., Cline, M. J., Metcalf, D., and Fox, C. F. (eds.). New York: Academic Press, pp. 241–259.

Nakeff, A. and S. Daniels-McQueen. 1976. *In vitro* colony assay for a new class of megakaryocyte precursor: Colony-forming unit megakaryocyte (CFU-M). *Proc. Soc. Exp. Biol. Med.* 151:587–590.

Nakeff, A., K. A. Dicke, and M. J. van Noord. 1975. Megakaryocytes in agar cultures of mouse bone marrow. *Ser. Haematol.* 8:4–21.

Odell, T. T. and D. A. Boran. 1977. The mitotic index of megakaryocytes of mice after acute thrombocytopenia. *Proc. Soc. Exp. Biol. Med.* 155:149–151.

Odell, T. T. and T. P. McDonald. Unpublished observations.

Odell, T. T., C. W. Jackson, T. J. Friday, and D. E. Charsha. 1969. Effects of thrombocytopenia on megakaryocytopoiesis. *Br. J. Haematol.* 17:91–101.

Odell, T. T., J. R. Murphy, and C. W.Jackson. 1976. Stimulation of megakaryocytopoiesis by acute thrombocytopenia in rats. *Blood* 48:765–775.

Odell, T. T., T. P. McDonald, C. Shelton, and R. Clift. 1979. Stimulation of mouse megakaryocyte endomitosis by plasma from thrombocytopenic rats. *Proc. Soc. Exp. Biol. Med.* 160:263–265.

Weiner, M. and S. Karpatkin. 1972. Use of the megathrombocyte to demonstrate thrombopoietin. *Thromb. Diathes. Haemorr.*28:24–30.

Weintraub, A. H., and S. Karpatkin. 1974. Heterogeneity of rabbit platelets. II. Use of the megathrombocyte to demonstrate a thrombopoietin stimulus. *J. Lab. Clin. Med.* 83:896–901.

Williams, N., H. Jackson, A. P. C. Sheridan, M. J. Murphy, Jr., A. Elste, and M. A. S. Moore. 1978. Regulation of megakaryocytopoiesis in long-term murine bone marrow cultures. *Blood* 51:245–255.

Discussion

Bessmann: Since you show that the platelet size increases with the thrombopoietin administration, do you interpret that as evidence of youth of the platelet reflecting the greater size, or the stimulation of platelet production?

McDonald: I would like to look at these data as evidence that we stimulated the production of young platelets by thrombopoietin.

Jackson: Have you combined thrombopoietic active substances from platelet depleted mice and from the kidney cell conditioned medium, in the same *in vivo* assay system, to determine if you get additive or synergistic effects?

McDonald: Not yet.

Long: When you reported the small cell content, was your base line for small acetylcholinesterase-positive cells 10% of all marrow cells? That figure seems high to me. What was the sizing criteria for small cell estimates?

Kalmaz: It was subjective estimation.

Breton-Gorius: We have attempted to evaluate the number CFU-M and the level of thrombocytopenia in human subjects. We have tried to compare the number of CFU-M in normals and in patients with immune thrombocytopenia. In both bone marrow and peripheral blood cells from such patients, the number of CFU-M was similar to that observed in normal subjects. Would you comment on these results?

McDonald: We have no experience with human cells.

Mayer: Is TSF production from human embryonic kidney cells related to a certain cell line?

McDonald: Human embryonic kidney cells are grown for commercial urokinase production (Abbott Labs). The cells were grown for 4–6 weeks to obtain a large overgrowth and then cultured for an additional 5–6 weeks in a serum-free medium composed primarily of lactalbumin hydrolysate.

Levin: The TSF was given to mice that already had high thrombopoietin levels because of the irradiation.

McDonald: The animals are not thrombocytopenic until day 7, the day of assay.

Published 1981 by Elsevier North Holland, Inc.
Evatt, Levine, and Williams, Editors
MEGAKARYOCYTE BIOLOGY AND PRECURSORS:
IN VITRO CLONING AND CELLULAR PROPERTIES

The Separate Roles of Factors in Murine Megakaryocyte Colony Formation

N. Williams, H. M. Jackson, R. R. Eger, and M. W. Long

Sloan-Kettering Institute for Cancer Research, Rye, New York

In vitro cloning of hemopoietic precursor cells with megakaryocyte maturation of the colony cells has facilitated the study of factors necessary for the regulation of megakaryocyte proliferation, nuclear endoreduplication and cytoplasmic maturation to the point of platelet shedding (Metcalf et al., 1975; Nakeff and Daniels-McQueen, 1976; McLeod et al., 1976; Williams et al., 1978a). Previous studies have shown that megakaryocyte precursor cells (CFU-MK) undergo 1–8 cell divisions (average colony size: 16-32 cells) (Metcalf et al., 1975; Williams and Jackson, 1978). The growth of colonies containing megakaryocytes depends on the addition of conditioned medium derived from stimulated spleen cells (Metcalf et al., 1975; Nakeff and Daniels-McQueen, 1976; Metcalf and Johnson, 1978), erythropoietin (McLeod et al., 1976) or WEHI-3 cell-conditioned medium (Williams et al., 1978a). The activities in these various preparations have been termed megakaryocyte colony-stimulating factor (MK-CSF). The growth of megakaryocyte colonies is enhanced if supernatant from either bone marrow cells (Williams et al., 1978a) or peritoneal macrophages (Williams, 1979) is also present in the cell cultures. Addition of these supernatants to the bioassay permits the detection of greater numbers of megakaryocyte colonies (Williams et al., 1978a); Williams et al., 1980). While it is possible that this activity, termed megakaryocyte potentiating activity (MK-Potentiator), increases the cloning efficiency of the megakaryocyte progenitor cell, it does not influence the number of granulocyte-macrophage precursor cells that clone in semi-solid agar (Williams et al., 1978a; Williams et al., 1980). It has been suggested that the MK-Potentiator may primarily influence megakaryocyte maturation (Williams et al., 1979).

This manuscript investigates the hypothesis that two factors are involved in megakaryocyte colony development, one being required for the proliferation and maturation of diploid precursor cells, the other being involved in the differentiation of developing megakaryocytes.

Materials and Methods

Cells. Bone marrow cells were obtained from C57BL/6 mice (Cumberland Farms) by flushing the marrow cavity with McCoy's medium containing 5% fetal calf serum (FCS). A single cell suspension was obtained by repeated pipetting, and viable cell counts were obtained by exclusion of trypan blue dye.

In vitro growth of colonies. For megakaryocyte colony growth, McCoy's modified 5A medium with 20% FCS (optimum concentration for growth), 10^{-4} M 2-mercaptoethanol and 3×10^{-7} M prostaglandin E_2 were used (Williams, 1979). Medium was brought to approximately 40°C before adding one-tenth volume of 2.5% agar. The cells were then added to the above medium-agar mix to give a concentration of 5×10^4 to 10^5 cells/ml. The growth factors (100 μl of concentrated WEHI-3 CM and 200 μl of peritoneal cell supernatant) were added directly to the culture dishes followed by 1 ml of cells suspended in agar-McCoy's. The cultures were placed at 4°C for 5 min before incubating cells in a humidified incubator with 7% CO_2 in air for 6 or 7 days at 37°C. PGE_2 specifically inhibits growth of macrophage colonies, and since the majority of remaining colonies contain granulocytes, detection of megakaryocyte colonies is easier (Williams, 1979).

Granulocyte-macrophage colony-forming cells were grown using the same conditions except that 0.3% agar was used and the peritoneal cell supernatant (MØSN) and the prostaglandin E_2 were omitted. Cultures were scored for colonies (> 50 cells) at day 7 at 25× magnification using a dissecting microscope.

Analysis of megakaryocyte colonies. Megakaryocyte colonies were generally detected directly in the culture dishes at 40× under a dissecting microscope. WEHI-3 CM stimulates the growth of both megakaryocyte and macrophage-granulocyte colones (Williams et al., 1978a); however, the cells comprising the megakaryocyte colonies were markedly larger than the cells comprising either granulocyte or macrophage colonies. Each colony scored contained at least 3 large recognizable megakaryocytes. Approximately 10–25 colonies per 10^5 cells were observed with this procedure. A less subjective method is to dry the cell cultures *in situ* and stain for acetylcholinesterase. Between 25 and 45 megakaryocyte colonies per 10^5 cells can be confidently scored using this method. Acetylcholinesterase is specific for megakaryocytes among murine hemopoietic cells (Nakeff and Daniels-McQueen, 1976) and can be used to recognize a class of small cells, thought to be the earliest recognizable megakaryocytes (Jackson, 1973; Long and Henry, 1979; Long and Williams, 1980). This procedure was used for experiments where the number of cells per colony were estimated and where total plate analyses (i.e., single megakaryocytes, two cell groups and megakaryocyte colonies) were performed.

Procedure for drying agar cultures and acetylcholinesterase staining. Stock solutions: O.1 M sodium phosphate buffer, pH 6.0, 0.1 M sodium citrate, 30 mM copper sulphate ($CuSO_4 \cdot 5H_2O$), 5 mM potassium ferricyanide, Harris' hemotoxylin, lithium carbonate (saturated solution) and 95% ethanol.

Whatman No. 1 filter paper discs (33 mm) were carefully placed over the agar layer in the 35 mm Petri dish. The dish was then placed on a heating plate (approximately 45–50°C), and the agar dried by blowing warm air across the plate from a hair dryer positioned slightly above and 12–15 inches from the dish. When the filter paper appeared dry (15–20 min), it was gently removed using fine forceps, leaving behind the colonies embedded in a fine agar film.

The Petri dishes were then placed on a slide warmer (45–50°C) for 30–60 min, prior to staining for 4 hr at room temperature in the following manner: (1) Preparation of acetylcholine substrate: dissolve 10 mg of acetylthiocholine iodide in 15 ml phosphate buffer, and add the following, in order, with constant mixing: 1.0 ml sodium citrate, 2.0 ml copper sulphate, and 2.0 ml potassium ferricyanide; (2) add between 1.5–2.0 ml of the above substrate to each dish; (3) after 4 hr fix in 95% ethanol (10 min); (4) air dry at room temperature; (5) counterstain in Harris' hematoxylin 30 sec; (6) rinse gently with tap water; (7) ''blue'' in saturated lithium carbonate solution 30 sec; and (8) rinse in gently running tap water.

Acetylcholinesterase-positive colonies were scored at 40 × –94.5 × magnification with surface plates covered with a thin film of water. A standard microscope with a 4× or 6.3× long-working length objective was used.

Conditioned medium preparation. Cells of a murine myelomonocytic leukemic cell line (WEHI-3) were conditioned in serum-free McCoy's medium (SKI) for 3–4 days at 37°C before harvesting by centrifugation and subsequent clarification through a 0.45 micron filter. The conditioned medium (CM) was stored at −20°C. Megakaryocyte potentiator activity (MØSN) was derived from the supernatants of macrophages comprising the adherent cell population of a peritoneal cell exudate obtained from C57BL/6 mice (Williams, 1979). Conditioned media from various tissues and cell lines were prepared as described elsewhere (Williams et al., 1980).

Column chromatography. One liter of thawed or freshly collected WEHI-3 CM was applied to a Sephadex G-50 medium column (10 × 93 cm) equilibrated with 2 mM sodium phosphate (pH 8.0) in 0.001% polyethylene glycol 6000 (PEG). The fraction comprising the void volume (>25,000 daltons) and containing both the MK-potentiator and the MK-CSF was collected, concentrated by lyophilization and stored at −70°C. Concentrates of this fraction which gave optimal colony growth (× 10–30) were used for routine assays of megakaryocyte and granulocyte-macrophage colonies.

The freeze-dried fraction was resolubilized to 10–50 times the concentration of the original CM with glass-distilled water and dialyzed 6–8 hr at 4°C against 2 mM sodium phosphate (pH 8.0) in PEG. The sample was applied to a DEAE Sephadex

A-25 column (1.6 × 14 cm), washed with 3–4 column volumes of phosphate buffer (pH 8.0) in PEG. The breakthrough fraction containing 20–30% of the protein and greater than 90% of MK-stimulating activity recovered from the column was lyophilized. The bound molecules were eluted with 1 M NaCl in the equilibrating buffer with 85–95% of all protein recovered, based on the method of Lowry et al. (1951). The breakthrough fraction was resolubilized with glass-distilled water to a concentration 200 times that of the original CM and 1.5–2.0 ml was applied to a Sephadex G-150 column (1.8 × 90 cm) pre-equilibrated with 0.15 M NaCl in 10 mM sodium phosphate (pH 7.0) with 0.01% PEG. Fractions of 4.6 ml were collected and filtered for testing in the megakaryocyte colony-forming bioassay. Sephadex was purchased from Pharmacia. All chromatographic procedures were performed at 4°C.

DNA content. Feulgen staining of isolated megakaryocytes derived from colonies was carried out after methanol fixation for 10 min. Colonies of megakaryocytes were transferred onto albumin-coated glass slides using a drawn-out Pasteur pipette. Hydrolysis of the cells was performed with 1 M HCl for 40 min at room temperature before reacting with Schiff reagent (Leuchtenberger, 1958; Levine et al., 1980). Quantitation of DNA content was determined using an Artek Model 800 Image Analyser (Artek Systems Co., Farmingdale, NY). The range about the mean for a given population was ± 30%. Polymorphonuclear neutrophils were used as a standard source of 2N cells.

Results and Discussion

The respective dose responsiveness of CFU-MK to the two activities being investigated are described in Figure 1. The levels of MK-CSF in WEHI-3 CM were determined in the presence of optimal levels of macrophage supernatant (MK-Potentiator, upper panel). The MK-CSF was titrated over a 50-fold dilution from optimal levels. The MK-Potentiator activity was determined in the presence of maximal levels of MK-CSF in WEHI-3 CM. Without the addition of potentiator activity, these MK-CSF levels allowed only 5–10 megakaryocyte colonies per 10^5 bone marrow cells to be formed (lower panel). The MK-Potentiator activity in adherent peritoneal macrophage supernatant (MØSN) was present over a 5-fold dilution from optimal amounts.

Production of Megakaryocyte Colony Stimulating Activities by Various Organs, Peritoneal Macrophages and Hemopoietic Cell Lines

Conditioned medium from various organs was prepared and tested for activities that either directly stimulate megakaryocyte colony growth or potentiate the number of megakaryocyte colonies formed in the presence of optimal levels of WEHI-3 CM (Table 1). No organ reproducibly produced MK-CSF, although some activity was obtained from both lung and bone. No other tested organs elaborated MK-CSF. Lung, bone and heart produced GM-CSA in all experiments. Megakaryocyte po-

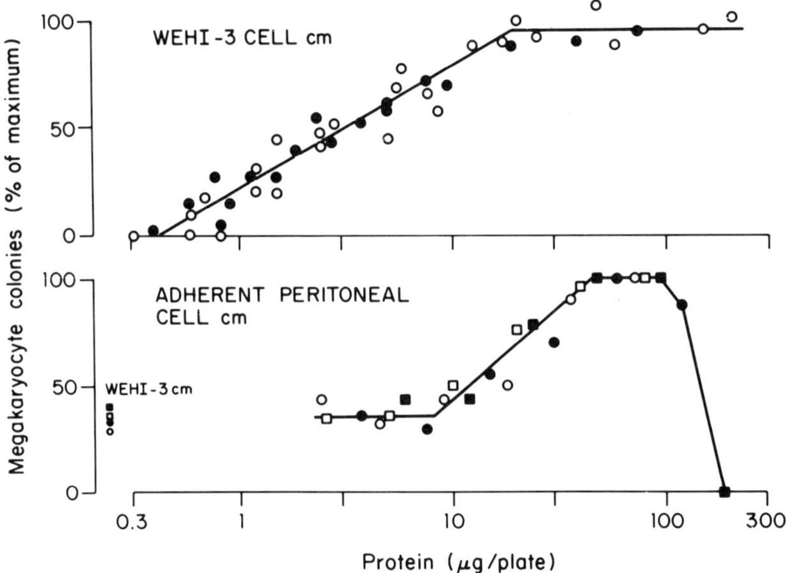

Fig. 1. The dose responsiveness of megakaryocyte progenitor cells to factors influencing megakaryocyte colony growth. The upper panel shows the response of CFU-MK to MK-CSF present in WEHI-3 CM. Cultures were made optimal for MK-Potentiator by the additon of adherent peritoneal cell supernatant. The normalized distributions are from two batches of WEHI-3 CM (conditioned 50- and 36-fold, respectively). The data are generated from several experiments with each preparation. The lower panel shows the response of CFU-MK to MK-Potentiator in 4 separate batches of adherent peritoneal cell supernatants. One preparation (■) was concentrated 5-fold before assay. Optimal amounts of MK-CSF were added in the form of WEHI-3 CM. Maximal numbers of megakaryocyte colonies ranged from 18–40 colonies per 10^5 cells.

tentiating activity could be reproducibly obtained from lung, bone, both total and adherent peritoneal exudate cells and, to a lesser extent, bone marrow. Spleen, heart, thymus and kidney conditioned media did not enhance CFU-MK levels (Table 1).

Only two cell lines tested, other than WEHI-3, produced MK stimuli. Both types had to be stimulated with mitogen for activity to be detected. A macrophage cell line (P388D1) was exquisitely sensitive to lipopolysaccharide (LPS). As little as 0.01 μg LPS per ml enabled P388D1 to produce small amounts of MK-CSF and high levels of MK-potentiating activity (Table 1). A T-lymphocyte cell line (EL4-BU), when stimulated with 30 μg/ml Con A, produced small but detectable levels of MK-CSF, but not MK-Potentiator. In contrast to MK-CSF production, only macrophage cell lines (J774 and P388D1) produced MK-potentiating activity. No activity was obtained from J774 cells unless they were cultured at cell concentrations about 1 × 10^6 cells/ml.

Table 1. Sources of Stimuli of Megakaryocyte Colony Formation.

		MK-CSF	MK-Potent.	GM-CSF
Tissues:	Stim. Spleen CM	+++	++	+++
	Spleen	–	+/–	–
	Lung CM	+	++++	++
	Bone CM	+/–	+++	+
	Bone Marrow CM	–	+	–
	Peritoneal Cell CM	–	++	–
	Heart	–	–	+
	Kidney	–	–	–
Cell Lines:	WEHI-3CM	++	+	–
	P388D1 + LPS	+	+++	++
	EL4 BU + Con A	+	–	+

Requirement for More Than One Factor for Megakaryocyte Colony Formation

There are several postulates to explain the development *in vitro* of mature megakaryocytes from diploid precursor cells. First, both proliferation and maturation may be governed by MK-CSF alone. Second, the serum batch chosen to support colony growth may contain MK-Potentiator (Metcalf, 1977). Third, the generation of mature megakaryocytes in colonies may result from the MK-CSF being added exogenously while the MK-Potentiator is produced by a population of bone marrow cells (Williams et al., 1980). Fourth, both MK-CSF and MK-Potentiator may be present in the exogenous stimulus used, of which one, MK-Potentiator, is present at suboptimal levels.

Experiments were performed to determine conditions where CFU-MK could be separated from the endogenous potentiator producing marrow cells, and then determine if the CFU-MK population depleted of potentiator cells could respond to a source of MK-CSF (WEHI-3 CM). Megakaryocyte progenitor cells free of potentiator cells were obtained among cells that sediment between 3.5 and 4.5 mm hr^{-1} (Fig. 2). These CFU-MK did respond to WEHI-3 CM (Fig. 2), indicating that either maturation occurred in the presence of MK-CSF alone or that MK-Potentiator could be derived from the WEHI-3 CM or the serum.

Concentrates of WEHI-3 CM were fractionated and assayed for MK-CSF and MK-Potentiator using plateau levels of MØSN and WEHI-3 CM as outlined in Figure 1. Evidence was obtained to indicate that more than one factor was required for megakaryocyte colony formation after Sephadex G-150 fractionation of a component which was nonadsorbed on DEAE-Sephadex A-25. No loss of megakaryocyte colony stimulatory capacity was observed after DEAE-Sephadex chromatography. The number of megakaryocyte colonies was extremely low in all fractions from the Sephadex G-150 column when added alone to the bioassay (Fig. 3).

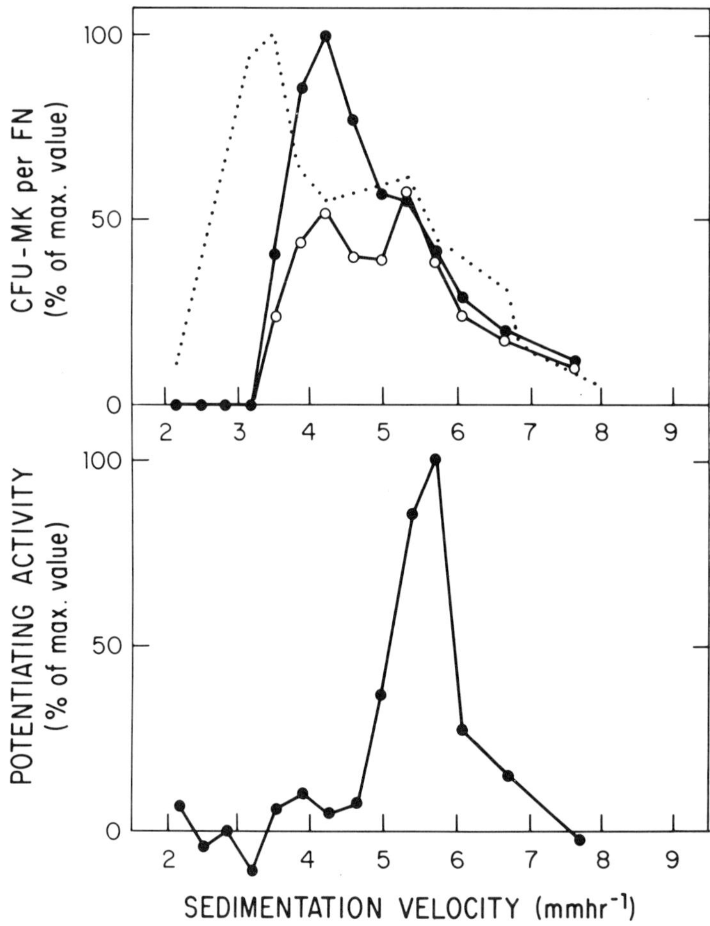

Fig. 2. The sedimentation velocity characteristics of megakaryocyte progenitor cells and bone marrow cells producing megakaryocyte potentiating activity. The upper panel shows the distribution profiles of total nucleated cells (••••••) and CFU-MK determined in the presence of WEHI-3 CM alone (○) and with both WEHI-3 CM and adherent peritoneal cell supernatant (●). The lower panel shows the potentiator activity determined by mixing 3×10^5 irradiated (750R) fractionated cells with 5×10^4 unirradiated unfractionated target cells (Williams et al., 1980).

However, extensive amounts of MK-CSF were present in fractions 23–28 when assayed in the presence of MØSN (middle plot). Colony numbers were increased 5 to 10-fold by the active fractions. Determinations of MK-Potentiator activity levels in the fractions were performed in the presence of MK-CSF from a pool of fractions 24 and 25. Potentiating activity was found in fractions 16–18 (MW 10,000) and did not overlap with those fractions containing MK-CSF.

Separate Roles of Factors in Megakaryocyte Colony Formation

The fractionation studies of WEHI-3 CM showed that two factors were necessary for colony formation as determined by the recognition of mature megakaryocytes in each colony. Experiments were performed to ascertain if the level of detection of the assay was altered as a result of using the fractionated material. Cultures stimulated by the various fractions were dehydrated and the colonies stained for acetylcholinesterase. Results similar to that shown in Figure 1 were obtained. Fewer than three acetylcholinesterase—positive colonies were observed in cultures stimulated by any of the fractions unless mixing experiments were performed. This

Fig. 3. The separation of MK-CSF from MK-Potentiator in WEHI-3 CM by filtration through Sephadex G-150. The distribution of protein is given in the upper panel. Fractions were assayed for MK-CSF (middle panel) alone (○) and in the presence of optimal levels of adherent peritoneal cell supernatant (MØSN) (●). MØSN alone did not stimulate colony formation. MK-Potentiator (lower panel) was assayed in the presence of fraction 25. All fractions contained 4.6 ml. The column was calibrated with Blue Dextran (V_0 , bovine serum albumin (BSA), ovalbumin (OA), chymotrypsinogen A (CHY), and ribonuclease A (RNAase).

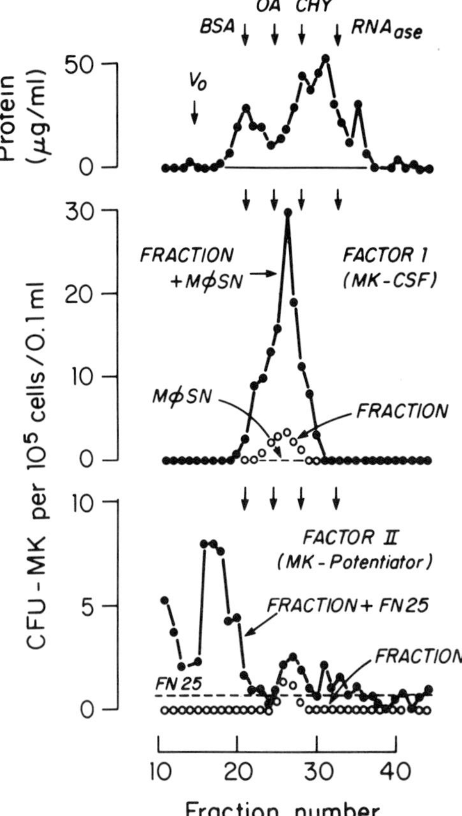

result suggests that detection was not influenced but rather that both factors were required for and prior to megakaryocyte development. Therefore, to study the relative roles of the two factors, it was necessary to use conditions for colony growth where optimal amounts of MK-CSF, but suboptimal levels of MK-Potentiator (WEHI-3 CM) were compared to conditions where both factors were present at maximal levels (WEHI-3 CM + MØSN). Colony size (Fig. 4) was not influenced by these two types of culture conditions. The ploidies of the colony megakaryocytes were affected, however. A bias towards higher values was obtained when optimal levels of both activities were present in the cultures. Figure 5 shows the range of ploidy values obtained. Cells of 64N were obtained under both sets of conditions, indicative that full ploidization was possible. Analysis of the proportion of cells in each ploidy group shows, however, that 68% of colony megakaryocytes obtained ploidy values of 16N and greater, when optimal culture conditions were employed, compared to 55% of megakaryocytes being 8N and less in the cultures with suboptimal levels of MK-Potentiator (Fig. 6).

A consequence of this finding was that, if MK-Potentiator preferentially enhanced nuclear endoreduplication, then megakaryocyte progenitor cells may respond to form single megakaryocytes if placed in the correct environment. Matching cultures were set up and stimulated with WEHI-3 CM, MØSN, optimal levels of both, and just medium alone. After culturing for 7 days, the plates were dehydrated, stained

Fig. 4. The effect of optimal levels of MK-Potentiator (MØSN) on megakaryocyte colony size. The range in colony size is shown on the abscissa (e.g., 3/5 is 3–5 acetylcholinesterase-positive cells per colony). Colony size was estimated by counting acetylcholinesterase-positive cells obtained either directly in the culture dish or staining after transferring the colonies to albumin-coated glass slides. No difference between the two methods was observed. The data are the results of 7 experiments. The number of colonies examined is given in brackets for both groups. Some large colonies (>40 cells) are estimates rather than true values due to overlap of colony cells.

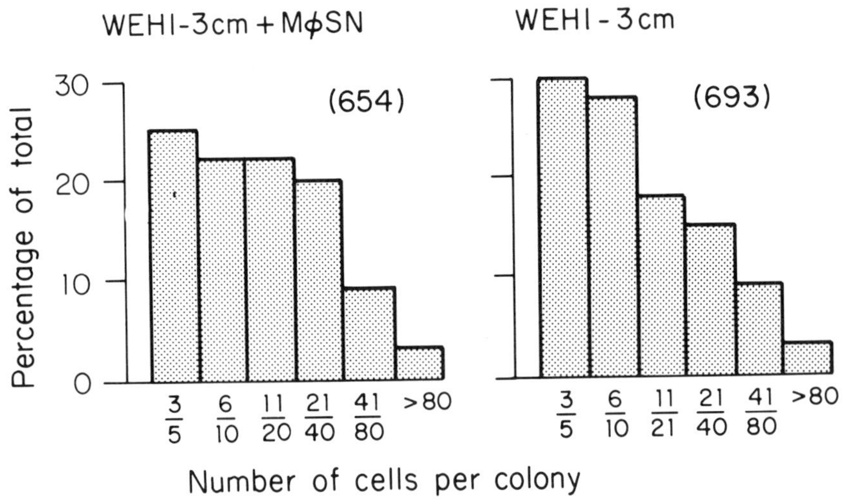

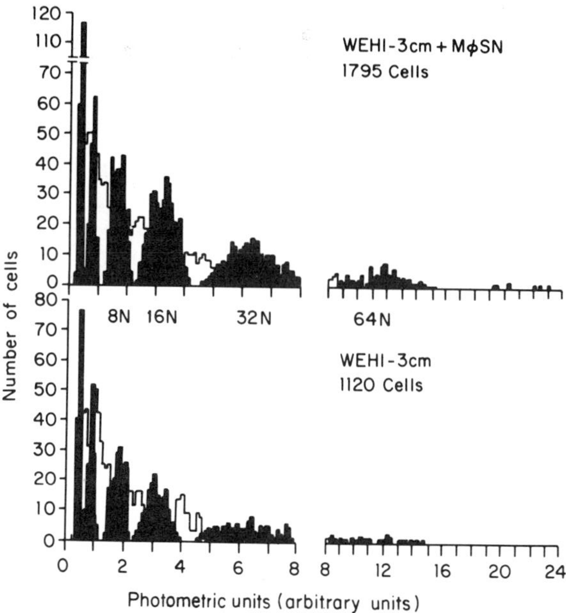

Fig. 5. The DNA content of megakaryocyte colony cells measured after Feulgen staining. The number of cells is plotted against arbitrary photometric units measured by microdensitometry. The filled-in regions represent the distributions of the various ploidy values of the colony cells (mean ± 30%). The culture conditions used for colony growth are shown in the upper right-hand corner of each plot, together with the number of cells studied in each group. Cells were taken from cultures after 6 or 7 days.

for acetylcholinesterase and assayed for single megakaryocytes, two cell aggregates and megakaryocyte colonies (>3 cells). By drying the cell cultures immediately after plating, the number of megakaryocytes put into the plates was obtained. Approximately 30 single megakaryocytes were found in each plate; together with about 2 two-cell aggregates, and a very low but detectable number of colonies (Table 2). The latter two observations presumably reflect coincidental megakaryocytes and clumps of cells present in the cell suspension used for plating. After culturing the cells for 7 days, surprisingly, no decrease in the number of single megakaryocytes was observed in any cultures, except for those cultures containing MØSN, where a significant decrease (P = 0.05) was observed. Kinetic experiments show that the number of single megakaryocytes observed was not a rebound phenomenon, a result which would suggest stimulation of precursor cells to form single megakaryocytes (data not shown). The results suggest that megakaryocyte progenitor cells do not form megakaryocytes directly, but require some further commitment involving cell division before endomitosis and cytoplasmic maturation occurs.

The results show that two separable factors can be obtained which differentially influence megakaryocyte formation from a diploid progenitor cell. Both factors can be obtained from *in vivo* sources as conditioned media and may have physiologic

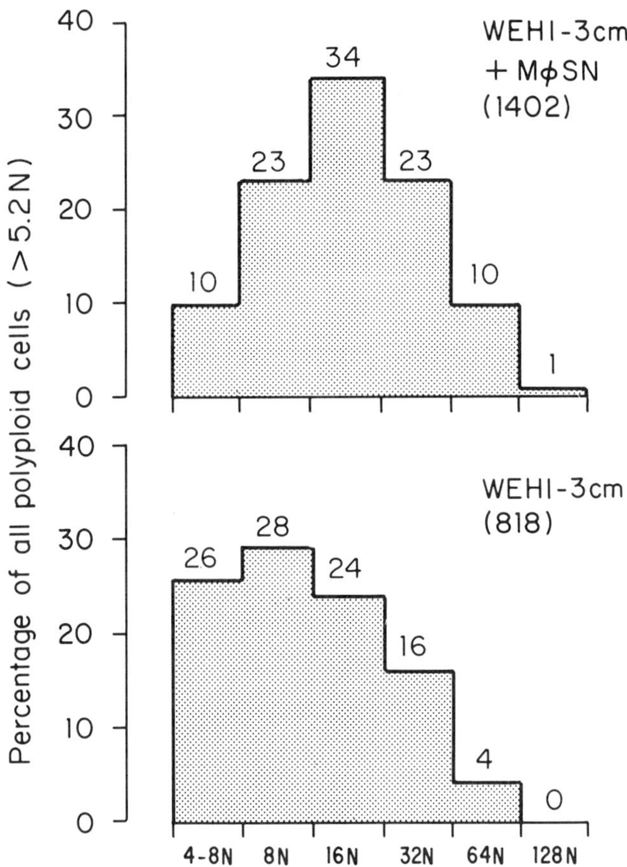

Fig. 6. The effect of optimal levels of MK potentiator (MØSN) on the ploidy distribution of colony megakaryocytes. The ploidy distribution of cells grown in the presence of WEHI-3 CM and MØSN (upper) is compared to the distribution of cells grown with WEHI-3 CM alone (lower). The percent of cells in each ploidy class is shown. 4N cells were excluded since some colonies included polymorphic neutrophils and macrophages and may have included immature 4N cells of those cell lineages. The number of cells in each group with ploidy values of 5.2N or greater is given in brackets. 5.2N represents the ploidy at the limit of the distribution of 4N cells (see Fig. 5).

significance. Parallels have been drawn between megakaryocyte potentiator and Thrombopoietic Stimulating Factor (TSF) (Williams et al., 1979). Like other CSFs, MK-CSF appears to be elaborated by both macrophages and T cells when appropriately stimulated (Ralph et al., 1977). Both factors are present in WEHI-3 CM and can be separated by Sephadex G-150 column chromatography. The factors appear to act prior to megakaryocyte formation since acetylcholinesterase-positive colonies are not detected unless both factors are present in the cultures. The bioassay, however, is limited by the intensity of the acetylcholinesterase stain and the mag-

Table 2. The Effect of WEHI-3CM and MØSN on Total Megakaryocyte Activity.

Culture time	Addition	Number of experiments	Single megakaryocytes	Two megakaryocytes	Megakaryocyte colonies
Day 0	Medium	10	31.7 ± 2.0	2.0 ± 0.2	0.2 ± 0.1
Day 7	Medium	5	30.5 ± 2.9	2.5 ± 0.3	0.1 ± 0.1
	MØSN	7	22.0 ± 1.6	2.0 ± 0.5	0.5 ± 0.3
	WEHI-3CM	8	26.1 ± 2.3	6.2 ± 0.9	21.7 ± 1.5
	WEHI-3CM + MØSN	8	28.6 ± 2.3	5.3 ± 0.6	36.5 ± 3.2

10^5 cells were cultured in triplicate per group per experiment.

The results are shown as mean values ± SEM.

nification used. Generally, colonies were analyzed at 60-100 × magnification. Colonies comprised of small acetylcholinesterase cells would not be detected, since these cells are difficult to observe below 200 × magnification (Long, unpublished observation).

A two-stage differentiation process for megakaryocyte development from precursor cells is suggested similar to that hypothesized for erythroid development (Iscove, 1978). In that scheme, a molecule (burst-promoting activity or BPA) is required for the commitment of cells to the erythroid series, and erythropoietin influences the final cell divisions and the induction of hemoglobin synthesis. The data presented here and elsewhere suggest that two factors also govern separate phases of megakaryocytopoiesis, with MK-CSF influencing cell division and MK-potentiator influencing nuclear endoreduplication and cytoplasmic maturation. Cell division may be a necessary requirement for a precursor cell to differentiate and form a megakaryocyte. A working hypothesis is shown in Figure 7. It is suggested that early cells restricted to the megakaryocyte line are highly sensitive to MK-CSF, and following several cell divisions and the beginning of maturation, there is a loss of response to the MK-CSF and a concomitant increase in sensitivity to MK-Potentiator.

The concept of regulation of megakaryocytopoiesis at different stages raises several important and unanswered issues. The interrelationships between the factors described here to thrombopoietin and other CSFs, their physiological roles, and their specificity within the entire hemopoietic differentiation process remains central to the understanding of megakaryocytopoiesis.

Fig. 7. Scheme for two factor stimulation of megakaryocyte production from diploid precursor cells. MK-CSF: Megakaryocyte colony stimulating factor. CFU-MK: Colony forming unit-megakaryocyte. SACHE: small acetylcholinesterase-positive cell.

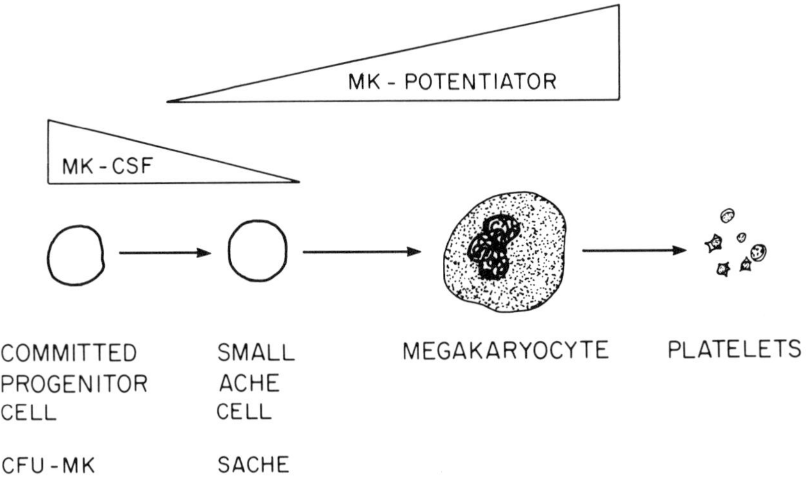

Summary

The *in vitro* cloning technique was employed to determine if two factors were involved in megakaryocyte formation from clonable precursor cells. Murine megakaryocyte colonies were detected *in situ* by dehydration of the cell cultures and subsequent staining of the cells for acetylcholinesterase. WEHI-3 cell-conditioned medium with the capacity to stimulate megakaryocyte colony formation was fractionated by Sephadex G-150 column chromatography. The development of colonies containing megakaryocytes was observed when mixing experiments were performed. Individual fractions did not support megakaryocyte colony growth. The two activites were obtained from a variety of tissue sources and cell lines. Megakaryocytes colony formation depended on the presence of one factor (megakaryocyte colony-stimulating factor) in the cell cultures. Colony size was not influenced by the concentration of a second factor (megakaryocyte potentiating factor), but in the presence of optimal levels of this factor increased ploidy values of the colony megakaryocytes was observed. Maturation of precursor cells into single megakaryocytes was not observed under the culture conditions employed. It is suggested that these two factors independently regulate cell division of megakaryocyte progenitor cells and maturation of their progeny.

ACKNOWLEDGMENTS

The authors wish to acknowledge the help, goodwill and hospitality of Drs. Richard Levine and Paul Bunn of the Veterans Administration Hospital of Washington, D.C., for instruction and access to the Artex Image Analyzer. N. Williams is a Leukemia Society of America Scholar. M. Long is a Leukemia Society of America Fellow. The work was funded by grants HL 22451 from the National Institutes of Health, CH-3C from the American Cancer Society, CA 0874B from the National Cancer Institute and the Gar Reichman Foundation.

References

Iscove, N. N. 1978. Erythropoietin-independent stimulation of early erythropoiesis in adult marrow cultures by conditioned media from lectin-stimulated mouse spleen cells. In *Hematopoietic Cell Differentiation*. Golde, D. W., Cline, M. J., Metcalf, D., and Fox, C. F., eds. New York: Academic Press. pp. 37–52.

Jackson,C. W. 1973. Cholinesterase as a possible marker of early cells of the megakaryocyte series. *Blood* 42:413–421.

Leuchtenberger, C. 1958. Quantitative determination of DNA in cells by Feulgen microspectrophotometry. In *General Cytological Methods* (Vol. 1). Danielli, J. F., ed. New York: Academic Press. pp. 220–277.

Levine, R. F., P. A. Bunn, Jr., K. C. Hazzard, and M. L. Schlam. 1980. Flow cytofluorometric analysis of megakaryocyte ploidy. Comparison with Feulgen microdensitometry. Validation of method and discovery that 8N is the predominant ploidy class in guinea pig and monkey marrow. *Blood* 56:210–217.

Long, M. W. and R. L. Henry. 1979. Thrombocytosis-induced suppression of small acetylcholinesterase-positive cells in bone marrow of rats. *Blood* 54:1338–1346.

Long, M. W. and N. Williams. 1980. Relationship of small acetylcholinesterase positive cells to megakaryocytes and clonable megakaryocytic progenitor cells. In *Megakaryocytes In Vitro*, Evatt, B. L., Levine, R. F., and Williams, N. T., eds. New York: Elsevier North Holland, pp. 293-298.

Lowry, O. H., N. J. Rosebrough, A. L. Farr, and R. J. Randall. 1951. Protein measurement with the folin phenol reagent. *J. Biol. Chem.* 193:265–275.

McLeod, D. L., M. M. Shreeve, and A. A. Axelrad. 1976. Induction of megakaryocytes colonies with platelet formation *in vitro*. *Nature* 261:492–493.

Metcalf, D. 1977. In *Hemopoietic Colonies*. New York: Springer-Verlag. p. 156.

Metcalf, D. and G. R. Johnson. 1978. Production by spleen and lymph node cells of conditioned medium with erythroid and other hemopoietic colony-stimulating activity. *J. Cell. Physiol.* 96:31–42.

Metcalf, D., H. R. MacDonald, N. Odartchenko, and B. Sordat. 1975. Growth of mouse megakaryocytes colonies *in vitro*. *Proc. Nat. Acad. Sci. USA* 72:1744–1748.

Nakeff, A. and S. Daniels-McQueen. 1976. *In vitro* colony assay for a new class of megakaryocyte precursor: Colony-forming unit megakaryocyte (CFU-M). *Proc. Soc. Exp. Biol. Med.* 151:587–590.

Ralph, P., H. E. Broxmeyer, M. A. S. Moore, and I. Nakoinz. 1978. Induction of myeloid colony stimulating activity in murine monocyte tumor cell lines by macrophage activators and in a T cell line by concanavalin A. *Cancer Res.* 38:1414–1419.

Williams, N. 1979. Preferential inhibition of murine macrophage colony formation by prostaglandin E. *Blood* 53:1089–1094.

Williams, N. and H. Jackson. 1978. Regulation of the proliferation of murine megakaryocyte progenitor cells by cell cycle. *Blood* 52:163–170.

Williams, N., H. Jackson, P. Ralph, and I. Nakoinz. 1980. Cell interactions influencing murine marrow megakaryocytes: Nature of the potentiator cell in bone marrow. *Blood* 57:157–163.

Williams, N., R. R. Eger, M. A. S. Moore, and N. Mendelsohn. 1978a. Differentiation of mouse bone marrow progenitor cells to neutrophil granulocytes by an activity separated from WEHI-3 cell conditioned medium. *Differentiation* 11:59–63.

Williams, N., T. P. McDonald, and E. M. Rabellino. 1979. Maturation and regulation of megakaryocytopoiesis. *Blood Cells* 5:43–55.

Williams, N., H. Jackson, A. P. C. Sheridan, M. J. Murphy, A. Elste, and M. A. S. Moore. 1978b. Regulation of megakaryocytopoiesis in long-term murine bone marrow cultures. *Blood* 51:245–255.

Discussion

Levine: *In situ* megakaryocytes often occur in colonies of 2–4 or up to 8 cells, so that depending on the effectiveness of your marrow disruption procedure, it might not be surprising to find a few colonies at the beginning of the cultures. Single megakaryocytes might persist in culture without changes for several days.

Evatt: Have you looked at the specificities of the potentiator?

Williams: The macrophage colony-forming cell is not affected by the presence of potentiator.

Levine: Could your potentiator be the same as thrombopoietin?

Williams: We need to reserve judgment of that possibility. We need to have better purification and higher concentrations to inject into animals before we can answer that.

Evatt: Our data on differentiated megakaryocytes are consistent with Neil's model of a potentiator role for thrombopoietin.

Levin: One comment on the use of tumor cell supernatants. I had the opportunity to use another WEHI-3 cell line which, unfortunately, did not support the growth of megakaryocytes. I am not challenging your data, but just wish to point out that these tumor lines can change. These cell lines are not a universal source of stimulator. You very conservatively used plus/minus (±) for thrombopoietin stimulation of megakaryocyte colony formation. Now we are going to hear, and have heard a lot already, about the ability of thrombopoietin to stimulate CFU-M. I was a little surprised by your conservatism, since you have published that, in fact, there is a dose-response relationship between thrombopoietin and CFU-M.

Williams: What we showed was that it was uncertain that thrombopoietic stimulating factor (TSF) could directly stimulate megakaryocyte colony formation. However, our data did show that TSF can potentiate colony formation in the presence of a preparation of megakaryocyte CSF (*Blood Cells* 5:43, 1979). The work of Dr. Kalmaz showed that you can get some colonies with TSF, but that these are markedly smaller than those obtained when a source of megakaryocyte colony stimulating factor (MK-CSF) is employed. That is one of the reasons why I suggested this gradation of responsiveness: a progenitor cell might have some degree of responsiveness to TSF, but it is extremely responsive to the megakaryocyte CSF. I think they are extremely interesting and important questions, but the bioassays and the molecular probes that we have now are quite unsuitable to address those issues. We clearly are going to need purified molecules, radioimmunoassays, and monoclonal antibodies before we have discrete and meaningful answers to those questions.

Petursson: We used pokeweed mitogen stimulated spleen cell-conditioned medium as our standard source of megakaryocyte colony stimulating activity in the methylcellulose system. I just want to comment that when we tested TSF supplied by Dr. McDonald, we got no megakaryocyte colony formation. Likewise, when we used thrombocytopenic mouse serum we also found no megakaryocyte colonies, but utilizing these two substances in conjunction with the suboptimal amounts of mitogen-stimulated spleen cell-conditioned medium we got synergy suggesting that these two substances do appear to be acting at separate levels of megakaryocyte production.

Paulus: Under our conditions using erythropoietin in plasma clots, we have data showing at least two ways in which the ploidy of megakaryocytes may become decreased. First, in mixed erythroid-megakaryocyte colonies the average ploidy is far lower than that obtained with pure colonies. Second, with pure colonies there is an inverse relationship between the average number of endoreduplications per colony and the number of doublings achieved by the colony cells. That might relate to your data and the need for potentiator which might become rate limiting in large colonies.

Williams: It is possible. What I didn't show in the DNA content distribution were the 2N–4N cells and the 4N cells. These two groups were very pronounced in the underpotentiated cultures. I deliberately omitted them from the data, in case they did belong to another cell lineage. I do believe, however, that the majority of them are megakaryocytic. I think that your point is well taken.

Jackson: By varying the concentrations of your two activities, have you been able to induce small colonies with high ploidy megakaryocytes or the reverse of that?

Williams: No, we haven't. We tried to look particularly at early time points for exactly that. When colonies can be first observed at 3 or 4 days of culture, there were no colonies comprised of small cells; we always saw at least one big megakaryocyte among the small cells. It can only mean that there is asynchrony in colony formation.

Bunn: I think it's very interesting that you have shown that the CSF affects the number and the potentiator affects the ploidy. Before we all accept the ploidy data, we need to do statistical analyses on the ploidies. It's going to be very important to show that these things are clearly different even though the ploidies look like they are shifted. We need statistical consultants to help us to show that.

Williams: I agree. With both stimulators, I think somewhere around 70% of all megakaryocytes were of 16N or greater. The reverse was true in the underpotentiated colonies; the majority of megakaryocytes were 8N and less.

Evatt: That is an important point. In these cultures, the younger or lower ploidy megakaryocytes may be the only ones responding to thrombopoietin or other potentiators. Statistical evidence will be important. Changes in ploidy patterns should be correlated with changes in thymidine incorporation to get a better perspective on the responses to these with factors.

McDonald: Richard Levine proposed that the potentiator might be thrombopoietin, but I'd like to point out that neither the spleen-conditioned media, the bone marrow-conditioned factor, nor WEHI-3 CM have stimulated platelet production in mice, as does the TSF from the kidney cell culture media.

Williams: Yes, I acknowledge that point. The dose of your kidney cell-conditioned medium that you inject into an animal is something like 30–300 times as great as what we have been adding to a plate. The amount of concentration that is necessary for us to inject *in vivo* may be quite considerable. We need about 3 liters of bone marrow-conditioned medium to be able to do that experiment, and we would certainly expect to see inhibition at those concentrations.

Published 1981 by Elsevier North Holland, Inc.
Evatt, Levine, and Williams, Editors
MEGAKARYOCYTE BIOLOGY AND PRECURSORS:
IN VITRO CLONING AND CELLULAR PROPERTIES

The Effects of Thrombopoietin on Megakaryocytopoiesis of Mouse Bone Marrow Cells *In Vitro*

G. D. Kalmaz and T. P. McDonald

University of Tennessee College of Medicine, Knoxville Unit, Department of Medical Biology, Memorial Research Center, Knoxville, Tennessee

Previous studies have shown that a plasma clot tissue culture system will support the proliferation and differentiation of mouse megakaryocytes (Freedman et al., 1977; Nakeff, 1977). This technique has allowed for the study of "stem cells" or "committed pools" of megakaryocyte progenitors (CFU-M) and will in all probability be a useful assay for the thrombocytopoiesis-stimulating factor (TSF or thrombopoietin). It was previously shown that TSF from both human embryonic kidney cell culture medium (Freedman et al., 1977) and plasma from thrombocytopenic mice (Nakeff, 1977) will stimulate an increase in CFU-M. In these previous studies megakaryocytic colonies were used as a measure of increased differentiation of the megakaryocytic compartment. However, because of the low number of cells per colony due to limited endomitotic properties of megakaryocytes and difficulties in counting colonies, it seemed possible that the total number of megakaryocytes might also be a good measure of megakaryocytopoiesis (Nakeff et al., 1975). In the studies to be reported we have cultured mouse bone marrow by use of the plasma clot system and have shown that TSF stimulated megakaryocytopoiesis as evidenced by increased numbers of megakaryocytes, megakaryocytic colonies, and megakaryocytes per colony. Also, in this study, the total number of megakaryocytes was compared with the number of megakaryocytic colonies after stimulation by TSF; the results show a high degree of correlation between these two indices.

Methods and Materials

Mouse bone marrow cells were cultured by use of a modified plasma clot culture system described previously (Nakeff and Daniels-McQueen, 1976). For bone marrow donors, 7–8 weeks-old C_3H male mice were used; marrow from each femur was flushed from the bones with 1 ml phosphate-buffered saline (PBS), pH 7.4, and a single cell suspension was prepared by mixing. The number of nucleated cells was determined from the final cell suspension by diluting 0.02 ml of marrow into buffered saline (1:500); Hematall (Fisher Scientific) was added to lyze RBC and the cells were counted with an electronic cell counter.

The marrow cultures were prepared in plastic tubes using the following quantities per treatment group: (a) 0.1 ml beef embryo extract (GIBCO) diluted 1:4 in Leibovitz (L-15) tissue culture medium (GIBCO); (b) 0.2 ml fetal calf serum (FCS); (c) 0.2 ml of production control medium (PMC) or TSF-rich medium in varying concentrations (0.05–0.2 mg protein) diluted into L-15 medium; (d) 0.3 ml of L-15 medium; (e) 1×10^5 bone marrow cells in 0.1 ml PBS, and (f) 0.1 ml of bovine citrated plasma. The materials were mixed gently by inversion and 0.1 ml was added to each well (8 per treatment group) of disposable microtiter plates (Cooke Engineering Co., Alexandria, VA) and allowed to clot. Before adding the medium and cells, the disposable microtiter plates were cut into sections with eight wells each, placed into a 15×100 mm plastic dish, and sterilized overnight by UV-irradiation. The cells were cultured for varying lengths of time in a water-jacketed CO_2 incubator (National Appliance Company) at 37°C in an atmosphere of high humidity with 5% CO_2 in air. In order to maintain a high humidity environment, a 10×35 mm Petri dish was filled with distilled water and placed alongside each microtiter plate.

TSF used in these studies was production culture medium from human embryonic kidney cells (Lot 68-041) which had been previously shown to contain high levels of TSF (McDonald et al., 1975). Prior to storage, the production medium was concentrated (300 ×) with an Amicon ultrafiltration cartridge. The concentrate was then clarified by centrifugation and stored at −76°C until used. Control production medium (medium without kidney cell growth) was treated in a similar manner.

Cultures were examined for megakaryocytes and megakaryocytic colonies at daily intervals for 3–7 days. At these times, plasma clots were rimmed and the clots transferred to glass slides. The clots were partially dehydrated by placing pieces of filter paper on their surfaces. A second piece of filter paper was applied to the first piece of paper and allowed to remain long enough for the paper to become moist. With the aid of forceps, the top piece of filter paper was removed and 5–7 drops of 5% gluteraldehyde (in 0.01 M phosphate buffer, pH 7–7.2) were placed on the remaining filter paper with a Pasteur pipette. The filter paper moistened with the gluteraldehyde was left for 10 min. After removing the remaining filter paper, the slides with fixed plasma clots were then rinsed in 0.1 M sodium phosphate for 1 min and stained by the use of the "direct-coloring" thiocholine method (Karnovsky and Roots, 1964) for acetylcholinesterase. For staining, the clots were incubated for 3 hr at room temperature in a solution consisting of: (a) 30 mg of acetylthiocholine iodide, (b) 45 ml of 0.1 M sodium phosphate, pH 6, (c) 3 ml of 0.1 M sodium citrate, (d) 6 ml of 30 mM copper sulfate, and (e) 6 ml of 5 mM potassium ferricyanide. Following a 1 min rinse in 0.1 M sodium phosphate, postfixation in absolute methanol for 10 min, and 50% methanol for 30 sec, the cells were counterstained in Harris' Hematoxylin "blued" stain for 2–3 min and then clarified in running tap water. Final preparations were examined microscopically at magnifications of 430 × for acetylcholinesterase-positive megakaryocytes. A colony was defined as 2 or more morphologically recognizable megakaryocytes in close proximity.

Student's *t* test, least squares analysis, and correlation coefficient analysis were used for determination of statistical significance.

Results and Discussion

Figure 1 shows the results of comparing the number of megakaryocytes in cultures with the number of megakaryocytic colonies. Some of these cultures were stimulated with thrombopoietin. There was a high degree of correlation between the two values. These data support the hypothesis that the total number of megakaryocytes in plasma clots can be used to estimate the number of megakaryocytic colonies and is, therefore, a good measure of megakaryocytopoiesis.

Table 1 compares the average number of megakaryocytes in plasma clot cultures of 1 × 10^4 bone marrow cells after incubation with L-15 medium, PMC and TSF-

Fig. 1. The number of megakaryocytes versus the number of megakaryocytic colonies in plasma clot cultures ($y = 7.75 + 8.38 \times$). The correlation coefficient was: $r = 0.94$, $P < 0.0005$. A total of 51 cultures was utilized in this study; some of the cultures were stimulated by thrombopoietin.

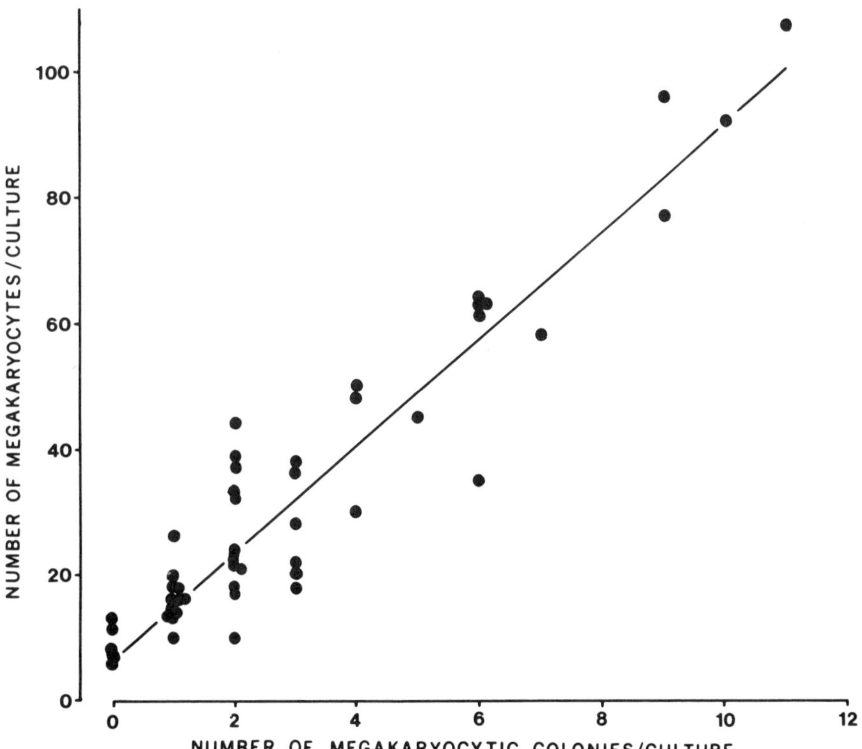

Table 1. Average Number of Megakaryocytes in Plasma Clot Cultures on Various Days After Culture With Control Production Medium (PMC) and TSF-Rich Kidney Cell Culture Medium.

Days of culture	L-15 medium	Dose (Mg Protein/Culture)							
		PMC[b]				TSF[b]			
	0[a]	.005	0.01	0.015	0.02	0.005	0.01	0.015	0.02
3	4.47 ±.56	4.50 ±.71	4.38 ±.68	4.80 ±.65	4.40 ±.65	18.63[c,e] ±1.24	22.30[c,e] ±2.24	28.80[c,d] ±6.70	17.70[c,d] ±3.80
4	4.86 ±.60	4.12 ±.52	4.90 ±.81	4.12 ±.90	4.12 ± .80	22.14[c,e,x] ±1.45	24.83[c,e] ±2.77	33.50[c,e] ±4.76	19.83[c,e] ±2.02
5	5.46 ±.69	4.12 ±.67	5.17 ±.94	5.00 ±.78	5.20 ±1.2	26.40[c,e,y] ±2.00	30.00[c,e] ±4.40	43.40[c,e,x] ±3.20	23.80[c,e] ±2.20
6	5.94[x] ±.53	4.28 ±.68	5.50 ±1.12	4.50 ±.65	5.00 ±.82	31.40[c,e,y] ±3.14	39.20[c,e,y] ±2.90	54.80[c,e,y] ±3.50	33.80[c,e,y] ±1.22
7	6.70[x] ±.93	4.00 ±1.30	3.12 ± .77	4.28 ±1.00	4.40 ±.81	22.80[c,e] ±2.60	25.10[c,e] ±3.20	33.90[c,e] ±2.40	24.00[c,e] ±2.00

PMC–Production Medium, Control.

[a]Average number of megakaryocytes in 16 cultures ± SE.

[b]Average number of megakaryocytes in 7-8 cultures ± SE.

Values were significantly different from control: medium-treated control (o) versus PMC or TSF, $^{c}P < 0.0005$; PMC control values versus TSF, $^{d}P < 0.005$, $^{e}P < 0.0005$; values found on day 3 of same treatment versus values found on days 4-7, $^{x}P < 0.05$, $^{y}P < 0.005$.

rich kidney cell culture medium. As shown, both the L-15 medium (zero control) and the different levels of PMC gave essentially the same average number of megakaryocytes per culture on various days of culture. However, those cultures incubated with TSF-rich medium showed a marked increase in the number of megakaryocytes per clot with peak numbers occurring on day 6 after addition of 0.015 mg of TSF/culture. No colonies were seen in cultures containing PMC.

Table 2 compares the results of another experiment in which L-15 medium and TSF were added to cultures and the average number of megakaryocytic colonies and megakaryocytes per colony determined. As shown, TSF increased the number of colonies as well as the number of megakaryocytes per colony in cultures. In agreement with the previous table, the highest number of megakaryocytic colonies and megakaryocytes/colony was found after 6 days of culture with 0.015 mg of TSF/culture.

The results from two additional experiments in which various doses of TSF were added to plasma clots are shown in Figure 2. The total number of megakaryocytes showed a linear increase with dose (Fig. 2A), reaching a peak at 0.015 mg of TSF. Higher doses resulted in a decreased number of megakaryocytes. This same pattern was also shown when the results are expressed as megakaryocytes per colony (Fig. 2B) or megakaryocytic colonies per culture (Fig. 2C). Therefore, TSF stimulated not only the total number of megakaryocytes but also stimulated megakaryocytic colonies and the size of the colonies (megakaryocytes per colony).

In this study, recognizable megakaryocytes begin to increase in number on day 2 and by day 6 mature megakaryocytes were numerous. Megakaryocyte colonies, however, first became apparent after 5 days of culture; their number peaked at day 6. With the addition of TSF, mitotic activity in "young" megakaryocytes was observed at day 3; mitosis was rarely seen after day 5. By day 5, the larger cells contained large amounts of cytoplasm, lobulated foamy nuclei, granules, and formed spreading pseudopods. There appeared to be platelet shedding. The length and number of pseudopods increased with additional time in culture. By day 6, platelets were seen at the periphery of many of the pseudopods of megakaryocytes. However, at this time, only a few megakaryocytes were seen shedding platelets. Degenerative changes, such as loss of granules and irregularity in shape, were observed in some megakarycytes by day 7 and these changes increased in frequency with additional time in culture. The results of this study demonstrate that TSF stimulates megakaryocytic growth, differentiation and cytoplasmic maturation, leading to platelet production.

Differentiation of mouse marrow megakaryocyte progenitors (CFU-M) has been studied previously *in vitro* by use of plasma clot colony assays (Freedman et al., 1977; Nakeff and Daniels-McQueen, 1976; Mizoguchi et al., 1979), and with semisolid medium techniques (Metcalf et al., 1975; McLeod et al., 1976; Williams and Jackson, 1978). These techniques differed in their capacity to support growth of megakaryocytes. For example, plating efficiencies for CFU-M per cell cultured of 1:5000 (Nakeff, 1977), 1:4000 (McLeod et al., 1976), or 1:2000 (Mizoguchi et al., 1979) have been found. The present study used 1×10^4 cells per culture with a plating efficiency of 1:1500.

Table 2. Megakaryocytic Colonies and Megakaryocytes Per Colony in Plasma Clot Cultures After Various Days of Culture and Stimulation With TSF-Rich Kidney Cell Culture Medium.

	Dose of TSF (Mg protein/culture)				
Days	**0**	**0.005**	**0.01**	**0.015**	**0.02**
5	0.75 ± 0.25 (2.17 ± 0.17)	1.00 ± 0.22 (2.00 ± 0.00)	1.33 ± 0.21 (2.12 ± 0.12)	2.43 ± 0.20[c] (3.41 ± 0.26)[a]	1.17 ± 0.17 (2.00 ± 0.00)
6	1.10 ± 0.02 (2.00 ± 0.00)	1.14 ± 0.46 (2.13 ± 0.12)	2.50 ± 0.22[c] (2.40 ± 0.13)[a]	5.00 ± 0.53[c] (3.51 ± 0.22)[c]	2.57 ± 0.20[c] (2.56 ± 0.14)[b]
7	1.12 ± 0.23 (2.11 ± 0.11)	0.80 ± 0.38 (2.25 ± 0.25)	1.29 ± 0.29 (2.33 ± 0.17)	1.75 ± 0.25 (2.40 ± 0.13)	1.38 ± 0.33 (2.27 ± 0.14)

Numbers in ( ) are the average number of megakaryocytes in colonies ± SE.

Values were significantly different from control: medium-treated control versus TSF treated cultures [a]$P < 0.05$, [b]$P < 0.005$, [c]$P < 0.0005$.

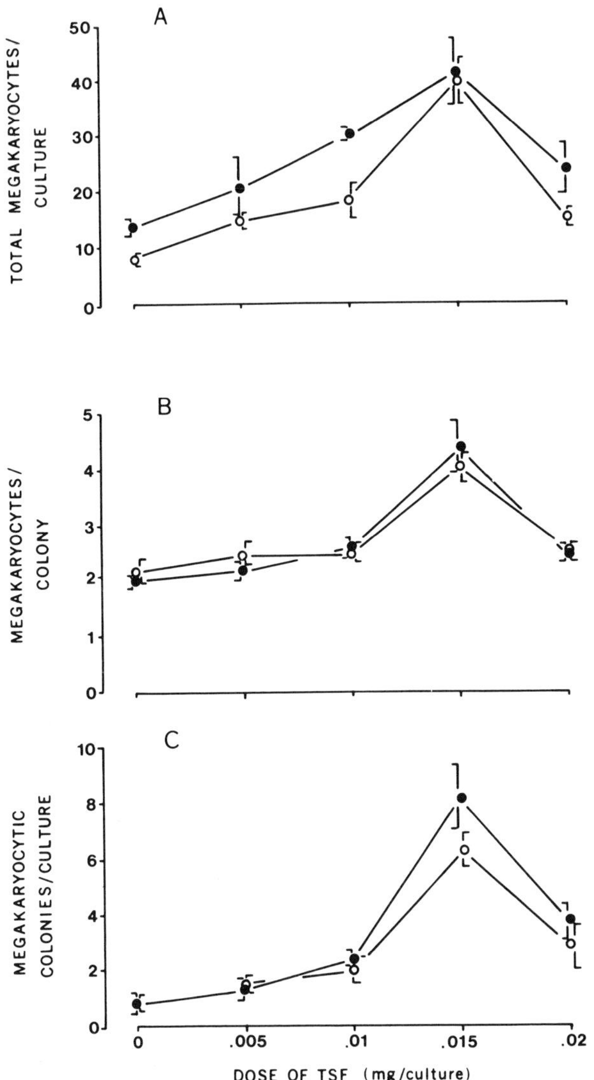

Fig. 2. The average number of megakaryocytes (A), megakaryocytes per colony (B), and megakaryocytic colonies (C) per culture of plasma clots after stimulation with various doses of TSF and 6 days of culture. Two separate experiments are shown (o and ●). Each point is the average of 5–6 cultures and the vertical lines indicate the standard errors.

It is well known that megakaryocytic colonies are difficult to count, primarily because of the diffuse nature of the colony and the small numbers of megakaryocytes in colonies. Therefore, it would be easier to count only the total number of megakaryocytes in plasma cultures as a measure of "megakaryocytopoiesis." Since megakaryocytes are believed to undergo 3–5 endoreduplications before platelet shedding, a high ploidy megakaryocyte could be considered to be a colony itself. We compared the number of megakaryocytes in a plasma clot versus the number of colonies and found excellent agreement (Fig. 1). Therefore, it seems possible that by counting the total number of megakaryocytes a good measure of colonies is found; thus, a measure of megakaryocytopoiesis is achieved. This simplification in technique will be useful for future studies.

The number of megakaryocytes in colonies after addition of PMC or L-15 medium was almost identical (Table 1). The cultures treated with PMC did not contain colonies; cultures were composed of small megakaryocytes and platelets were not being shed from these small megakaryocytes. It may be possible that an additional role of TSF, in addition to stimulation of CFU-M, is to stimulate cytoplasmic maturation and platelet shedding from megakaryocytes. As shown, a significant increase in megakaryocytopoiesis (Tables 1 and 2; Fig. 2) was found in cultures stimulated with TSF. This finding points out that kidney cell culture medium contains a stimulator of megakaryocytopoiesis.

In agreement with the results of the present study, other work using TSF from human embryonic kidney cell culture medium showed an increase in the number of megakaryocytic spleen colonies in irradiated-bone marrow transplanted mice (McDonald et al., 1978) and an increase in the number of megakaryocytic colonies *in vitro* (Freedman et al., 1977). In addition, numerous similarities between TSF from kidney cell culture medium and TSF from animal sources have been demonstrated (McDonald, Chapt 4, this volume). The present work showed that TSF from kidney cell culture medium stimulates CFU-M *in vitro*. Nakeff (1977) presented preliminary evidence that the addition of serum from thrombocytopoietic mice will also stimulate the production of megakaryocytic colonies *in vitro*. These data point out additional similarities in the two TSF sources.

In agreement with previous work by Freedman et al. (1977), the present work showed that TSF induced a dose response increase in CFU-M. Why 20μg of culture medium gave significantly less total numbers of megakaryocytes, megakaryocyte colonies, and megakaryocytes per colony is unexplained, but the decrease could be due to toxic substances associated with the higher dose. Other explanations are also possible.

Other workers have shown that erythropoietin will stimulate CFU-M growth *in vitro* (Freedman et al., 1977; Metcalf et al., 1975; McLeod et al., 1976). However, the TSF used in the present work contained no erythropoietin activity (McDonald et al., 1975). Other tissue culture studies have shown that the cruder the erythropoietin the greater the CFU-M response (Freedman et al., 1977; McLeod et al., 1976) and the greater the platelet production *in vivo* (McDonald and Clift, 1979). It, therefore, follows that the humoral control of megakaryocytic differentiation might not be due to erythropoietin itself but possibly to a contaminant.

The results of the present work appear to be consistent with the hypothesis that TSF from kidney cell culture media, like TSF from serum of thrombocytopenic animals, has a regulatory role on megakaryocyte differentiation at the stem cell level *in vitro*. The studies presented herein form the basis for the development of an *in vitro* assay for the detection of TSF and will be useful for further studies of the humoral control of thrombocytopoiesis at the stem cell level.

Summary

Previous studies have shown that a plasma clot culture system will support the proliferation and differentiation of megakaryocytes. This technique may be useful for the assay of a thrombocytopoiesis-stimulating factor (TSF or thrombopoietin). Therefore, in the studies reported, we cultured mouse bone marrow cells with various amounts of TSF. Bone marrow cells were prepared from 8-week old C_3H mice, cultured for 6–7 days in a standard plasma clot medium, and the megakaryocytes were stained for acetylcholinesterase (AChE). The results showed that TSF stimulated megakaryocytes in cultures as evidenced by increased numbers of megakaryocytes, megakaryocytic colonies, and megakaryocytes per colony. The number of megakaryocytes and megakaryocytic colonies were directly related to the amount of TSF added. In addition, the total number of megakaryocytes per culture was directly related to the number of megakaryocytic colonies observed (correlation coefficient: $r=0.94$, $P < 0.005$). This finding supports the hypothesis that the number of megakaryocytic colonies can be estimated by counting the total number of megakaryocytes in plasma clot cultures. This *in vitro* technique may be useful for the assay of TSF from patients with platelet production disorders.

ACKNOWLEDGMENTS

The authors thank Marilyn Cottrell and Rose Clift for technical support, Pat Taylor for stenographic aid, and Drs. Neil T. Williams and Alexander Nakeff for helpful advice. This investigation was supported by the National Heart, Lung, and Blood Institute, Grant No. HL 14637; G. D. Kalmaz is the recipient of a National Research Service Award, Fellowship Number HL 06072.

References

Freedman, M. H., T. P. McDonald, and E. F. Saunders. 1977. The megakaryocyte progenitor (CFU-M): *In vitro* proliferative characteristics. *Blood* 50 (Suppl. 1):146.

Karnovsky, M. J., and L. Roots. 1964. A "direct-coloring" thiocholine method for cholinesterases. *J. Histochem. Cytochem.* 12:219–221.

McDonald, T. P. 1980. Thrombopoietin and its control of thrombocytopoiesis and megakaryocytopoiesis. In *Megakaryocytes In Vitro,* Evatt, B. L., Levine, R. F., and Williams, N. T., eds. New York: Elsevier North Holland. pp. 39-57.

McDonald, T. P. and R. Clift. 1979. Effects of thrombopoietin and erythropoietin on platelet production in rebound-thrombocytotic and normal mice. *Am. J. Hematol.* 6:219–227.

McDonald, T. P., R. Clift, R. D. Lange, C. Nolan, I. I. E. Tribby, and G. H. Barlow. 1975. Thrombopoietin production by human embryonic kidney cells in culture. *J. Lab. Clin. Med.* 85:59–66.

McDonald, T. P., M. Cottrell, C. C. Congdon, O. Walasek, and G. H. Barlow. 1978. Stimulation of megakaryocytic spleen colonies in mice by thrombopoietin. *Life Sci.* 22:1853–1857.

McLeod, D. L., M. M. Shreeve, and A. A. Axelrad. 1976. Induction of megakaryocyte colonies with platelet formation *in vitro*. *Nature* (London) 261:492–494.

Metcalf, D., H. R. MacDonald, N. Odartchenko, and B. Sordat. 975. Growth of mouse megakaryocyte colonies *in vitro*. *Proc. Nat. Acad. Sci. USA* 72:1744–1748.

Mizoguchi, H., K. Kubota, Y. Miura, and F. Takaku. 1979. An improved plasma culture system for the production of megakaryocyte colonies *in vitro*. *Exp. Hematol.* 7:345–351.

Nakeff, A. 1977. Colony-forming unit, megakaryocyte (CFU-M): Its use in elucidating the kinetics and humoral control of the megakaryocytic committed progenitor cell compartment. In *Experimental Hematology Today*. Baum, S. J., and Ledney, G. D., eds. New York: Springer-Verlag. pp. 111–123.

Nakeff, A. and S. Daniels-McQueen. 1976. *In vitro* colony assay for a new class of megakaryocyte precursor: Colony-forming unit megakaryocyte (CFU-M). *Proc. Soc. Exp. Biol. Med.* 151:587–590.

Williams, N., and H. Jackson. 1978. Regulation of the proliferation of murine megakaryocyte progenitor cells by cell cycle. *Blood* 52:163–170.

Discussion

Dukes: I would like to make a few comments on the single megakaryocytes versus megakaryocyte colony observations. We reported last year (*Exp. Hematol.* 7 (Suppl. 6): 64, 1979) that there is a good correlation between these two types of entities at day 7. But this correlation does not continue when one makes observations at day 10 or day 12. In other words, what we had found was that the colonies did not increase in number after day 7, whereas the single recognizable megakaryocytes did increase, suggesting that the precursors of these two entities are different cells. Further, interferon was much more inhibitory to the expression of the precursor of megakaryocyte colonies than the precursors of the single recognizable megakaryocytes in culture.

Leven: I'd like to know if, with thrombopoietic stimulating factor (TSF) in your cultures, you see any increase in any particular morphological stage of megakaryocytes? In particular, did you see any effect on this pseudopod formation that you've mentioned?

Kalmaz: Maturation depends on time in culture. For example, on day 6 we see more mature megakaryocytes than immature or blasts forms.

Leven: Do you see any predominance of any one of these particular stages with TSF?

Kalmaz: No.

Published 1981 by Elsevier North Holland, Inc.
Evatt, Levine, and Williams, Editors
MEGAKARYOCYTE BIOLOGY AND PRECURSORS:
IN VITRO CLONING AND CELLULAR PROPERTIES

Modulation of Megakaryocyte Colony Formation by Protein Fractions from the Urine of Normal and Anemic Individuals

Peter P. Dukes, Edward Gomperts, and Parvin Izadi

Departments of Pediatrics and Biochemistry, University of Southern California School of Medicine and Division of Hematology-Oncology, Childrens Hospital of Los Angeles, Los Angeles, California

Megakaryocytic colony formation in plasma clot cultures of mouse bone marrow cells can be stimulated by erythropoietin (EPO) derived either from sheep plasma or from human urine (McLeod et al., 1976). Recent work in this laboratory (Dukes et al., 1979) suggested that anemic human urine contained a protein fraction with erythroid burst promoting activity. In this communication we report that this urinary protein fraction also has megakaryocytic colony stimulating activity and that this activity does not seem to be additive to the activity of urinary EPO.

Methods and Materials

Urinary proteins were separated by chromatographic procedures which have previously been shown to concentrate erythroid burst promoting activity in a fraction, code named RP (Dukes, et al., 1979).

Three batches of RP were used in this study. RP (H7-A-7) was derived from pooled urine of patients with iron deficiency anemia, RP (4-53-3-13) was prepared from the urine of a patient with acquired aplastic anemia, and RP (PD1-3-4) from the urine of a normal individual. A hot water extract of an iron deficiency anemia urine protein concentrate obtained by the benzoic acid method (Espada and Gutnisky, 1970) was partially desalted by chromatography on Sephadex G-25. The aplastic anemia urine and the normal urine were processed by combined concentration and dialysis with an Amicon hollow fiber system DC10 employing an H10P10 cartridge. Amounts of the concentrates which contained 0.5–1.0 g protein were dissolved in 0.005 M sodium phosphate buffer, pH 8.0 (Buffer 1), and the solutions were dialyzed against two changes of 50 volumes of the same buffer. Dialysands were applied to 2.6 × 40 cm columns packed with QAE Sephadex A-50 (Fig. 1). The columns were developed with Buffer 1 at a flow rate of 20 ml/hr.

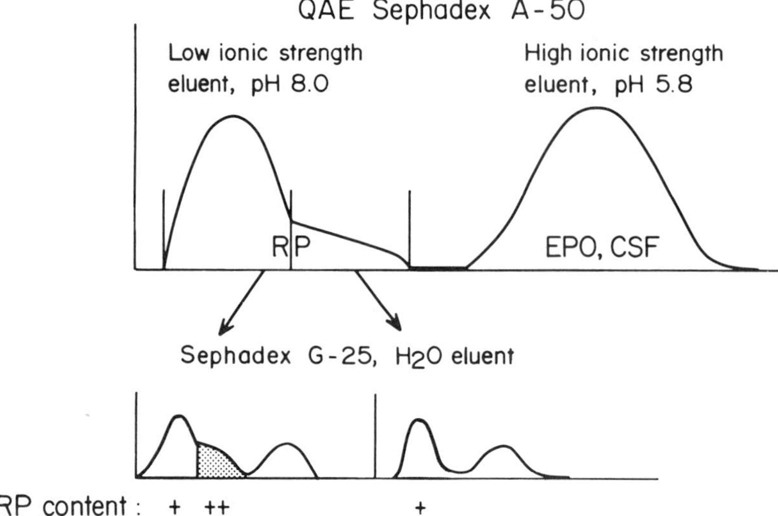

Fig. 1. Chromatographic separation of urinary protein.

This resulted in the appearance of a major protein peak, followed by a second smaller peak of trailing material. The first peak contained more than 90% of the total RP activity recovered, the remainder of the RP was found in the second peak. Proteins which had remained bound to the QAE Sephadex were subsequently eluted with 0.25 M sodium phosphate buffer – 1 M NaCl, pH 5.8. These included EPO and granulocyte-macrophage colony stimulating factors (GM-CSF) active on mouse cells.

Chromatography of the first peak material on Sephadex G-25 with H_2O as the developing agent resulted in the appearance of a significant portion of the input protein in fractions trailing the excluded peak. Fractions containing unretarded and retarded material, respectively, were pooled. It was found that the retarded protein contained most of the RP from the first QAE Sephadex peak. All other proteins recovered from QAE Sephadex were fully excluded by Sephadex G-25.

The human urinary EPO used in this study was kindly provided by the National Heart, Lung, and Blood Institute (preparation OPR 148.1 U/mg protein). It had been collected and concentrated by the Department of Physiology, University of the Northeast, Corrientes, Argentina, and was further processed, including the removal of the RP fraction, in our laboratory under NIH Grant HL 10880. The *in vivo* potency of EPO was estimated by the exhypoxic polycythemic mouse assay (Cotes and Bangham, 1961) as modified in our laboratory (Dukes et al., 1969).

Mouse marrow cell plasma clot cultures were performed based on the procedure of McLeod et al. (1976) as adapted in our laboratory. Suspensions of marrow cells from seven-week-old female C57BL/6 mice were prepared in supplemented Eagles' minimal essential medium with Hanks' salt mixture (H MEM) containing 2% fetal calf serum (FCS). For each ml of final clot volume desired, the following were combined: 0.1 ml L-asparagine (2 mg/ml)-$CaCl_2$(280 μg/ml) in NCTC 109 medium, 0.2 ml FCS, 0.1 ml 10% bovine serum albumin (detoxified with BioRad AG-501-X8 treatment) in H MEM, 0.1 ml diluting medium for optional additions (one part H MEM, three parts NCTC 109 medium), 0.1 ml cells in H MEM (4×10^6 nucleated cells/ml), 0.3 ml NCTC 109 medium containing 100 U Penicillin–100 μg Streptomycin/ml and 0.1 ml bovine citrated plasma (BCP). Human urinary EPO and RP were added as indicated. The mixture was plated in 0.5 ml aliquots into 2 ml wells of plastic Disposo trays (Flow Laboratories). The clots were incubated at 37°C for 7 days in a fully humidified atmosphere of 5% CO_2 in air. At the end of the incubation period, the clots were scooped out and fixed with 5% glutaraldehyde on glass slides. Megakaryocytes were identified by acetylcholinesterase stain (Jackson, 1973) and morphology. Closely spaced groups of two or more megakaryocytes were scored as colonies. Student's t-test was used to determine the significance of differences between means.

Results and Discussion

It was found that when H MEM was substituted for FCS the system became much more sensitive to EPO (Table 1). However, this held true only when BCP was present; the increase in sensitivity was abolished when mouse plasma was used instead of BCP in the FCS-free system.

The effect of adding RP to the system with increased sensitivity was tested (Table 2). RP derived from the urine of anemic patients caused megakaryocyte colony formation in a dose dependent fashion. RP from normal urine showed comparatively less activity than that from the patient's urine. It seemed that the population of cells which forms colonies in response to RP was not identical with that stimulated by EPO since in two experiments with an almost threefold different response to 1.5 U EPO, the responses to the RP preparations remained constant. A difference

Table 1. Effect of Substitution of 20% FCS with H MEM on Number of Megakaryocyte Colonies Formed per Clot in Response to 1.5 U EPO.

	Mean ± SE (4 replicates)		
Exp.	**FCS – BCP**	**H MEM – BCP**	**% Difference**
1	7 ± 1	29 ± 4	+ 314
2	12 ± 3	55 ± 6	+ 358
3	5 ± 1	20 ± 5	+ 300
4	5 ± 1	24 ± 8	+ 380

Mean difference: + 338. $0.05 > p > 0.02$ (paired t-test).

Table 2. Effect of RP on Megakaryocyte Colony Formation.

	Colonies/clot Mean ± SE (3-8 replicates)	RP dose			
		10 μg	20 μg	40 μg	80 μg
Exp. 1					
1.5 U EPO	27 ± 3				
RP (H7-A-7)		7 ± 2	10 ± 1	12 ± 2	–
RP (4-53-3-13)		–	6 ± 1	8 ± 1	12 ± 2
Exp. 2					
1.5 U EPO	81 ± 9				
RP (H7-A-7)		–	10 ± 5	12 ± 2	–
RP (4-53-3-13)		–	4 ± 1	10 ± 2	–
Exp. 3					
0.75 U EPO	47 ± 11				
RP (PD1-3-11)		–	2 ± 1	2 ± 1	–

Controls without EPO or RP contained < 1 colony/clot.

between target cells is further supported by the observation that the presence of RP in the system reduced the effect of large doses of EPO (Table 3). The number of colonies elicited by small doses of EPO (<0.75 U) was not diminished by RP (data not shown). These data suggest that a human urinary protein fraction, different from EPO, has megakaryocyte colony stimulating activity. It remains to be determined whether this activity, the erythroid burst promoting activity, and the CFU-S protecting activity found in the same RP protein fraction (Dukes, et al., 1979; Dukes, 1980) are properties of the same or of different molecules.

Summary

Mouse bone marrow cells were cultured in a plasma clot system containing no fetal calf serum. Megakaryocyte colonies were scored after seven days incubation. Human urinary erythropoietin (EPO) was used as the standard stimulator of colony formation. Urinary proteins were separated by chromatography on QAE-Sephadex, by a procedure which has been shown previously to concentrate erythroid burst promoting activity in a nonretained fraction, code named RP (EPO was retained). An RP preparation from a normal individual, which contained no detectable EPO activity, stimulated megakaryocyte colonies at 20 μg/clot (5 ± 2, mean ± SE of seven experiments). Higher doses caused no further increase in colony formation. Addition of 20 μg of this RP to increasing doses of EPO augmented the number of colonies only at low EPO concentration and had no effect or decreased the number of colonies at higher EPO levels. Data from a representative experiment were as follows (Mean ± SE of quadruplicates):

Table 3. Effect of Simultaneous Addition of EPO and RP.

	Addition	Colonies/clot Mean ± SE	Difference from EPO Control	
Exp. 1				
	1.5 U EPO	27 ± 3		
	40 μg RP (H7-A-7)	12 ± 2		
	1.5 U EPO + 40 μg RP (H7-A-7)	12 ± 1	−56%	$0.02 > p > 0.01$
	40 μg RP (4-53-3-13)	8 ± 1		
	1.5 U EPO + 40 μg RP (4-53-3-13)	20 ± 3	−26%	NS
Exp. 2				
	1.5 U EPO	81 ± 9		
	20 μg RP (H7-A-7)	10 ± 5		
	1.5 U EPO + 20 μg RP (H7-A-7)	41 ± 6	−49%	$p < 0.01$
	20 μg RP (4-53-3-13)	4 ± 1		
	1.5 U EPO + 20 μg RP (4-53-3-13)	70 ± 15	−14%	NS
Exp. 3				
	0.75 U EPO	47 ± 11		
	20 μg RP (PD1-3-11)	2 ± 1		
	0.75 U EPO + 20 μg RP (PD1-3-11)	11 ± 3	−77%	$0.02 > p > 0.01$

	No EPO	0.37 U EPO	0.75 U EPO	1.5 U EPO
No RP	0	5 ± 1	14 ± 3	28 ± 7
+ 20 μg RP	2 ± 1	14 ± 3	18 ± 4	7 ± 1

RP preparations from the urine of patients with iron deficiency or aplastic anemia were also found to be active. They contained more stimulatory activity per mg, but otherwise behaved like "normal" RP in this system.

ACKNOWLEDGMENTS

We thank P. Cheung, K. Harris, and C. Polk for expert technical assistance and C. Bogdan for skillful secretarial assistance. This work was supported in part by USPHS Grants HL 24629, HL 10880, and AM 26500.

References

Cotes, P. M. and D. R. Bangham. 1961. Bio-assay of erythropoietin in mice made polycythaemic by exposure to air at reduced pressure. *Nature*, 191:1065–1067.

Dukes, P. P. 1980. A urinary protein factor which increases CFU-S in bone marrow cultures. *Exp. Hematol.* 8 (Suppl 7):40.

Dukes, P. P., D. Hammond, and N. A. Shore. 1969. Comparison of erythropoietin preparations yielding different dose-response slopes in the exhypoxic polycythemic mouse assay. *J. Lab. Clin. Med.* 74:250–256.

Dukes, P. P., A. Ma, D. Meytes, J. A. Ortega, and N. A. Shore. 1979. Erythroid burst promoting activity in the urine of anemic patients. *Conference on "Aplastic Anemia – A Stem Cell Disease"*. San Francisco. In press.

Espada, J., and A. Gutnisky. 1970. A new method for concentration of erythropoietin from human urine. *Biochem. Med.* 3:475–484.

Jackson, C. W. 1973. Cholinesterase as a possible marker for early cells of the megakaryocytic series. *Blood* 42:413–421.

McLeod, D. L., M. M. Shreeve, and A. A. Axelrad. 1976. Induction of megakaryocyte colonies with platelet formation *in vitro*. *Nature* 261:492–494.

Discussion

Hoffman: Do you believe that erythropoietin is an important regulator in the proliferation and differentiation of the CFU-M? These data would suport that.

Dukes: It is my impression that erythropoietin does stimulate megakaryocyte colony formation. It may not be an obligatory factor and it is not likely to me that it would be important.

Hoffman: Is erythropoietin an important factor in megakaryocyte differentiation *in vivo*?

Dukes: I would guess that erythropoietin cross-reacts with other stimulators that have receptors on CFU-M.

Hoffman: Is there any immunological similarity between erythropoietin and a megakaryocytic-promoting factor?

Dukes: Not that I know of.

Evatt: In an hypoxic animal, platelet production is not stimulated. That does not mean that erythropoietin and thrombopoietin do not have similarities and certainly they co-purify with many biochemical techniques. Other hormones also have cross-reactivity at high enough doses. One and half units of erythropoietin per clot culture is a high dose, probably a much higher dose than that present *in vivo*.

Hoffman: It is true that all of these *in vitro* systems are somewhat artificial when dealing with the levels of a specific stimulator, but we have in our clinic, for instance, many patients who have secondary erythrocytosis. They have astronomically high levels of erythropoietin and associated erythrocytosis. These patients do not have elevated platelet counts. The polycythemia vera study group used elevated platelet levels as a major criterion to distinguish this myeloproliferative disorder from secondary erythrocytosis. I would question the relevance of the erythropoietin data to the *in vivo* situation.

Dukes: *In vivo* the spleen may have some normal inhibitory role on the number of platelets in the circulation, especially in situations where you have enlarged spleens.

Jackson: Does the dose of the urinary fraction (RP) inhibit or enhance BFU-C formation?

Dukes: These are the same doses that enhance erythroid colony formation, by preincubation and then challenging the cultures with erythropoietin. Addition of both substances simultaneously diminishes the result.

Published 1981 by Elsevier North Holland, Inc.
Evatt, Levine, and Williams, Editors
MEGAKARYOCYTE BIOLOGY AND PRECURSORS:
IN VITRO CLONING AND CELLULAR PROPERTIES

Megakaryocyte Progenitor Cells in Nude Mice Transplanted with Colony-Stimulating Factor-Producing Tumor Tissue

Hideaki Mizoguchi, Yasusada Miura, Toshio Suda, Kazuo Kubota, and Fumimaro Takaku

Division of Hemopoiesis, Institute of Hematology, and First Department of Medicine, Jichi Medical School, Tochigi, Japan

Extreme granulocytosis in mice bearing certain transplantable tumors was first studied by Bateman (1951), then by others (Asano et al., 1977; Burlington et al., 1977; Reincke et al., 1978). It has been demonstrated that colony-stimulating factor (CSF) is produced by tumor tissue.

Recently, we experienced a patient with a squamous cell carcinoma of the lung who developed marked granulocytosis. This tumor was successfully transplanted into nude mice of the BALB/c strain. These mice developed marked granulocytosis as the transplanted tumor grew.

It was confirmed that CSF was produced by the transplanted tumor, resulting in marked granulocytosis in the nude mice. The present study was designed to investigate the changes that occurred in the hemopoietic progenitor cells of these nude mice.

Methods and Materials

Pieces of tissue taken from a tumor removed from a 59-year-old female with lung cancer and granulocytosis ($9 \times 10^4/mm^3$) were transplanted into nude mice. Ten weeks after the transplantation, the mice showed marked granulocytosis of more than $6 \times 10^5/mm^3$ and consisting mainly of mature neutrophils. These mice are referred to as G-mice. It was determined that the conditioned medium of the tumor cells and the homogenate of the tumor tissue contained CSF for both human and mouse bone marrow cells.

After the G-mice were sacrificed, CFU-S, CFU-C, CFU-E, and CFU-M in the femoral marrow and in the spleen were assayed. CFU-S were assayed using the spleen-colony assay method reported by Till and McCulloch (1961). CFU-C were determined by the method of Robinson et al. (1967), CFU-E were assayed using the plasma culture system of McLeod et al. (1974) and CFU-M were assayed by the plasma culture method reported by us (Mizoguchi et al., 1979).

Results and Discussion

The mean spleen weight of the G-mice was about 13 times as much as that of the control nude mice. On the other hand, the mean number of nucleated cells in the femur of the G-mice did not show any significant increase as compared with that of the control mice (Table 1).

The tumor transplantation markedly increased the total number of CFU-C and CFU-S in the spleen, although there was no difference between the total number of CFU-C and CFU-S in the femurs of the G-mice and those of the control mice. The total number of CFU-E and CFU-M in the spleens of the G-mice increased markedly in comparison with those of the control mice. However, the total number of CFU-E and CFU-M in the femurs of the G-mice decreased markedly.

When the total number of all the hemopoietic progenitor cells in the whole body was calculated on the assumption that one femoral shaft was 8.5% of total marrow, the CFU-C, CFU-E, CFU-M, and CFU-S of the G-mice were found to have increased to 7.1 times, 8.1 times, 6.0 times and 38.7 times as much as those of control mice, respectively.

Several possible explanations can be put forward to account for these results. One is that the CSF continuously produced by the transplanted tumor induced replication of CFU-C which may then have stimulated proliferation of CFU-S, resulting in an increase in the inflow of CFU-S into CFU-E and CFU-M. Since Metcalf reported recently that pure CSF had direct effects on CFU-mix, the latter perhaps identical to CFU-S, and induced their differentiation (Metcalf et al., 1980), there is a possibility that the CSF produced by our tumor worked directly on CFU-S.

The other possibility is that the tumor produced not only CSF but also stimulators for CFU-E or CFU-M. When the marrow cells of BALB/c mice were cultured with the sera of the G-mice, however, only granulocyte-macrophage colonies were formed. No erythroid or megakaryocyte colonies were formed by the stimulation of the sera from the G-mice. Moreover, while there was an increase in CFU-E or CFU-M, there was no comparable increase in RBC or platelets in the peripheral blood of the G-mice (Table 2). These results may indicate that stimulators for CFU-E or CFU-M were not produced in the G-mice.

It still remains unclear why the hemopoietic progenitors accumulated in the spleens but not in the marrows of the G-mice. Similar findings on CFU-C and CFU-S were observed by Reincke et al. (1971) in mice transplanted with granulocytosis-inducing tumor tissue (mammary carcinoma of mice).

The number of progenitor cells per organ depends on their self-reproduction, the influx rate from more immature progenitors, the rate of differentiation, emigration, and immigration. It is precisely because there are so many variables that the mechanisms which lead to the accumulation of progenitor cells have not been clarified. One of the possibilities is that the cycling fraction of the progenitor cells increased

Table 1. Spleen Weight, Number of Nucleated Cells in the Femur and Number of Hemopoietic Progenitor Cells.

	G-mice (n = 6)		Control nude mice (n = 6)		
	$\bar{x}$	sd	$\bar{x}$	sd	p value
Spleen weight, mg	1,456	± 31	112	± 25	$p < 0.001$
Number of nucleated cells per femur, × 10^6	15.3	± 5.4	13.2	± 1.9	ns[b]
CFU-C					
per femur, × 10^4	2.26	± 1.29	2.36	± 1.07	ns[b]
per spleen,[a] × 10^4	172	± 115	0.42	± 0.29	$p < 0.01$
CFU-E					
per femur, × 10^3	0.36	± 0.23	30.3	± 11.7	$p < 0.01$
per spleen,[a] × 10^4	325	± 128	4.39	± 1.68	$p < 0.01$
CFU-M					
per femur, × 10^3	1.55	± 0.87	5.98	± 0.70	$p < 0.01$
per spleen,[a] × 10^4	41.53	± 16.61	0.14	± 0.05	$p < 0.01$
CFU-S					
per femur, × 10^3	1.81	± 1.64	4.05	± 1.50	ns[b]
per spleen,[a] × 10^4	222	± 156	1.03	± 0.96	$p < 0.02$

[a]The total number of CFU-C, CFU-E, CFU-M, and CFU-S in the spleen was calculated by multiplying the concentration of hemopoietic progenitor cells by the spleen weight, assuming that 1 mg spleen contains 1.5 × 10^6 nucleated cells.

[b]ns: not significant.

Table 2. Peripheral White Blood Cell Counts, Red Blood Cell Counts and Platelet Counts.

	G-mice (n = 6)		Control nude mice (n = 6)		
	$\bar{x}$	sd	$\bar{x}$	sd	p value
WBC, $\times$ $10^3/\mu l$	977 ±	290	13 ±	5	$p < 0.001$
RBC, $\times$ $10^3/\mu l$	7400 ±	1830	9480 ±	700	ns[a]
Platelet, $\times$ $10^3/\mu l$	1400 ±	460	1220 ±	200	ns[a]

[a] ns: not significant.

more in the spleens of the G-mice than in their marrows. In preliminary experiments, however, no significant deviations from normal were noted in the DNA-synthesizing fraction of any of the hemopoietic progenitors. In order to clarify the role played by the spleen in the G-mice, similar studies using splenectomized G-mice are in progress.

Summary

A 59-year-old female with a lung cancer and leukocytosis (90,000/mm^3) was operated and pieces of tumor tissue were transplanted to nude mice. Ten weeks after the transplantation, the mice showed marked leukocytosis up to 6 $\times$ $10^5/mm^3$ consisting mainly of mature neutrophils. It was determined that the conditioned medium of the tumor cells and the homogenate of the tumor tissue contained colony-stimulating factor (CSF) for both human and mouse bone marrow cells. After the nude mice with leukocytosis (more than 6 $\times$ $10^5/mm^3$) induced by the transplantation of the tumor tissue (G-mice) were killed, CFU-S, CFU-C, CFU-E, and CFU-M in the femoral marrow and in the spleen were assayed. As a result, the average spleen weight and the number of nucleated cells in a femur increased to 13$\times$ and 1.5$\times$ those of control nude mice, respectively. The number of any progenitor cells in the marrow decreased but that in the spleen markedly increased. The total number of CFU-S, CFU-C, CFU-E, and CFU-M in the whole body of the G-mice increased to 38.7$\times$, 7.1$\times$, 8.1$\times$, and 6.0$\times$ that of control nude mice, respectively. The serum of G-mice had only CSF. It had stimulatory activity neither on CFU-E nor on CFU-M. From these findings, it could be considered that CFU-M was increased by the proliferation of CFU-S. CSF may have stimulated proliferation and differentiation of CFU-C resulting in the proliferation of CFU-S. Further it was shown that no competitive proliferation occurred between CFU-M and other progenitor cells such as CFU-E and CFU-C.

References

Asano, S., A. Urabe, T. Okabe, N. Sato, Y. Kondo, Y. Ueyama, S. Chiba, N. Ohsawa, and K. Kosaka. 1977. Demonstration of granulopoietic factor(s) in the plasma of nude mice transplanted with a human lung cancer and in the tumor tissue. *Blood* 49:845–852.

Bateman, J. C. 1951. Leukemoid reactions to transplant mouse tumors. *J. Nat. Cancer Inst.* 11:671–687.

Burlington, H., E. P. Cronkite, J. A. Laissue, U. Reincke, and R. K. Shadduck. 1977. Colony-stimulating activity in cultures of granulocytosis-inducing tumor. *Proc. Soc. Exp. Biol. Med.* 154:86–92.

McLeod, D. L., M. M. Shreeve, and A. A. Axelrad. 1974. Improved plasma culture system for production of erythrocytic colonies *in vitro*: Quantitative assay method for CFU-E. *Blood* 44:517–534.

Metcalf, D., G. R. Johnson, and A. W. Burgess. 1980. Direct stimulation by purified GM-CSF of the proliferation of multipotential and erythroid precursor cells. *Blood* 55:138–147.

Mizoguchi, H., K. Kubota, Y. Miura, and F. Takaku. 1979. An improved plasma culture system for the production of megakaryocytes colonies *in vitro*. *Exp. Hematol.* 7:345–351.

Reincke, U., H. Burlington, A. L. Carsten, E. P. Cronkite, and J. A. Laissue. 1978. Hemopoietic effects in mice of a transplanted, granulocytosis-inducing tumor. *Exp. Hematol.* 6:421–430.

Robinson, W., D. Metcalf, and T. R. Bradley. 1967. Stimulation by normal and leukemic mouse sera of colony formation *in vitro* by mouse bone marrow cells. *J. Cell. Physiol.* 69:83–91.

Till, J. E., and E. A. McCulloch. 1961. A direct measurement of the radiation sensitivity of normal mouse bone marrow cells. *Rad. Res.* 14:213–222.

Discussion

Williams: Have you tested the effect *in vivo* of a lung carcinoma that is not CSF-producing?

Mizoguchi: Yes, several tumors, including other lung carcinomas, have been used and no increase in the progenitor cell compartments was observed.

Megakaryocyte Colonies

Published 1981 by Elsevier North Holland, Inc.
Evatt, Levine, and Williams, Editors
MEGAKARYOCYTE BIOLOGY AND PRECURSORS:
IN VITRO CLONING AND CELLULAR PROPERTIES

Megakaryocyte Progenitor Cells *In Vitro*

Neil Williams

Sloan-Kettering Institute for Cancer Research, Rye, New York

Observations of the *in vitro* growth of colonies of megakaryocytes from precursor cells are contributing significantly to the understanding of megakaryocytopoiesis. The work of Metcalf (1975) and Nakeff (1975, 1976) have been paramount in the development of this cell culture technique. A megakaryocyte precursor cell is defined as an undifferentiated cell capable of responding *in vitro* to stimuli to produce colonies comprised of megakaryocytes. Such cells can be induced to colony formation from both mouse and human tissues, in a variety of culture media (Metcalf et al., 1975; Nakeff and Daniels-McQueen, 1976; Vainchenker et al., 1979; Fauser and Messner, 1979). Megakaryocytes in murine colonies can be readily recognized by their cell size and by the presence of acetylcholinesterase, an enzyme specific for the megakaryocytic series in mouse hemopoietic tissue (Nakeff and Daniels-McQueen, 1976). The recognition of colonies of human megakaryocytes has proven more difficult, but specific markers are being developed which should allow easy discrimination of these cells (Rabellino et al., 1979; Mazur et al., 1980). The most commonly used term to describe the cell giving rise to megakaryocyte colonies is the colony-forming unit-megakaryocyte (CFU-M or CFU-MK) (Nakeff and Daniels-McQueen, 1976). This nomenclature is consistent with that used for the other hemopoietic cell lineages. The term does not imply that the assay is clonal, nor does it suggest that a homogeneous or a well-defined cell population is being monitored.

Characteristics and Distribution of Megakaryocyte Progenitor Cells in the Mouse

All stages of normal megakaryocytopoiesis have been observed in the *in vitro* culture system. Megakaryocyte progenitor cells have been shown to be diploid cells giving rise to colonies of 2–300 acetylcholinesterase cells (Metcalf et al., 1975; Nakeff, 1977; Williams and Jackson, 1978). The most mature colony megakaryocytes increase in area from days 4–7 of culture (Mizoguchi et al., 1979), and later release "platelet-like entities" (McLeod et al., 1976; Williams et al., 1978). These have the appearance of large platelets with dense granules distributed throughout the cytoplasm (Williams et al., 1978). No functional studies have been performed on these "platelets".

Some of the properties of megakaryocyte progenitor cells are outlined in Table 1. The progenitor cells have characteristics in common with precursor cell populations of other hemopoietic cell lineages and these immature cells are, therefore, thought to represent maturation stages between the pluripotent stem cell and the megakaryocyte. Most megakaryocyte colonies are comprised only of cells containing acetylcholinesterase. A small proportion of colonies, however, contain both megakaryocytes and neutrophils (Metcalf et al., 1975). Various investigators have employed different criteria for enumerating megakaryocyte colonies. Kalmaz and McDonald (1980) considered a single megakaryocyte to be a colony on the basis that a megakaryocyte in its less differentiated state can be classed as a precursor cell as it undergoes nuclear endoreduplication as well as cytoplasmic maturation. The cells giving rise to single megakaryocytes in culture appear to be small and

Table 1. Some Properties of Mouse Megakaryocyte Progenitor Cells.

Frequency in bone marrow	5-50/10^5 cells	Metcalf et al., 1975; Nakeff and Daniels-McQueen, 1976; Mizoguchi et al., 1979; Burnstein et al., 1979
Frequency in spleen	1-5/10^6 cells	Metcalf et al., 1975; Burnstein et al., 1979
Frequency in blood	8/ml	Burnstein et al., 1979
Frequency in 12-day fetal liver	300/10^5 cells	Williams (unpublished)
Buoyant Density	1.070-1.076 g cm^{-3}	Nakeff, 1977; Williams et al., 1980
Sedimentation rate	3.5-7.0 mm hr^{-1}	Metcalf et al., 1975; Burnstein et al., 1979
Adherence	nonadherent	Williams et al., 1979a
Fc receptors	negative	Williams et al., 1979b
Ia receptors	negative	Williams et al., 1979b
Probable ploidy	2N	Metcalf et al., 1975; Nakeff, 1977
Strain differences	Yes	Nakeff, 1977
Radiation sensitivity	D_O = 128 rads	Nakeff et al., 1979
Cycling Status *in vivo*	10-20% in DNA syn.	Nakeff and Bryan, 1978; Williams and Jackson, 1978; Burnstein et al., 1979
Cycling Status *in vitro*	~10% in DNA syn.	Burnstein et al., 1979
	30-50% in DNA syn.	Nakeff and Bryan, 1978; Williams and Jackson, 1978; Mizoguchi et al., 1978

acetylcholinesterase positive (Long and Williams, 1980). These cells differ from progenitor cells capable of cell division in their acquisition of acetylcholinesterase and in their physical properties (Nakeff and Floeh, 1976; Nakeff, 1977). Nakeff and Daniels-McQueen (1976) defined a colony as being four or more recognizable megakaryocytes. This number omits the possibility that coincidental megakaryocytes remaining in the dish over the culture period would be scored as colonies, but requires a progenitor cell to undergo at least two cell divisions prior to nuclear endoreduplication and cytoplasmic maturation.

The number of megakaryocyte progenitor cell subpopulations that can be induced to form colonies is uncertain. Morphologically, a large spectrum of cell and colony types can be grown, ranging from colonies comprised of only a few mature cells to very large colonies with few mature megakaryocytes and many small cells. Burstein et al. (1979) concluded that megakaryocyte progenitor cells comprised a single population based on the shape of their sedimentation velocity profile. Progenitor cells with increased sensitivity to high specific activity tritiated thymidine were found to sediment with a value theoretically equivalent to cells in the G_2 portion of the cell cycle. These findings are consistent with that of a single cycling cell population (Miller, 1973). This is illustrated in Figure 1. Data averaged from five experiments are shown on the upper graph as a linear plot to show the mean progenitor cell distribution, and on a semilog plot in the lower graph for comparison with the theoretical curve for a dividing cell population (Miller, 1973). It can be estimated on the basis of such theoretical considerations that approximately 30% of the clonable megakaryocyte progenitor cells are in DNA synthesis (Fig. 2, upper panel). These data are consistent with that obtained from incubating marrow cells *in vitro* in the presence of S phase-specific drugs (Nakeff, 1978; Williams and Jackson, 1978; Mizoguchi et al., 1979). These estimates are, however, quite incompatible with *in vivo* estimates of the proportion of megakaryocyte progenitor cells in DNA synthesis. Only 10–20% of all megakaryocyte progenitor cells have been found to be sensitive to cell cycle-specific drugs injected into mice (Nakeff and Byran, 1978; Williams and Jackson, 1978; Burstein et al., 1979).

How can these data be reconciled? It is unlikely that the drugs are metabolized *in vivo* since other control cell populations were sensitive to the same agents (Nakeff and Byran, 1978; Williams and Jackson, 1978). Similarly, the sedimentation profile cannot be explained by poor resolution of the separation technique. A single noncycling population of fetal monkey granulocyte-macrophage precursor cells is symmetrically distributed about a mean sedimentation rate; the spread about the peak value is determined by the resolution of the method (Williams and Moore, 1973). Two possible interpretations are shown in Figure 2. The upper panel shows the distribution as a single population of precursor cells in cell cycle. In order to accommodate the *in vivo* cell cycle estimates, the majority of these precursor cells would have to be prevented from progression through the cell cycle by some marrow cell or factor. The lower panel shows one possible distribution for two overlapping cell populations; one subclass of progenitor cells being a noncycling population of 4 mm hr^{-1} (population A); the second subclass being a cycling cell population with

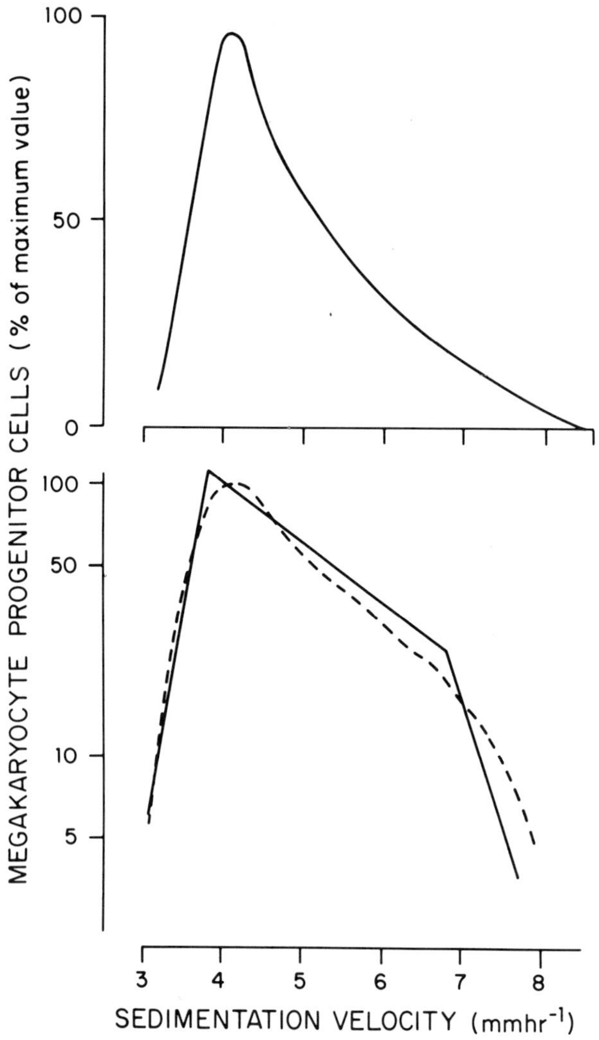

Fig. 1 Sedimentation rate distribution of megakaryocyte progenitor cells from mouse bone marrow. Top panel: The average profile from five separate experiments is shown. Each distribution was derived from 15–20 fractions and each fraction was assayed in triplicate. The data are expressed as a percentage of the peak value. Lower panel: The drawn line represents the theoretical distribution of an exponentially growing homogeneous cell population in which the G_1 cells sediment at a rate of 4.0 mm hr^{-1}. The curve is derived as outlined by Miller (1973). The dots show the plot presented in the upper panel.

a G_1 value of approximately 5 mm hr^{-1} (population B). This explanation is similar to that suggested for the pluripotent stem cell (Visser et al., 1977). Another possibility for two progenitor cell subclasses (not shown) is that population A is slowly cycling and that population B is noncycling (and has a sedimentation rate of .5 mm

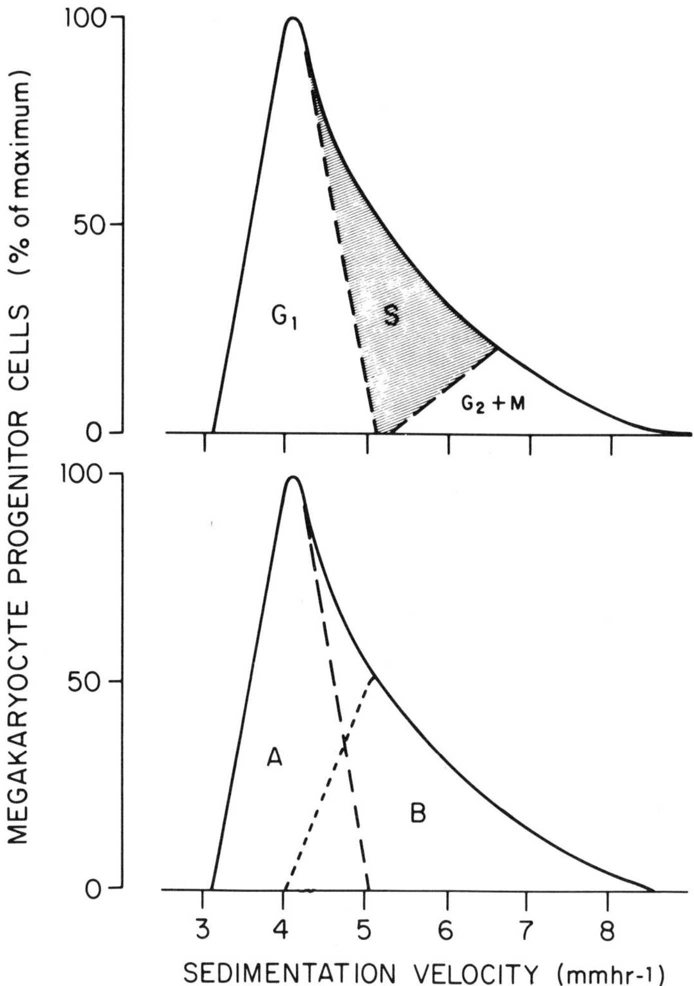

Fig. 2 Analysis of sedimentation rate distributions of megakaryocyte progenitor cells. The upper panel shows the conventional distribution of a single homogeneous population of cells separated by cell cycle status. The areas of G_1 and G_2 portions are governed by the resolution of the method (Miller, 1973). The lower panel shows two overlapping cell populations with A being a noncycling precursor cell pool and B representing a precursor cell population in active cell cycle.

hr^{-1}). Clearly, the spectrum of clonable progenitor cells and the cycle characteristics of each population warrants further investigation.

The Clonal Origin of Megakaryocyte Progenitor Cells

To date, the clonal origin of megakaryocyte progenitor cells remains unproven. In initial studies, the number of colonies detected was not found to be a linear function of the number of cultured cells (Metcalf et al., 1975; Williams et al., 1978; Mizoguchi et al., 1979). This suggests that cellular interactions within the cultures may augment cloning efficiency and leave open the possibility that colonies arise from aggregates. However, when culture conditions were made optimal by the addition of bone marrow conditioned medium, a linear and quantitative assay was obtained (Williams et al., 1978). Under these circumstances, the assay may be clonal, but formal proof of the clonal origin of mouse megakaryocytes by karyotypic analysis is necessary, especially for colonies containing both megakaryocytes and cells of other lineages of hemopoiesis (Metcalf et al., 1975; Vainchenker et al., 1979). Fauser and Messner (1979) have shown by mixing human cells from both sexes that the cells comprising mixed colonies and including megakaryocytes contained chromosomes of only one sex. However, colonies are not necessarily always of unicellular origin. Results with a high proportion (15–75%) of hemopoietic cell colonies derived from more than one precursor cell have been reported (Singer et al., 1979).

The Relationship Between Megakaryocyte Progenitor Cells and Pluripotent Hemopoietic Stem Cells

Megakaryocyte progenitor cells represent one or more differentiation states between the pluripotent hemopoietic stem cell and megakaryocytes. Based on several properties of the cells and on the resultant colony types, the progenitor cell appears to be closely associated with the *in vivo* hemopoietic stem cell (colony-forming unit-spleen, CFU-s). This cell class differs from all other clonable precursor cell populations in its extensive self-renewal capacity (Siminovitch et al., 1964). Both the CFU-s and the megakaryocyte progenitor cell populations (measured *in vivo*) are slowly cycling, and they are very similar in their physical properties (Becker et al., 1965; Worten et al., 1969; Metcalf et al., 1975; Nakeff, 1977; Nakeff and Byran, 1978; Williams and Jackson, 1978; Burstein et al., 1979). The colony types which comprise both *in vivo* and *in vitro* assays are also similar. Two classes of colonies are observed in both systems, one being a small group of mature cells, the second being foci of few mature cells and many small cells (Curry and Trentin, 1967; Williams and Jackson, 1978). The megakaryocyte progenitor cell that grows *in vitro* differs from its *in vivo* counterpart, however, in its irradiation sensitivity (Nakeff et al., 1979). These observations imply that the *in vitro* assay may detect a subpopulation of the pluripotent stem cell pool which is biased by the culture conditions towards megakaryocyte development. A second possibility is that the *in vitro* assay may monitor the cell that gives rise to the megakaryocytic spleen colonies and that it is committed and restricted primarily to megakaryocytopoiesis. Such stem cell populations have been reported for both the granulocytic and lymphocytic cell lineages (Abramson et al., 1977; Jones-Villeneuve and Phillips, 1980). A third possibility is that the colonies are derived from different precursor cells which differ in their degree of commitment, but that the cells are closely related.

Cells in murine marrow which *in vitro* form mixed colonies of erythroid cells, neutrophils, eosinophils, macrophages and megakaryocytes are thought to be derived from the stem cell compartment (Metcalf et al., 1979). If stem cells and megakaryocyte progenitor cells have a parent-progeny relationship as suggested in the third possibility outlined above, then it would be anticipated that the megakaryocyte colonies should be obtained when cells from the mixed colony cells are replated. In such experiments no megakaryocytic colonies were observed (Metcalf et al., 1979). Either this latter possibility is unlikely or the marrow colony-forming cell that gives rise to mixed colonies is restricted in its differentiation capacity and biased away from megakaryocytopoiesis.

Human Megakaryocyte Colony Growth

There are two reports describing the growth of human colonies which contain megakaryocytes (Vainchenker et al., 1979a,b; Fauser and Messner, 1979). They differ in the media, serum and stimuli used, as well as the numbers and types of colonies scored (Table 2). Fauser and Messner (1979) claimed that the differences between their findings and those of Vainchenker et al., (1979a,b) resulted from culture conditions influencing the promotion of the growth of certain hemopoietic cell lineages, rather than growing colonies of cells with different potentials for differentiation. This conclusion has been made for both mouse and human granulocyte-macrophage progenitor cells, and that compartment has been conclusively shown to be comprised of several cell populations (Johnson et al., 1977; Williams and Eger, 1978). Thus it is not obvious that the various systems do assay similar populations and further experience comparing and contrasting the colonies containing megakaryocytes is required.

Summary

Factors necessary for megakaryocyte colony formation and megakaryocyte maturation were obtained from WEHI-3 cell CM, adherent peritoneal cell supernatant

Table 2. Comparison of Conditions of Megakaryocyte Colony Growth From Human Bone Marrow.

	Vainchenker et al.	**Fauser and Messner**
Semisolid culture	Plasma clot	Methyl cellulose
Stimuli	Erythropoietin	PHA-leukocyte CM + erythropoietin
Serum	Human AB	Fetal calf serum
Numbers of mixed colonies containing megakaryocytes	$95/10^6$ cells	$1.5/10^6$ cells
Numbers of spontaneous colonies	$34/10^6$ cells	N.R.
Numbers of megakaryocyte colonies	$190/10^6$ cells	N.R.
Minimum colony size	2 cells	N.R.

N.R. = not reported.

and a human embryonic kidney cell line (*in vivo* Thrombopoietic Stimulating Factor; TSF). The WEHI-3 CM contained two essential factors (MK-CSF and MK potentiator) for megakaryocyte colony formation, one of which (MK potentiator) is always present in low concentrations. Neither MK-CSF or MK potentiator influenced the number of single megakaryocyte in bone marrow cultures, indicating that these entities did not influence the maturation of the clonable precursor cells into single megakaryocytes. Colony megakaryocytes were influenced differentially by the two factors. Megakaryocytes of higher ploidy values were obtained when saturating levels of MK potentiator were used (36%, 16N; 11%, 64N) compared to the ploidy of megakaryocytes grown in suboptimal levels of the MK potentiator (25%, 16N; 3%, 64N). The concentration of MK potentiator did not influence the rate of appearance of colonies or colony size. The factors studied also differentially influenced the cell intermediate between the CFU-MK and the megakaryocyte, the small acetylcholinesterase positive cell. Isolated small acetylcholinesterase cells formed single megakaryocytes by day 3 of culture in the presence of MK potentiator or TSF, but were not influenced by MK-CSF. A differentiation model is postulated where maturing cells differ in the degree of sensitivity to stimuli, with MK-CSF primarily stimulating cell division of precursor cells, and MK potentiator and TSF influencing the development and maturation of MK from the small acetylcholinesterase cell.

ACKNOWLEDGMENTS
N. Williams is a Leukemia Society of America Scholar. This work was funded by Grants HL 22451 from the National Institutes of Health, CA 08748 from the National Cancer Institute and the Gar Reichman Foundation.

References

Abramson, S., R. G. Miller, and R. A. Phillips. 1977. The identification in adult bone marrow of pluripotent and restricted stem cells of the myeloid and lymphoid systems. *J. Exp. Med.* 145:1567–1579.

Becker, A. J., E. A. McCulloch, L. Siminovitch, and J. E. Till. 1965. The effect of differing demand for blood cell production on DNA synthesis by haemopoietic colony-forming cells of mice. *Blood* 26:296–308.

Burstein, S., J. W. Adamson, D. Thorning, and L. A. Harker. 1979. Characteristics of murine megakaryocyte colonies *in vitro*. *Blood* 54:169–179.

Curry, J. L. and J. J. Trentin. 1967. Hemopoietic spleen colony studies. I. Growth and differentiation. *Devel. Biol.* 15:395–413.

Fauser, A. A. and H. A. Messner. 1979. Identification of megakaryocytes, macrophages and eosinophils in colonies of human bone marrow containing neutrophilic granulocytes and erythroblasts. *Blood* 53:1023–1027.

Johnson, G. R., C. Dresch, and D. Metcalf. 1977. Heterogeneity in human neutrophil, macrophage and eosinophil progenitor cells demonstrated by velocity sedimentation separation. *Blood* 50:823–832.

Jones-Villeneuve, E. and R. A. Phillips. 1980. Potentials for lymphoid differentiation by cells from long term cultures of bone marrow. *Exp. Hematol.* 8:65–76.

Kalmaz, G. D. and T. P. McDonald. 1980. The effects of thrombopoietin on megakaryocytopoiesis of mouse bone marrow cells *in vitro*. In *Megakaryocytes In Vitro*. Evatt, B. L., Levine, R. F., and Williams, N., eds. New York: Elsevier North Holland.

Long, M. W. and N. Williams. 1980. Relationship of small acetylcholinesterase positive cells to megakaryocytes and clonable megakaryocytic progenitor cells. In *Megakaryocytes In Vitro*. Evatt, B. L., Levine, R. F., and Williams, N., eds. New York: Elsevier North Holland.

McLeod, D. L., M. M. Shreeve, and A. A. Axelrad. 1976. Induction of megakaryocyte colonies with platelet formation *in vitro*. *Nature* 261:492–493.

Mazur, E., R. Hoffman, E. Bruno, S. Marchesi, and J. Chasis. 1980. Identification of two classes of human megakaryocyte progenitor cells. In *Megakaryocytes In Vitro*.

Metcalf, D., C. R. Johnson, and T. E. Mandel. 1979. Colony formation in agar by multipotential hemopoietic cells. *J. Cell. Physiol.* 98:401–420.

Metcalf, D., H. R. MacDonald, N. Odartchenko, and B. Sordat. 1975. Growth of mouse megakaryocyte colonies *in vitro*. *Proc. Nat. Acad. Sci. USA* 72:1744–1748.

Miller, R. G. 1973. Separation of cells by velocity sedimentation. In *New Techniques in Biophysics and Cell Biology* (Vol. I). Pain, R. H. and Smith, B. J., eds. London: John Wiley and Sons. pp. 87–112.

Mizoguchi, H., K. Kubota, Y. Miura, and F. Takaku. 1979. An improved plasma culture system for the production of megakaryocyte colonies *in vitro*. *Exp. Hematol.* 7:345–351.

Nakeff, A. 1977. Colony-forming unit, megakaryocyte (CFU-M): Its use in elucidating the kinetics and humoral control of the megakaryocytic committed progenitor cell compartment. In *Experimental Hematology Today*. Baum, S. J., and Ledney, D. G., eds. New York: Springer-Verlag. pp. 111–123.

Nakeff, A. and J. E. Byran. Megakaryocyte proliferation and its regulation as revealed by CFU-M analysis. In *Hematopoietic Cell Differentiation*. Golde, D. W., Cline, M. J., Metcalf, D., and Fox, C., eds. New York: Academic Press. pp. 241–259.

Nakeff, A. and S. Daniels-McQueen. 1976. *In vitro* colony assay for a new class of megakaryocyte precursor: Colony-forming unit megakaryocyte (CFU-M). *Proc. Soc. Exp. Biol. Med.* 151:587–590.

Nakeff, A. and D. P. Floeh. 1976. Separation of megakaryocytes from mouse bone marrow by density gradient centrifugation. *Blood* 48:133–138.

Nakeff, A., K. A. Dicke, and M. J. van Noord. 1975. Megakaryocytes in agar cultures of mouse bone marrow. *Ser. Haematol.* 8:1–21.

Nakeff, A., W. L. McLellan, J. Byran, and F. A. Valeriote. 1979. Response of megakaryocyte, erythroid, and granulocyte-macrophage progenitor cells in mouse bone marrow to gamma-irradiation and cyclophosphamide. In *Experimental Hematology Today*. Baum, S. J., and Ledney, D. G., eds. New York: Springer-Verlag. pp. 99–104.

Rabellino, E. M., R. L. Nachman, N. Williams, R. J. Winchester, and G. D. Ross. 1979. Human megakaryocytes. I. Characterization of the membrane and cytoplasmic components of isolated marrow megakaryocytes. *J. Exp. Med.* 149:1273–1287.

Siminovitch, L., J. E. Till, and E. A. McCulloch. 1963. The distribution of colony-forming cells among spleen colonies. *J. Cell. Comp. Physiol.* 63:327–336.

Singer, J. W., P. J. Fialkow, L. W. Dow, C. Ernst, and L. Steinmann. 1979. Unicellular or multicellular origin of human granulocyte-macrophage colonies *in vitro*. *Blood* 54:1395–1399.

Vainchenker, W., J. Guichard, and J. Breton-Gorius. 1979a. Growth of human megakaryocyte colonies in culture from fetal, neonatal and adult peripheral blood cells. Ultrastructural analysis. *Blood Cells* 5:25–42.

Vainchenker, N., J. Bouquet, J. Guichard, and J. Breton-Gorius. 1979b. Megakaryocyte colony formation from human bone marrow precursors. *Blood* 54:940–945.

Visser, J., G. van den Engh, N. Williams, and D. Mulder. 1977. Physical separation of the cycling and noncycling compartments of murine hemopoietic stem cells. In *Experimental Hematology Today*. Baum, S. J., and Ledney, D. G., eds. New York: Springer-Verlag. pp. 21–27.

Williams, N. and R. R. Eger. 1978. Purification and characterization of clonable murine granulocyte-macrophage precursor cell populations. In *Hematopoietic Cell Differentiation*. Golde, D. W., Cline, M. J., Metcalf, D., and Fox, C., eds. New York: Academic Press. pp. 385–398.

Williams, N. and H. M. Jackson. 1978. Regulation of the proliferation of murine megakaryocyte progenitor cells by cell cycle. *Blood* 52:163–170.

Williams, N. and M. A. S. Moore. 1973. Sedimentation velocity characterization of the cell cycle of granulocytic progenitor cells in monkey hemopoietic tissue. *J. Cell. Physiol.* 82:81–92.

Williams, N., H. Jackson, and P. Meyers. 1979a. Isolation of pluripotent hemopoietic stem cells and clonable precursor cells of erythrocytes, granulocytes, macrophages and megakaryocytes from mouse bone marrow. *Exp. Hematol.* 7:524–534.

Williams, N., T. P. McDonald, and E. M. Rabellino. 1979b. Maturation and regulation of megakaryocytopoiesis. *Blood Cells* 5:43–55.

Williams, N., H. Jackson, A. P. C. Sheridan, M. J. Murphy, Jr., A. Elste, and M. A. S. Moore. 1978. Regulation of megakaryocytopoiesis in long term murine bone marrow cultures. *Blood* 51:245–255.

Worton, R. A., E. A. McCulloch, and J. E. Till. 1969. Physical separation of hemopoietic stem cells from cells forming colonies in culture. *J. Cell. Physiol.* 74:171–182.

Published 1981 by Elsevier North Holland, Inc.
Evatt, Levine, and Williams, Editors
MEGAKARYOCYTE BIOLOGY AND PRECURSORS:
IN VITRO CLONING AND CELLULAR PROPERTIES

Examination of Culture Conditions for Optimizing the Growth of Megakaryocyte Colonies

Alexander Nakeff, Ph.D.

Section of Cancer Biology, Division of Radiation Oncology, Mallinckrodt Institute of Radiology, Washington University School of Medicine, St. Louis, Missouri

The mitotic activity associated with megakaryocyte proliferation resides in a class of progenitor cells (CFU-M) which are detected by their ability to form colonies of megakaryocytes in either plasma (Nakeff and Daniels-McQueen, 1976; McLeod et al., 1976; Mizoguchi et al., 1979; Vainchenker et al., 1979) or agar (Metcalf et al., 1975; Williams et al., 1978; Burstein et al., 1979) culture systems. Since the application of clonogenic cell culture techniques to the study of megakaryocyte proliferation is relatively new, culture conditions remain to be defined further, particularly in light of the low incidence of CFU-M in marrow (approximately 1 CFU-M/10^4 bone marrow cells), the small average colony size (<50 megakaryocytes/colony), the presence of numerous single megakaryocytes among colonies and the stimulation of CFU-M by prerequisite conditioned media (CM) that are relatively non-specific.

It remains to be determined whether these properites of CFU-M reflect the limited proliferative capacity of megakaryocytic progenitors *in vivo* or less than optimal culture conditions. We have approached this question by summarizing published data from a number of different laboratories on the culture of CFU-M, then analyzing some of these variations in our own laboratory, including the age and sex of mice as marrow donors, plasma versus agar as semi-solid media, horse serum lot variations and lastly, lectin-stimulated mouse spleen cell CM in terms of different mitogens, spleen cell sub-populations, lengths of culture and storage time and CM batch variations.

Culture Conditions for CFU-M

In vitro growth of megakaryocyte colonies from mouse and, more recently, human hematopoietic tissue has utilized two quite different types of semi-solid supporting media, namely, agar and plasma as described originally by Metcalf et al. (1975) and Nakeff and Daniels-McQueen (1976), respectively. In addition, there are numerous differences in other aspects of the culture conditions which make comparisons among various techniques difficult. This problem is compounded as culture conditions have become more diverse with subsequent modifications being made by each new laboratory establishing a CFU-M assay.

A comparison of the various techniques used to culture CFU-M is presented in Table 1 and clearly illustrates numerous, potentially important differences. Included are those involving different mouse strains from which bone marrow is obtained for CFU-M culture which are known to influence CFU-M yield under identical culture conditions (Nakeff, 1977). In addition, five different culture media have been used satisfactorily with different sources (fetal calf, horse, and human AB) and concentrations of serum for supporting CFU-M growth.

The presence of a stimulatory substance(s) is a prerequisite for colony formation; in this respect, conditioned media have been obtained from 2-mercaptoethanol (2-ME) or pokeweed mitogen (PWM)-stimulated mouse spleen cell cultures, as well as a mixture obtained from long-term cultures of mouse bone marrow cells and a myelomonocytic leukemia (WEHI-3) cell line. Human urinary (NIH) and sheep (Connaught) erythropoietin (Ep) have also been used successfully. Generally, the number of cells plated in primary culture is either 7.5×10^4 or 1×10^5 cells except for human bone marrow where relatively higher numbers (5×10^5) must be used. Under these conditions, mouse bone marrow CFU-M normalized to 10^5 nucleated cells vary almost ten-fold, from 6 to 50 CFU-M/10^5 cells plated. The fraction of CFU-M in human bone marrow stimulated by Ep is a factor of 2 to 3-fold less than that in mouse marrow, which may reflect differences in culture technique rather than physiological differences. Interestingly, Burstein et al., (1979) have shown that under the same culture conditions, spleen and blood contain 3 to 5-fold proportionately less CFU-M than that assayed in the bone marrow of C57B1 mice; this has been observed also for CFU-M in human peripheral blood (Vainchenker et al., 1979). It appears from these observations that CFU-M are present in all hematopoietically-active tissues in the mouse and circulate in the peripheral blood to seed from one organ site to the other.

In addition to the above, three other variables need to be considered. Firstly, the culture volumes vary considerably from 1 ml soft-agar in 35 mm plates (Metcalf et al., 1975; Williams et al., 1978; Burstein et al., 1979) to 0.4–0.5 ml plasma clots in microtiter plates (Mizoguchi et al., 1979; McLeod et al., 1976) to 0.1 ml microtiter plasma clots (Nakeff and Daniels-McQueen, 1976). Thus, the propensity for cell-cell contact and media utilization is greatly enhanced in the latter cultures as compared to the former, although the extent to which this affects CFU-M growth remains undetermined.

The use of acetylcholinesterase (AChE) as a secondary specific cytoplasmic marker to aid in identifying megakaryocytes within colonies varies with different laboratories. Those using plasma culture routinely stain all cultures for AChE prior to scoring CFU-M *in situ*. On the other hand, investigators using agar cultures

Table 1. CFU-M Culture Conditions.

Cell source	Culture medium	Serum	CM	Cells plated	CFU-M/10^5	
Plasma						
BDF1-BM	L-15	20% HS	MS-PWM	1×10^5	10	Nakeff and Daniels-McQueen, 1976
C57B1-BM	NCTC-109	20% FCS	Ep-human sheep	1×10^5	10	McLeod et al., 1976
Balb/c-BM	NCTC-109	15% HS	MS-PWM	1×10^5	50	Mizoguchi et al., 1979
Human BM	α-Med.	20% human	Ep-human	5×10^5	20	Vainchenker et al., 1979
Agar						
C57B1-BM	Eagle's (mod.)	15% FCS	MS-2ME	7.5×10^4	6	Metcalf et al., 1975
BDF1-BM	McCoy's 5A	15% FCS	$WEHI^{-3}$ (10x) + BM	7.5×10^4	20	Williams et al., 1978
C57B1-BM	Eagle's (mod.)	20% FCS	MS-PWM	1×10^5	24	Burstein et al., 1979
C57B1-spleen	”	”	”	”	5	”
C57B1-blood	”	”	”	”	8	”

generally score colonies using inverted phase or light microscopy with only occasional plucking of individual colonies and staining for AChE on a glass slide to confirm their megakaryocytic origin. The latter technique may have the limitation that colonies consisting of only small megakaryocytic cells that stain specifically for AChE may be missed under phase or light microscopy alone. An additional advantage of plasma over agar culture as used presently is that slide preparations of the former can be stored for long periods of time thus making it possible to review cultures at a later time in light of new findings and as a means to monitor culture conditions.

Lastly, the minimum number of cells that constitutes a megakaryocyte colony varies. Metcalf et al., (1975); Burstein et al., (1979); McLeod et al., (1976) and Vainchenker et al., (1979) all consider 2 cells to constitute a colony, whereas Williams et al. (1978) use 3 cells and Nakeff and Daniels-McQueen (1976) and Mizoguchi et al. (1979) use a lower limit of 4 cells. As shown in Figure 1, colony morphology may be tight (1A) or spread (1B) with larger-sized colonies being mainly spread (1C and 1D). This determination is important since a significant proportion of megakaryocytes in all these culture systems are present as single cells rather than in colonies (Fig. 1E). Thus, the frequency distribution of colony size is severely skewed towards smaller sizes making determination of these difficult. Clearly, it is imperative to be both conservative and consistent in enumerating CFU-M.

In summary, although differences exist in culture conditions among the various laboratories assaying for CFU-M, it is not clear which play a major, and which a minor, role in explaining satisfactorily the degree of variation among them.

We have assayed some of the above variables in our own culture system. As described by Nakeff and Daniels-McQueen (1976), each 0.1 ml of plasma culture contained the following proportion (v/v) of ingredients: 40% tissue culture medium (Leibovitz, L-15 from GIBCO); 20% horse serum; 10% pokeweed (1/320)-stimulated mouse spleen cell CM; 10% bovine embryo extract (BEE from GIBCO) diluted 1:4 with L-15; 10% BM cells (10^7/ml L-15) from B6D2F1 (C57 B1/6^0 × DBA^0) mice, 6–8 weeks of age and 10% citrated bovine plasma (BCP from GIBCO).

In order to compare plasma to agar as semi-solid substrates, we made up 0.1 ml agar microtiter cultures with the following proportion (v/v) of ingredients: 30% double-strength (2×) L-15; 20% horse serum; 10% BM cells (10^7/ml L-15); and 30% autoclaved agar or agarose (1%) in Travenol water kept at 42°C prior to mixing.

Both sets of cultures were incubated at 37°C in 7.5% CO_2 in air for 4 days (plasma) or 7 days (agar), then harvested and stained for AChE. CFU-M were counted by light microscopy at a total magnification of 125× and expressed both per 10^5 total nucleated cells (TNC) and as a percent of the total obtained in plasma culture. Of the agar cultures tested (Table 2), those composed of Noble agar supported the growth of about the same number of CFU-M as in plasma culture, with agarose slightly less efficient and bacto-agar being least efficient with less than half the values obtained in plasma culture. Bacto-agar is used most extensively in the

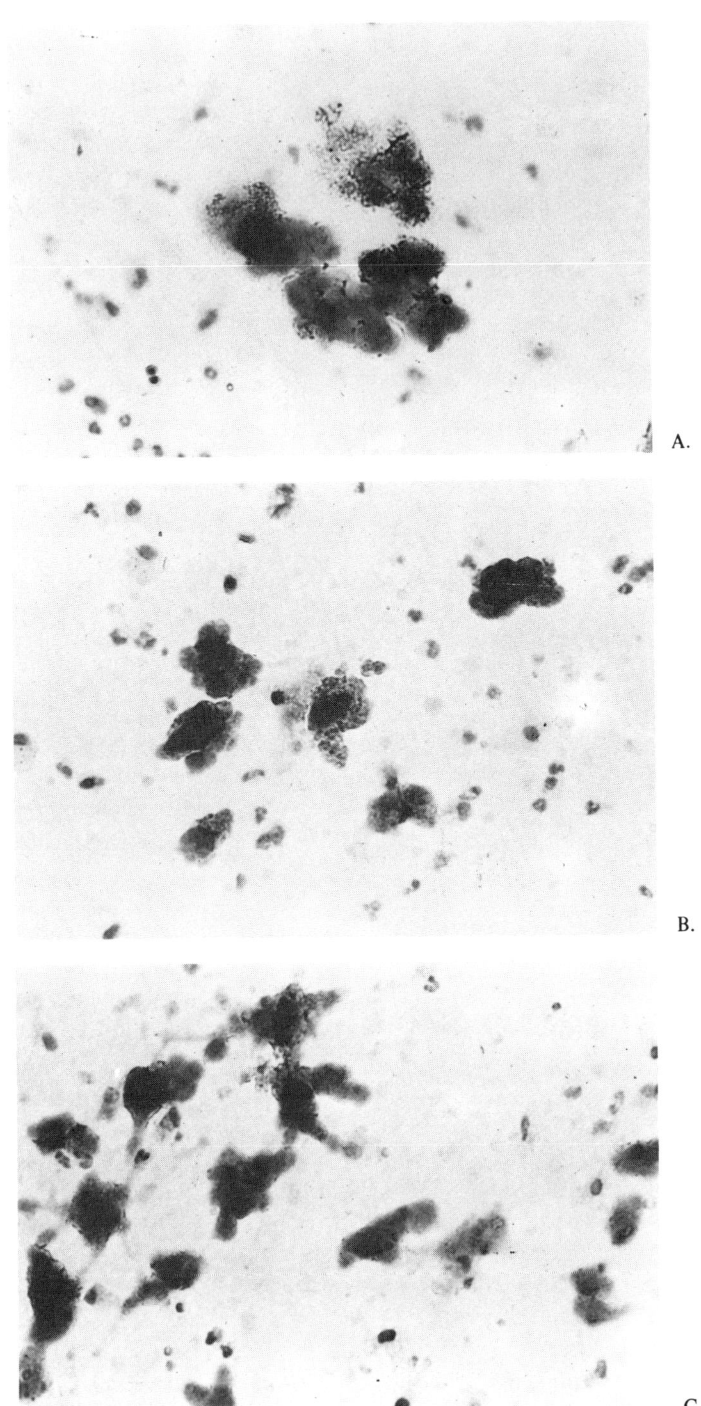
A.

B.

C.

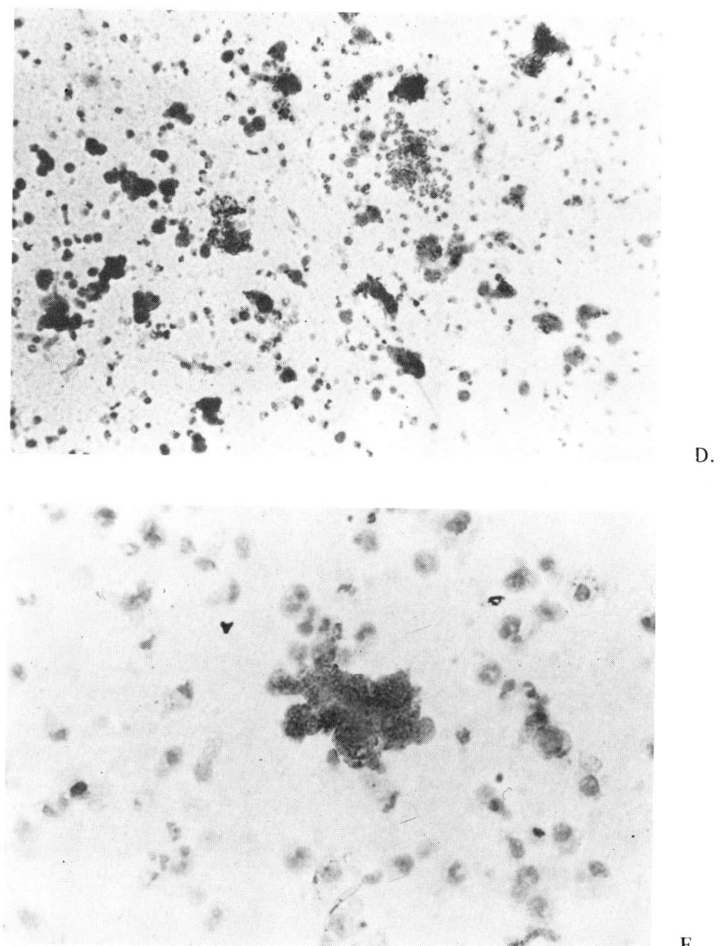

Fig. 1 Photomicrographs of megakaryocyte cultures illustrating different colony morphologies and the appearance of single megakaryocytes. A. Tight colony of 4 cells (mag. 400 ×). B. Spread colony of 6 cells (mag. 300 ×). C. Spread colony of 14–15 cells (mag. 250 ×). D. Spread colony of >30 cells (mag. 100 ×). E. Single megakaryocyte (mag. 400 ×).

culturing of CFU-GM (granulocyte-macrophage) while agarose is the most highly purified of the three with a single repeating sub-unit structure which might more easily be broken down by enzymatic digestion for those studies where isolated clones are desired that are free of suspending medium. Thus, two-fold differences may be obtained within a single culture system dependent on the type and quality of the semi-solid support used.

Since mouse strain differences for CFU-M have been reported (Nakeff, 1977) and mice of different sexes and ages may be used as marrow donors, we investigated

Table 2. Comparison of Semi Solid Media.

Suspending media	CFU-M/10^5 TNC	Control
Plasma	8.6 ± 1.4[a]	100 ± 16
Bacto Agar	3.7 ± 1.7	43 ± 19
Noble Agar	9.0 ± 2.7	105 ± 31
Agarose	6.8 ± 1.2	79 ± 14

[a]Mean ± I SE of 9 cultures from each of two experiments.

the latter to determine the degree of variability that may be obtained. As shown in Table 3 (for B6D2F1 mice), there was little difference in TNC/femur in mice at 6, 16 or 26 weeks of age (range 2.2–2.8 × 10^7) although generally fewer cells are obtained in female than in male mice. No consistent differences in CFU-M/10^5 TNC or CFU-M/femur were observed although there may be some sex difference at 26 weeks of age. Thus, within the age bracket studied, age and sex differences do not contribute significantly to differences in CFU-M at least in this one mouse strain.

As we reported earlier (Nakeff, 1977), a change in our serum source from fetal calf to horse resulted in more reproducible growth of CFU-M. As shown in Table 4, however, each new shipment of horse serum must be tested prior to its use against a control sample since we have noted differences of 2-fold or more from lot to lot. Of the six new lots tested, half of them were unsatisfactory for the assay of CFU-M.

Conditioned Media for CFU-M

As summarized in Table 5, many differences are noted in the types of CM that have been applied to CFU-M cultures and the methods used to obtain them. Cells from three different mouse strains have been used (B6D2F1, C57B1 and Balb/c)

Table 3. Effect of Age and Sex on B6D2F1 Mouse CFU-M.

Age in Weeks	Sex	TNC/femur[a] (x10^7)	CFU-M/10^5 TNC	CFU-M/femur (x10^3)
6	Male	2.5	12.9 ± 0.9[b]	3.2 ± 0.2
	Female	2.3	23.5 ± 1.0	3.2 ± 0.2
16	Male	2.8	11.2 ± 0.9	3.1 ± 0.2
	Female	2.2	12.0 ± 0.8	2.7 ± 0.2
26	Male	2.6	15.4 ± 1.1	4.0 ± 0.3
	Female	2.3	13.1 ± 0.9	3.1 ± 0.2

[a]BM from 2 groups, each pooled from 3 mice.

[b]Mean ± 1 SE from 9 cultures.

Table 4. Effect of Serum Lots on CFU-M Growth.

Serum lot number	CFU-M/10^5 TNC	Control
210 (control)	12.0 ± 1.3[a]	100 ± 11
212	11.2 ± 1.3	94 ± 11
603	12.0 ± 1.9	100 ± 16
305	12.7 ± 1.4	106 ± 12
095	8.4 ± 1.3	70 ± 11
042	6.8 ± 0.5	50 ± 4
043	8.3 ± 0.6	69 ± 5

[a]Mean ± SE from 9 cultures.

in addition to different cell sources (spleen and bone marrow) and use of normal and neoplastic (WEHI-3) cells. Thus, the target cells producing the CM are not well characterized and may be heterogeneous in the quantity and quality of CM they produce. Cell concentrations used in CM cultures vary almost 10-fold (0.3-2 $\times$ 10^6/ml); concentration of CM by Amicon filtration has also been used (Williams et al., 1978).

Many different inducers have been used to stimulate cell production of CM including mitogens (PHA and PWM), 2-ME, Ep of varying activity from human and sheep sources and thrombopoietin (thrombocyte stimulating factor(s), TSF) from human embryonic kidney cells (McDonald et al., 1975) and platelet-depleted sheep sera. Even considering just PWM, different concentrations have been used varying from dilutions of 1/250 to 1/350 as well as different concentrations of PWM-CM in the final CFU-M culture (either 10% or 20%). It would seem reasonable that the cell type producing the CM as well as the CM itself could be quite heterogeneous since in the case of PWM-CM, for example, there is evidence (Metcalf and Johnson, 1978) that other colony types including so called "mixed colonies" are stimulated *in vitro*. The length of time from addition of the stimulus to obtaining the CM varies from 1 day to >9 weeks (in long-term culture) and for one class of promoter (PWM) this may vary from 1 to 7 days.

Thus, with substantial differences in the types of stimuli used and even the use of the same stimulus in the hands of different investigators, it is difficult to compare CM, their potency and specificity from information in the literature.

Our data using identical CFU-M culture conditions and comparing PWM and PHA stimulation on both adherent and non-adherent mouse spleen cells is summarized in Table 6 in terms of both CFU-M and single megakaryocytes in cultures to which 10% CM was added. It can be seen that PWM supports the growth of slightly more CFU-M than PHA when both are used at their optimal concentrations in the cultures from which the CM was obtained. This effect was also independent of whether adherent or non-adherent cells were plated in the CM cultures. A ratio of about 1:10 was observed in the proportion of CFU-M to single megakaryocytes counted in each entire CFU-M culture under all the conditions tested. Thus, with

Table 5. CFU-M: Sources of Conditioned Medium.

CM cells					
Source	Concentration ($x10^6$ ml)	Promoters	Days of culture	Conc in CFU-M culture	Reference
B6D2F1 spleen	2	PWM-1/320 PHA-12.5 μg/ml	2	10%	Nakeff & Daniels-McQueen, 1976
C57BL spleen	1.25	2-ME (50 μM)	7-14	17%	Metcalf et al., 1975
		Ep (sheep = 2.1 u/mg) (human = 74.3 u/mg)		3 (u/ml)	McLeod et al., 1976
		Ep (sheep = 4 u/mg) (human = 75 u/mg)		3 (u/ml)	Freeman et al., 1977
		TSF (PDS-Sera)		5-320 (μg/ml)	
B5D2F1 BM	0.3-0.5		1-9 weeks	15.4%	Williams et al., 1979
+	+			+	
Balb/C WEHI-3 myelomono	1	2-ME (50 μM) (10x - 700 μg/ml)	7	7.7%	
C57BL spleen	2	PWM-1/300	7	20%	Burstein et al., 1979
C57BL spleen	1	PWM-1/250	1	10%	Mizoguchi et al., 1979
Human kidney		TSF-HEKC	> 6 weeks	18-500 (μg/ml)	Williams et al., 1979
+		+		+	
Balb/C		WEHI-3CM		7.7%	

Table 6. Effect of Mitogen and Spleen Cells on CM Production.

Mitogen	Spleen cells	CFU-M/10^5 TNC	MEGS/10^5 TNC
PHA (12.5 μg)	NA[a]	8.7 ± 1.3[b]	82 ± 13
	A	8.7 ± 1.3	96 ± 9
PWM (1/320)	NA	10.2 ± 2.2	107 ± 14
	A	11.0 ± 2.6	89 ± 10

[a]NA = nonadherent cells; A = adherent cells.
[b]Mean ± 1 SE of 9 cultures.

the conditions we used for adherence (8×10^7 spleen cells in 40 ml alpha-MEM with 10% fetal calf serum and incubation for 60 min at 37°C before rinsing off non-adherent cells), we could demonstrate no large difference in the ability of either adherent and non-adherent cells to produce CM or of PWM and PHA to stimulate CM production. It was clear from the large ratio of single megakaryocytes to CFU-M in the BM cultures that the minimum number of cells taken to constitute a colony and the proximity of megakaryocytes to each other within a single colony are important variables to be considered.

We have examined the time response of mouse spleen cells stimulated with PWM to produce CM. As presented in Figure 2, we have expressed CM activity as a percent of peak CM (day 2) and have determined it as a function of cell type (i.e., adherent vs non-adherent). For total spleen cells (2×10^6/ml in 40 ml per 75 cm^2 flask), incubation times either shorter or longer than 2 days resulted in CM with about 50% of the peak activity at day 2 with adherent and non-adherent cells accounting for 81 ± 21% and 72 ± 9%, respectively, of the day 2 activity. Although about equal amounts of CM were produced by adherent and non-adherent cells, their time response was somewhat different in that with the former high CM activity was obtained later in culture from day 2 to day 14 at which time about 3/4 of peak activity was still obtained, whereas with the latter, high levels of CM were obtained early in culture (during the first seven days) with only 1/3 of peak activity being obtained on day 14. This difference probably reflects not only the different cell types initially plated (i.e., macrophages, lymphocytes, etc.) but also those cell types present later in culture which either may have survived throughout 14 days of culture or may have been produced by mitogen-stimulated cell proliferation. The role of PWM in CM production remains unknown.

Since CFU-M cultures are dependent on CM whose activity has a finite shelf life, it is essential to be producing different lots of CM continuously, although they might vary in potency. An example of this is illustrated in Figure 3 where four lots of CM prepared sequentially over a two week period was compared to a known high potency lot, all tested at one time. Clearly, differences in lot potencies were obtained that may vary up to 3 to 4-fold in activity.

Our experience with storage lability at -20°C of a CM lot of high initial potency is presented in Figure 4. It can be seen that activity could be measured for up to

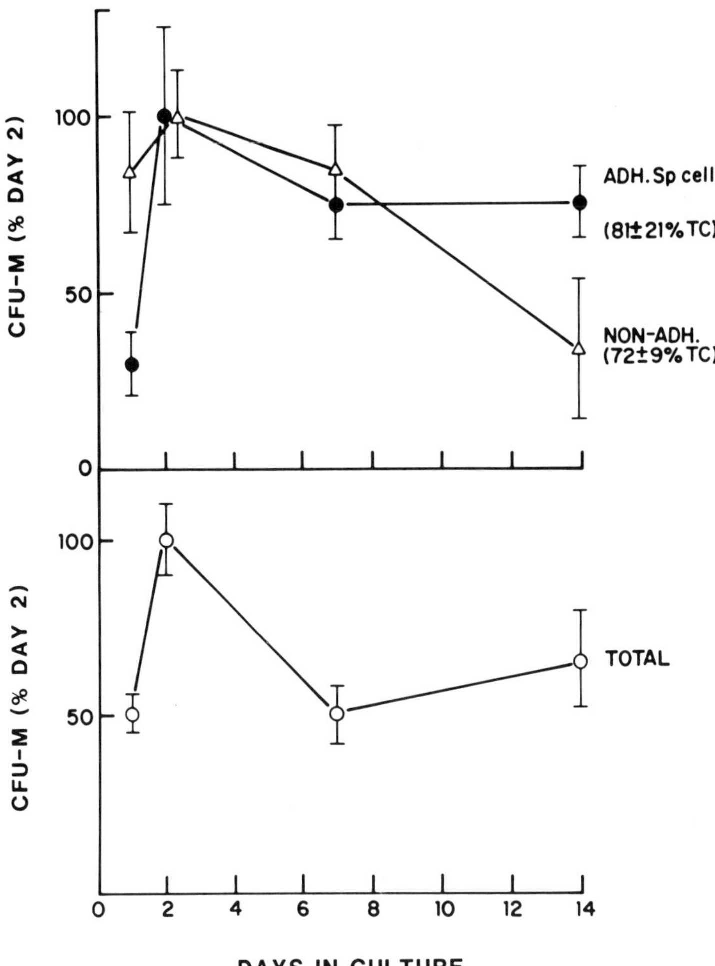

Fig. 2 Effect of time in culture on the activity of PWM-spleen CM to support growth of CFU-M in mouse spleen cell cultures comprised of either total, adherent or non-adherent cells following the addition of 1/320 concentration of PWM. Mean ± 1 SE of 9 cultures per point.

3 months by which time all activity was lost. Interestingly, it was clear that using this same batch of CM at different times throughout the three month period resulted in variations of almost two-fold in the number of CFU-M/10^5 TNC which were not a function of the length of storage time. To what extent this degree of variability reflects that inherent in the target cell or that in the CM itself is not known but must be recognized as a potential source of error.

This abbreviated examination of CFU-M culture condition variability has substantiated some of its many real and potential sources. The degree to which some

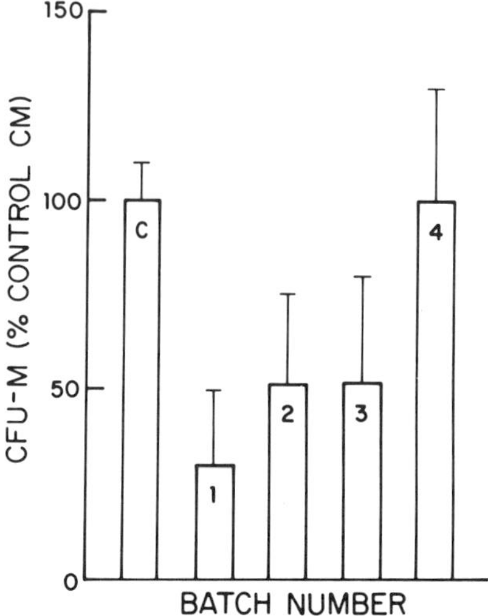

Fig. 3. Batch variation of day 2 PWM-spleen CM expressed as a percent of a known control batch of high activity on CFU-M. Errors shown are ± 1 SE of 9 cultures per point from 2 separate experiments.

of these may interact additively or synergistically remains to be determined. It is of obvious importance to determine and attempt to control as many of these as possible under the precise conditions used in each investigator's laboratory. The degree of inherent variability within our own laboratory precludes to some extent a meaningful comparison between laboratories on the basis of published data. Although the variability reflects, to some extent, the relatively early stage of development of this assay as compared to others (e.g., CFU-GM), some as yet undetermined component may reflect the inherently low proliferative, i.e., mitotic, potential of CFU-M. Furthermore, in distinction to that seen in other hematopoietic, clonogenic progenitors, megakaryocyte clonal development entails extensive DNA synthesis that is endomitotic and responsible for extensive polyploidy within colonies. The interplay of these DNA synthetic processes in megakaryocyte progenitors is not known but experimental approaches have been proposed to study this important area of regulation using flow cytometry and cell sorting.

In summary, an opportunity to test each culture technique in one location at one time using the same starting cell suspension and CM as has been attempted for CFU-GM (Rijswijk Workshop 1971) may be of some importance at this particular stage of development of the CFU-M assay and its ever-widening adoption by other laboratories.

Summary

The mitotic activity associated with megakaryocytic proliferation resides in a class of progenitor cells (CFU-M) which are detected by their ability to form colonies of megakaryocytes in either modified plasma or agar culture systems. Since the

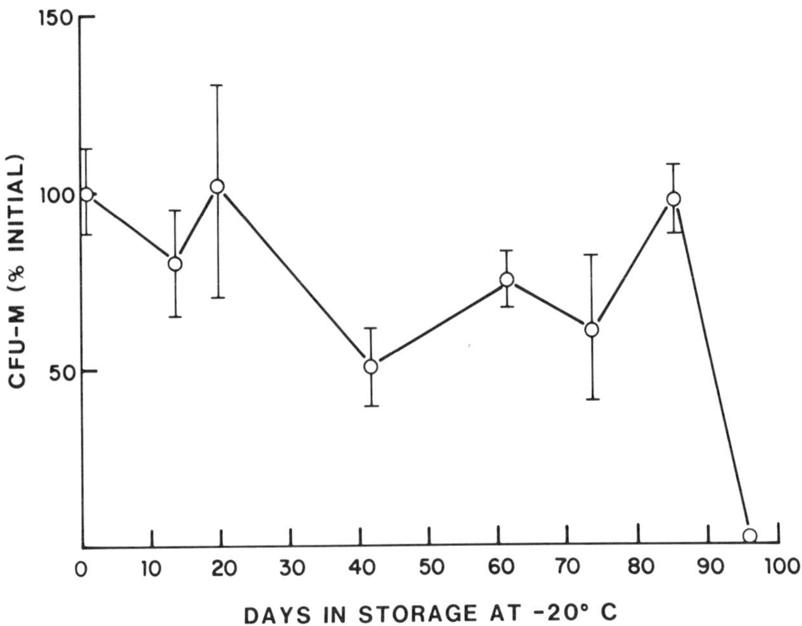

Fig. 4. Effect of storage time at -20°C on the activity of day 2 PWM-spleen CM to support CFU-M. Errors shown are ± 1 SE of 9 cultures per point.

application of clonogenic cell culture techniques to the study of megakaryocyte proliferation is relatively new, culture conditions remain to be defined further, particularly in light of the low incidence of CFU-M in marrow (1 CFU-M/10^4 bone marrow cells), the small average colony size (<50 megakaryocytes/colony), the presence of numerous single megakaryocytes among colonies and the stimulation of CFU-M by relatively nonspecific conditioned media (CM). It remains to be shown to what degree these properties of CFU-M reflect the limited proliferative capacity of megakaryocytic progenitors *in vivo* or non-optimal culture conditions. We have approached this question by analyzing a number of culture variables including lectin-stimulated mouse spleen CM in terms of different mitogens and spleen subpopulations, the source of serum including species and batch variations, the effect of marrow cell density in culture and, lastly, the use of morphologic and cytochemical colony markers such as acetylcholinesterase. The advantages and limitations in using *in vitro* culture systems to study initial events involving the induction of CFU-M and their role as physiologic target cells responsible for host megakaryocyte production is discussed.

References

Burstein, S. A., J. W. Adamson, D. Thorning, and L. A. Harker. 1979. Characteristics of murine megakaryocyte colonies *in vitro*. *Blood* 54:169–179.

Freedman, M. H., T. P. McDonald, and E. F. Saunders. 1977. The megakaryocyte progenitor (CFU-M): *In vitro* proliferative characteristics. *Blood* 50 (Suppl. 1):146.

McLeod, D. L., M. M. Shreeve, and A. A. Axelrad. 1976. Induction of megakaryocyte colonies with platelet formation *in vitro*. *Nature* 161:492–493.

McDonald, T. P., R. Clift, R. D. Lange, C. Nolan, I. I. E. Tibby, and G. H. Barlow. 1975. Thrombopoietin production by human embryonic kidney cells in culture. *J. Lab. Clin. Med.* 85:59–66.

Metcalf, D. and G. R. Johnson. 1978. Production by spleen and lymph node cells of conditioned medium with erythroid and other hemopoietic colony-stimulating activity. *J. Cell. Physiol.* 96:31–42.

Metcalf, D., H. R. MacDonald, N. Odartchenko, and B. Sordat. 1975. Growth of mouse megakaryocyte colonies *in vitro*. *Pro. Nat. Acad. Sci. USA* 72:1744–1748.

Mizoguchi, H., K. Kubota, Y. Miura, and R. Takaku. 1979. An improved plasma culture system for the production of megakaryocyte colonies *in vitro*. *Exp. Hematol.* 7:345–351.

Nakeff, A. 1977. Colony-forming unit megakaryocyte (CFU-M): Its use in elucidating the kinetics and humoral control of the megakaryocytic committed progenitor cell compartment. In *Experimental Haematology Today*. Baum, S. J., and Ledney, G. D., eds. New York: Springer-Verlag. pp. 111–123.

Nakeff, A. and S. Daniels-McQueen. 1976. *In vitro* colony assay for a new class of megakaryocyte precursor: Colony-forming unit megakaryocyte (CFU-M). *Proc. Soc. Exp. Biol. Med.* 151;587–590.

Nakeff, A. F. Valeriote, J. W. Gray, and R. J. Grabske. 1979. Application of flow cytometry and cell sorting to megakaryocytopoiesis. *Blood* 53:732–745.

Vainchenker, W., J. Bouquet, J. Guichard, and J. Breton-Gorius. 1979. Megakaryocyte colony formation from human bone marrow precursors. *Blood* 54;940–945.

Van Bekkum, D. W., and K. A. Dicke. 1971. Proceedings of a Workshop/Symposium on *In Vitro* Culture of Hemopoietic Cells. Rijswijk, The Netherlands.

Williams, N., H. Jackson, A. P. C. Sheridan, M. J. Murphy, Jr., A. Elste, and M. A. S. Moore. 1978. Regulation of megakaryocytopoiesis in long-term murine bone marrow cultures. *Blood* 51:245–255.

ACKNOWLEDGMENTS

The excellent secretarial aid of Ms. Donna Troeckler is gratefully acknowledged. This work is supported in part by Grant No. HL20826 and Research Career Development Award HL00440 awarded by the National Heart, Lung and Blood Institute.

Discussion

Ebbe: When you see these really diffuse, spread out megakaryocytes colonies, how do you know that that's one colony?

Nakeff: Beyond the borders of those large diffuse colonies there are no megakaryocytes that might be confused with other megakaryocyte colonies. One always has to bear in mind whether these are several colonies which have simply run into each other or not. The presence of such overlapping colonies is very low in the culture, but they are there.

Ebbe: Have megakaryocytes in the dishes been shown to migrate?

Nakeff: Yes, I believe that these cells can move; I don't know how to control for that.

Williams: I would agree with that. A colony at day 4 would be more confined that a colony at day 7.

Nakeff: If you are dealing with poor culture conditions there are often a lot of groups of small numbers of megakaryocytes. That leads to difficulty deciding what constitutes a colony. With good culture conditions the cells seem to form discrete colonies much more readily.

Weil: I am curious about the numbers of single megakaryocytes in the cultures on days 0 and 7. Are these input megakaryocytes or are these unrecognizable mononuclear cells that have no proliferative capacity, but with maturational capacity to form megakaryocytes?

Nakeff: In plasma clot cultures of marrow cells we plate about 30 megakaryocytes and this about doubles by day 7. We have used fractions containing progenitors in which there were no recognizable megakaryocytes. In this case, individual megakaryocytes appear by day 7. These are probably coming from an unidentifiable progenitor compartment that may be endoreduplicating.

Levin: Do you think that CFU-M could be polyploid? Both you and Burstein had sedimentation rate data that would indicate that CFU-M would have to be diploid cells.

Nakeff: I don't want to imply that CFU-M are diploid cells. I am uncertain. What we really want to be able to do is to separate out individual, unrecognizable progenitors on the basis of their DNA and subsequently put them into culture and see how able they are to produce colonies. Until we can do that experiment and show that tetraploid and diploid cells are different in their proliferative capacities, it is just speculative. However, one would think from the sedimentation velocity data that they are diploid.

Published 1981 by Elsevier North Holland, Inc.
Evatt, Levine, and Williams, Editors
MEGAKARYOCYTE BIOLOGY AND PRECURSORS:
IN VITRO CLONING AND CELLULAR PROPERTIES

Regulation of Murine Megakaryocytopoiesis

Samuel A. Burstein, John W. Adamson, Susan K. Erb, and Lawrence A. Harker

Department of Medicine (Hematology), Harborview Medical Center and the Hematology Research Laboratory and Hematology Section, Veterans Administration Medical Center, Seattle, Washington

The changes in marrow megakaryocytes which occur with experimentally induced thrombocytosis indicate that platelet production is subject to negative feedback regulation (Harker, 1968a; Ebbe, 1970). It has been suggested that a humoral factor analogous to erythropoietin is either produced or suppressed in response to platelet demand. The target of action of this putative regulator is unknown, but it has been postulated that a morphologically unidentifiable diploid megakaryocytic precursor is the cell upon which the hormone acts. In the past six years, clonal assays have been developed for a progenitor cell (CFU-M) which, under the appropriate conditions in culture, gives rise to colonies of morphologically identifiable megakaryocytes (Metcalf et al., 1975; Nakeff and Daniels-McQueen, 1976). Since this progenitor cell is committed to megakaryocytic differentiation, it is a candidate target cell of thrombopoietic regulation. In this communication, we present our hypotheses on the cellular anatomy and regulation of megakaryocytopoiesis and summarize evidence which suggests that the primary target cell for the day-to-day regulation of the platelet count is the megakaryocyte or a cell immediately preceding the megakaryocyte, and not the CFU-M.

Methods and Materials

Mice

Male C57B1/6 mice, six to eight weeks old and obtained from Jackson Laboratories (Bar Harbor, Maine) were used for all experiments.

Cell Counts

Platelet counts and hematocrits were determined using routine methods on blood obtained from the retro-orbital plexus.

Quantitation of Megakaryocytes

Femurs were cleaned and fixed in neutral formalin. Four micron sections were cut with a microtome and the sections decalcified and stained with hematoxylin and eosin. Megakaryocytes were enumerated in the central section and in the two adjacent sections, and corrected for multiple counting error (Harker, 1968b). Megakaryocyte diameters were measured with an optical micrometer in 250 randomly chosen megakaryocytes and volumes were determined using previously described methods (Harker, 1968b).

Quantification of CFU-M

CFU-M in the marrow and spleen were determined as previously described (Burstein et al., 1979).

Tritiated Thymidine (^{3}H-TdR) Suicide Studies

The percentage of CFU-M in DNA synthesis was determined by exposure of marrow cell suspensions to high specific activity ^{3}H-TdR and quantitating the effect on colony formation (Burstein et al., 1979).

Preparation of Antiplatelet Serum

To prepare platelets for antibody generation, mice were sacrificed by exposure to CO_2, and then bled in a sterile manner via the inferior vena cava. The blood was withdrawn into syringes containing acid citrate dextrose (ACD). Platelet rich plasma was produced by centrifugation at room temperature of the blood of ten donor mice. After the platelets were washed and resuspended in Ringer's citrate, three subsequent slow centrifugations were performed to remove the majority of mononuclear cells.

Antiplatelet serum was produced by four intravenous injections of 10^9 mouse platelets in 2 ml Ringer's citrate to New Zealand White rabbits over a period of one week. The rabbits were bled 21 days after the first injection. Subsequently, the blood was allowed to clot, the serum heat inactivated, and then absorbed three times with packed mouse red cells. No decrease in hematocrit or white blood cell count of recipient mice was noted after injection of the antiplatelet serum.

Platelet Preparation for Transfusion

Platelets were prepared as above with the omission of the slow centrifugations. Platelets from ten donor mice were used for each transfusion of a single recipient.

Experimental Design

A. Kinetics of Thrombocytopenia

To determine the effect of acute thrombocytopenia on megakaryocytopoiesis, mice were injected intravenously with 0.2 ml of antiplatelet serum. Control mice were treated with equal volumes of normal rabbit serum. At various times after antiplatelet serum injection, mice were sacrificed by cervical dislocation and the spleen, humeri, and femora were removed for quantitation of megakaryocyte numbers and volumes, and quantitation of CFU-M numbers and percent in DNA synthesis.

B. Effect of Platelet Hypertransfusion on Normal Animals

Normal mice were given transfusions of platelets two days apart. Each transfusion consisted of the platelets obtained from 10 donor mice. Control animals were given platelet buffer solution. Two days after the second platelet transfusion, mice were killed and megakaryocyte numbers and volumes, and CFU-M numbers and percent in DNA synthesis were determined.

C. Effect of Platelet Hypertransfusion on the Recovery of Megakaryocytopoiesis in Irradiated Animals

Mice were given 900 rads from a ^{37}Cs source and within four hr infused with 25 $\times$ 10^6 normal isogeneic marrow cells. Five and seven days later, experimental mice received platelet transfusions while control animals received platelet buffer solution. On day 8 after marrow infusion, mice were killed and megakaryocyte number and volume, and CFU-M number and percent in DNA synthesis were determined.

Results and Discussion

Figure 1 reveals the time-related effects of acute thrombocytopenia on megakaryocytes and on CFU-M. Following a single injection of antiplatelet serum, the platelet count fell within two hr to less than 10% of control values, remained low for 40

Fig. 1. The time-related response of megakaryocytes and CFU-M after the induction of acute thrombocytopenia by a single injection of rabbit antiplatelet serum. Results are expressed as a percent ($\pm$ SEM) of concurrent controls injected with normal rabbit serum.

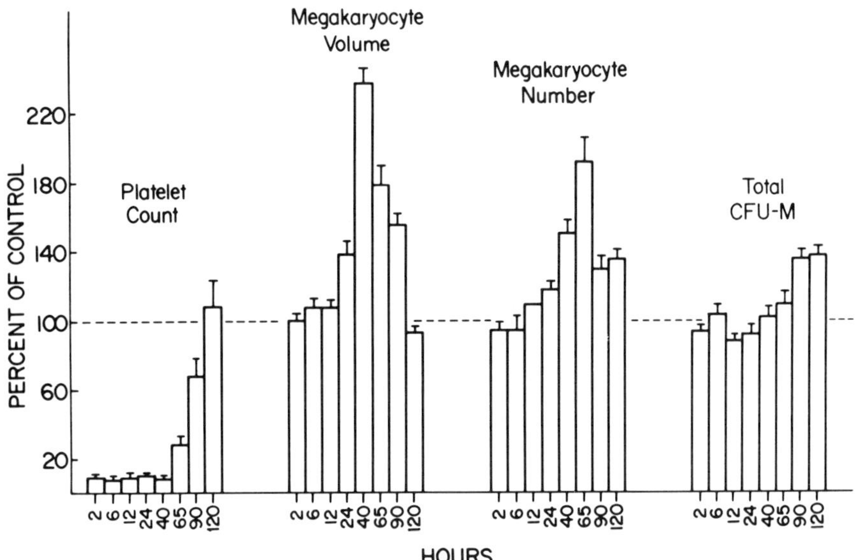

hr and then gradually returned to normal by 120 hr. The first increase in megakaryocyte volume occurred at 24 hr with a maximal increase of greater than twice the control volume seen at 40 hr. Similarly, the first increase in megakaryocyte numbers occurred between 12 and 24 hr and was maximal at 65 hr, at which time almost a two-fold increase was seen. At no time after antiserum injection was a decrease in megakaryocyte number noted. In contrast, the estimated number of total body CFU-M (the sum of marrow plus splenic CFU-M), was not changed at times when marrow megakaryocyte number and volumes were increased. Indeed, the first significant change in CFU-M number occurred at 90 hr after antiserum injection and was less than 40% above control values.

Figure 2 shows the effect of acute thrombocytopenia on the percent of CFU-M in DNA synthesis. At 24 hr, the first significant increase in ^{3}H-TdR suicide rate was noted, and at 40 hr, half of CFU-M were in DNA snythesis. By 90 hr, the suicide rate had returned to normal levels.

Figure 3 reveals the effect of platelet hypertransfusion on normal mice. The mean platelet count was increased to approximately 2 1/2 times normal at the time of sacrifice. This resulted in a significant decrease in the number of megakaryocytes per central marrow section and in megakaryocyte volume. However, no change was noted in the number of CFU-M per humerus or spleen, or in the percent of CFU-M in DNA synthesis.

The effect of platelet hypertransfusion on regenerating marrow is also shown in Figure 3. Eight days after marrow infusion, the mean platelet count of control animals was less than $4 \times 10^5/1$ while the mean platelet count of transfused animals

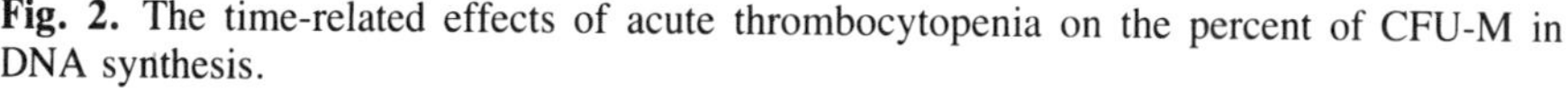

Fig. 2. The time-related effects of acute thrombocytopenia on the percent of CFU-M in DNA synthesis.

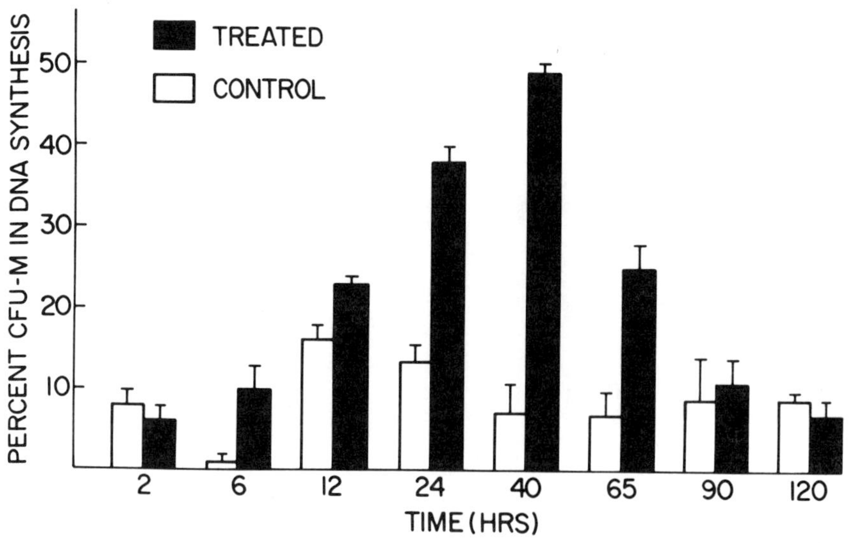

was > 2.5 × 10^6/1. At this time, megakaryocytes per central marrow section and their volume were markedly decreased in hypertransfused animals. However, the numbers of CFU-M per humerus and spleen were similar, and the percent of CFU-M in DNA synthesis was not suppressed, being over 40% in both groups.

Changes in the platelet count, bone marrow megakaryocytes, and CFU-M were studied in mice after the acute induction of thrombocytopenia or thrombocytosis. Through such studies, we hoped to learn the sequence of changes that occur under conditions of altered platelet demand, and thereby: (1) where regulatory factors worked, and (2) whether more than one mechanism of regulation could be shown. After induction of thrombocytopenia, the megakaryocyte mass increased before any changes were noted in the number of CFU-M. Four days after antiserum injection, small increases (<40%) were seen in CFU-M. After induction of thrombocytosis in either normal animals, or in mice with regenerating marrow, decreases in megakaryocytic mass were the only changes noted; CFU-M numbers were unaltered.

Following the onset of thrombocytopenia, the first change in the CFU-M was an increase in the ^{3}H-TdR suicide rate which occurred simultaneously with the increases seen in marrow megakaryocyte number and volume. Thus, although the number of CFU-M do not change until 90 hr after megakaryocytes increase, these results do not allow one to determine whether the primary response to acute throm-

Fig. 3. The response of megakaryocytes and CFU-M to platelet hypertransfusion in both normal mice and in animals with regenerating marrow. Results are expressed as the percent of normal controls. The open symbols indicate pre-transfusion values, while the closed symbols represent values at the time of sacrifice.

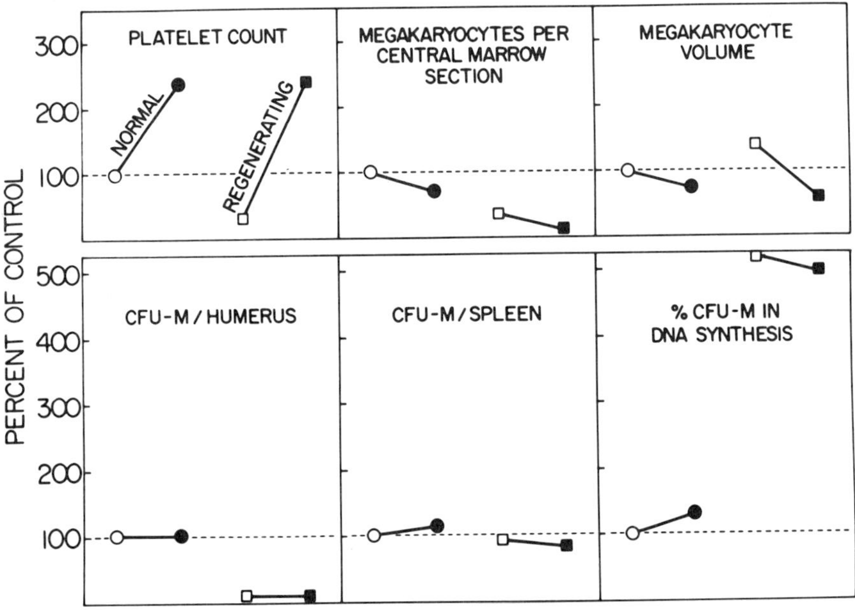

bocytopenia is an increase in megakaryocyte number and volume, an increase in the percent of CFU-M in DNA synthesis, or both.

To further examine these possibilities, we assumed that if a change in the cell cycle rate of CFU-M is the initial response to acute thrombocytopenia, then a decrease in cell cycle activity of this progenitor might be seen in platelet hypertransfused animals. However, the fact that the percent of CFU-M in DNA synthesis in both hypertransfused and control mice was similar does not settle the issue. Since the percent of CFU-M in DNA synthesis is normally low, with a range of 0–20% (Burstein et al., 1979), any suppressive effect of platelet transfusion on normal mice would not be discernible.

To resolve this problem, the platelet hypertransfusion experiments were repeated in mice under conditions of marrow regeneration. Progenitor cells, including CFU-M, have been shown to go into active cell cycle upon infusion into irradiated mice (Williams and Jackson, 1978). Therefore, in this model in which the baseline percent of CFU-M in DNA synthesis would be high, a suppressive effect of platelet transfusion would be readily detectable. As noted in Figure 3, however, this does not occur. A marked decrease in both megakaryocyte number and volume is seen in the platelet transfused group, while no change is seen in CFU-M number in marrow or spleen, or in the percent of CFU-M in DNA synthesis.

These experiments do not prove that a thrombocytopenic stimulus cannot directly affect the percent of CFU-M in DNA synthesis. It is possible that the mechanisms of regulation of CFU-M under normal conditions differ from conditions of increased platelet demand. The experiments do demonstrate, however, that platelet demand is not necessary for activation of CFU-M into cell cycle or for regeneration of the CFU-M population. We postulate, therefore, that the first detectable event in response to alterations in platelet demand is a change in megakaryocyte numbers and volume, and that the increases in the cell cycle rate of CFU-M and the number of CFU-M in the thrombocytopenic animals are secondary and tertiary events, respectively.

These results are analogous to those from studies of erythropoietic regulation, where three classes of colony-forming cells are recognized. A late erythroid progenitor, designated the erythroid colony-forming cell (CFU-E), gives rise in two days of culture to 1–2 small aggregates (8–64 cells) of hemoglobinized erythroblasts. Another progenitor, designated the erythroid burst-forming cell (BFU-E), gives rise to clusters of erythroid colonies of identical morphology to CFU-E-derived colonies after 7 days of culture. An intermediate progenitor, often referred to as a day 4 BFU-E, gives rise to a small number of clusters of erythroid colonies (Gregory and Eaves, 1978). Physiologic studies in mice have revealed that BFU-E and CFU-E are regulated independently. Induction of anemia or polycythemia results in marked changes in the CFU-E, minimal or no changes in BFU-E, and moderate changes in the intermediate progenitor cell (Iscove, 1977; Adamson et al., 1978).

Megakaryocytopoiesis appears similarly structured. In our formulation, CFU-M is an early progenitor, perhaps closely related to the pluripotent stem cell (CFU-S) and unresponsive in the short term to platelet demand. Other experimental evidence supports this hypothesis. CFU-M have a velocity sedimentation rate of 4.2 mm/hr, which is similar only to values for early hematopoietic progenitor cells, including CFU-S (Worton et al., 1969), BFU-E (Heath et al., 1976), and the granulocyte-macrophage progenitor colony forming cell (CFU-C) (Metcalf and MacDonald, 1975). In addition, CFU-M in the C57B1/6 mouse have a low proliferative rate (10 ± 5% in DNA synthesis by ^{3}H-TdR or hydroxyurea studies) (Burstein et al., 1979). Low or intermediate ^{3}H-TdR suicide rates are seen with CFU-S (Vassort et al., 1973), BFU-E (Gregory and Eaves, 1978), and CFU-C (Metcalf, 1972), whereas the later erythroid progenitor, CFU-E, has a higher proliferative rate (70–80%) (Gregory and Eaves, 1978). Finally, CFU-M circulate in the peripheral blood (Burstein et al., 1979) as do CFU-S (Barnes and Loutit, 1967), BFU-E (Hara and Ogawa, 1976), and CFU-C (Metcalf, 1977), whereas CFU-E do not.

Other cell(s) further along the megakaryocyte differentiation pathway appear to be acutely responsive to platelet demand. We suggest that this cell is a morphologically unidentifiable megakaryocyte of low ploidy, analogous to the CFU-E. Comparison of the proliferative capacity of the two cells types is consistent with this view. The CFU-E gives rise to colonies of 8–64 cells. A 2N megakaryocyte has the potential to undergo endomitosis and become at least a 64N cell. Thus, one might consider a 64N megakaryocyte equivalent to a 32-cell erythroid colony. A factor(s) analogous to erythropoietin may operate at these stages of megakaryocytopoiesis, affecting megakaryocyte cytoplasmic differentiation, ploidy, volume, and platelet release. However, it is possible that some or all of these processes are differentially regulated by separate factors.

Alternative hypotheses exist. Since megakaryocytic colonies contain from two to several hundred cells, it may be simplistic to regard all of these colonies as being derived from a homogeneous population of progenitors. Larger colonies may be derived from an earlier progenitor with marked proliferative capacity, whereas colonies with fewer cells might be derived from a more "differentiated" type of CFU-M, more analogous to CFU-E. If this hypothesis is correct, then the ratio of small to large colonies would increase in response to thrombocytopenia, and decrease in response to thrombocytosis. However, observation of the number and size of megakaryocytic colonies after experimental manipulation of the platelet count does not favor this interpretation. The proportion of two and three cell colonies does not appear to change with varying platelet demand.

Other physiologic mechanisms which influence CFU-M may be signaled by changes in the immediate precursors of megakaryocytes. Through unknown mechanisms, CFU-M may recognize an absence or a decrease in the number of this intermediate progenitor which may be analogous to the day 4 BFU-E. The increases noted in CFU-M number in the marrow four days after induction of thrombocytopenia, and in the spleen at 65 hr, might reflect this recognition. If this mechanism does operate, then chronic thrombocytopenia or thrombocytosis might elicit more profound changes in CFU-M numbers. Alternatively, it is possible that the changes which occur in CFU-M number or cycle characteristics are unrelated to megakaryocyte demand but, rather, result from an immunologic stimulus related to the rabbit antiserum-mouse platelet reaction. Such an antigen-antibody reaction may result in the *in vivo* production of cell-derived factors analogous to the factors produced when pokeweed mitogen is added to spleen cells *in vitro*.

Megakaryocytopoiesis may also be influenced by the modulating effects of hormones. Adrenergic and cholinergic agents and androgenic hormones have been shown to rapidly activate the CFU-S into cell cycle (Byron, 1972a,b). Cholinergic agonists have been shown to increase the number of CFU-C detected in culture (Kurland et al., 1977; Oshita et al., 1977), while thyroid hormones and beta adrenergic agonists increase the number of CFU-E detected in culture (Popovic et al., 1977). The presence of the enzyme acetylcholinesterase in the cytoplasm of mouse megakaryocytes led us to speculate that this enzyme is the vestige of a primitive cholinergic regulatory mechanism. We tested this hypothesis and found that the cholinergic agonist carbamylcholine increased the number of CFU-M-derived colonies detected in culture and triggered CFU-M into cell cycle. *In vivo* studies, using neostigmine, an acetylcholinesterase inhibitor, and thus a potentiator of endogenous acetylcholine, resulted in a marked increase in both the number of CFU-M and the percent in DNA synthesis (Burstein et al., 1980). These data suggest that cholinergic agonists may modulate the regulation of early megakaryocytopoiesis. It is of interest that CFU-S appear responsive to both adrenergic and cholinergic stimuli, CFU-C and CFU-M are responsive to cholinergic agonists, and erythroid progenitors are responsive to beta adrenergic agonists. It is possible that in the process of differentiation, the responses to particular classes of agonist become restricted.

If the hypothesis of a multi-tiered structure of megakaryocytopoietic regulation is correct, it is unlikely that this structure is rigidly compartmentalized with the influence of early regulation factors ending abruptly at one stage, such as the CFU-M, and the influence of a thrombopoietic hormone(s) beginning subsequently. Rather, since cell differentiation is a continuum, we hypothesize that the sensitivity to early regulatory factors may decline in parallel with the gradual acquisition of sensitivity to late regulatory factors. As has been proposed for erythropoietic differentiation (Iscove, 1978), it is possible that the CFU-M has receptors for early regulatory factors. With differentiation, the number or affinity of these receptors might decline, while receptors for late regulatory factors might appear and increase (Fig. 4). If this hypothesis is correct, it implies the presence of intermediate populations of cells which may be sensitive to both early and late regulatory influences.

In addition, this hypothesis generates several predictions. *In vitro* assays for factors which regulate early megakaryocytopoiesis will be based on enumeration of CFU-M derived colonies or determination of the percent of CFU-M in DNA synthesis. *In vitro* assays for late regulatory factors, however, would not be based on the number of megakaryocytic colonies, but rather on the volume or ploidy of individual megakaryocytes within a colony.

Summary

Physiologic studies in mice indicate that after induction of thrombocytopenia, the

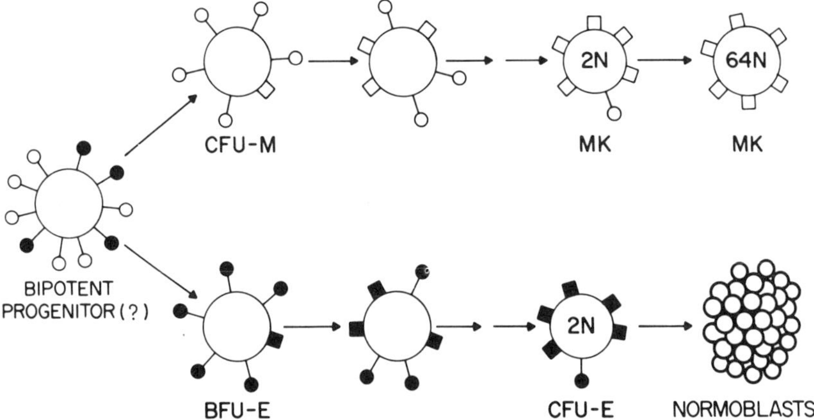

Fig. 4. A schematic outline of the cellular anatomy of megakaryocytopoiesis and erythropoiesis and their regulation. In this model, CFU-M and BFU-E are early progenitors possibly derived from a bipotent precursor (McLeod et al., 1976). An early megakaryocyte is analogous to the CFU-E, both having a similar capacity for nuclear replication. The symbols represent relative receptor density or avidity. The circles represent receptors for early regulatory factors, such as those present in conditioned media, while the squares represent receptors for late regulators—erythropoietin in the case of the erythroid series. As the cells mature, responsiveness to early regulatory influence declines while responsiveness to late regulatory factors increases.

number and volume of marrow megakaryocytes increases before any change is noted in the number of megakaryocytic colony-forming cells (CFU-M). Conversely, platelet hypertransfusion results are analogous to changes which occur in the erythroid lineage after induction of anemia or plethora. The late erythroid progenitor (CFU-E) responds acutely to changes in red cell demand, whereas the early progenitor, the erythroid burst-forming unit (BFU-E) does not.

Under certain conditions, CFU-M can be altered. A marked increase in the percent of CFU-M in DNA synthesis is observed in mice which have been lethally irradiated and transplanted with normal marrow. However, the number and cell cycle rate of CFU-M in regenerating marrow is the same in control as in platelet hypertransfused animals, and is thus independent of the platelet count.

We hypothesize that there are at least two levels of regulation of megakaryocytopoiesis. CFU-M is an early progenitor, closely related to CFU-S, and analogous to BFU-E and the granulocyte-macrophage colony-forming cell. At this level of megakaryocytopoiesis, regulatory mechanisms may include cell-derived products promoted by mitogen stimulation and the recognition of decreased marrow megakaryocytes. A later megakaryocytic progenitor cell, or perhaps the megakaryocyte itself is analogous to the CFU-E, being regulated by acute changes in platelet demand. A physiologic thrombopoietic regulator would act at this level of megakaryocytopoiesis, and not on the CFU-M.

ACKNOWLEDGMENTS
The authors appreciate the excellent technical assistance of Cathleen Wise and secretarial assistance of Linda Tufarolo. This work was supported by Grants HL24984, HL11775, and AM19410 from the National Institutes of Health, a Grant-in-Aid from the American Heart Association (77915), and the Veterans Administration.

References

Adamson, J. W., B. Torok-Storb, and N. Lin. 1978. Analysis of erythropoiesis by erythroid colony formation in culture. *Blood Cells* 4:89–103.

Barnes, D. W. H. and J. F. Loutit. 1967. Effects on irradiation and antigenic stimulation on circulating haemopoietic stem cells of the mouse. *Nature* 213:1142–1143.

Burstein, S. A., J. W. Adamson, D. Thorning, and L. A. Harker. 1979. Characteristics of murine megakaryocytic colonies *in vitro*. *Blood* 54:169–179.

Burstein, S. A., J. W. Adamson, and L. A. Harker. 1980. Megakaryocytopoiesis in culture: Modulation by cholinergic mechanisms. *J. Cell. Physiol.* 103:201–208.

Byron, J. W. 1972a. Comparison of the action of ^{3}H-thymidine and hydroxyurea on testosterone-treated hematopoietic stem cells. *Blood* 40:198–203.

Byron, J. W. 1972b. Evidence for an adrenergic receptor initiating DNA synthesis in hematopoietic stem cells. *Exp. Cell Res.* 71:228–232.

Ebbe, S. 1970. Megakaryocytopoiesis. In *Regulation of Hematopoiesis* (Vol. 2). Gordon, A. S., ed. New York: Appleton-Century-Crofts. pp. 1587–1610.

Gregory, C. J. and A. C. Eaves. 1978. Three stages of erythropoietic progenitor cell differentiation distinguished by a number of physical and biologic properties. *Blood* 51:527–537.

Hara, H. and M. Ogawa. 1976. Erythropoietic precursors in mice with phenylhydrazine-induced anemia. *Am. J. Hematol.* 1:453–458.

Harker, L. A. 1968a. Kinetics of thrombopoiesis. *J. Clin. Invest.* 47:458–465.

Harker, L. A. 1968b. Megakaryocyte quantitation. *J. Clin. invest.* 47:452–457.

Heath, D. S., A. A. Axelrad, D. McLeod, and M. M. Shreeve. 1976. Separation of the erythropoietin responsive progenitors BFU-E and CFU-E in mouse bone marrow by unit gravity sedimentation. *Blood* 47:777–792.

Iscove, N. N. 1978. Erythropoietin-independent stimulation of early erythropoiesis in adult marrow cultures by conditioned media from lectin-stimulated mouse spleen cells. In *Hematopoietic Cell Differentiation*. Golde, D. W., Cline, M. J., Metcalf, D., and Fox, C. F., eds. New York: Academic Press. pp. 37–52.

Iscove, N. N. 1977. The role of erythropoietin in regulation of population size and cell cycle of early and late erythroid precursors in mouse bone marrow. *Cell Tissue Kinet.* 10:323–334.

Kurland, J. I., J. W. Hadden, and M. A. S. Moore. 1977. Role of the cyclic nucleotides in the proliferation of committed granulocyte-macrophage progenitor cells. *Cancer. Res.* 37:4534–4538.

McLeod, D. L., M. M. Shreeve, and A. A. Axelrad. 1976. Induction of megakaryocyte colonies with platelet formation *in vitro*. *Nature* 261:492–494.

Metcalf, D. 1972. Effect of thymidine suiciding on colony formation *in vitro* by mouse hematopoietic cells. *Proc. Soc. Exp. Biol. Med.* 139:511–514.

Metcalf, D. 1977. Hemopoietic colonies: *In vitro* cloning of normal and leukemic cells. *Recent Results in Cancer Res.* 61:1–277.

Metcalf, D. and H. R. MacDonald. 1975. Heterogeneity of *in vitro* colony- and cluster-forming cells in the mouse marrow. Segregation by velocity sedimentation. *J. Cell. Physiol.* 85:643–654.

Metcalf, D., H. R. MacDonald, N. Odartchenko, and B. Sordat. 1975. Growth of mouse megakaryocyte colonies *in vitro*. *Proc. Nat. Acad. Sci. USA* 72:1744–1748.

Nakeff, A. and S. Daniels-McQueen. 1976. *In vitro* colony assay for a new class of megakaryocyte precursor: Colony-forming unit megakaryocyte (CFU-M). *Proc. Soc. Exp. Biol. Med. 15:587–590.*

Oshita, A. K., G. Rothstein, and G. Lonngi. 1977. cGMP stimulation of stem cell proliferation. *Blood* 49:585–591.

Popovic, W., J. E. Brown, and J. W. Adamson. 1977. The influence of thyroid hormones on *in vitro* erythropoiesis. *J. Clin. Invest.* 60:907–913.

Vassort, F., M. Winterholer, E. Frindel, and M. Tubiana. 1973. Kinetic parameters of bone marrow stem cells using *in vivo* suicide by tritiated thymidine or hydroxyurea. *Blood* 41:789–796.

Williams, N. and H. Jackson. 1978. Regulation of the proliferation of murine megakaryocyte progenitor cells by cell cycle. *Blood* 52:163–170.

Worton, R. G., E. A. McCulloch, and J. E. Till. 1969. Physical separation of hemopoietic stem cells from cells forming colonies in culture. *J. Cell. Physiol.* 74:171–182.

Discussion

Paulus: Your comparison of BFU-E and CFU-M is interesting and convincing. I would like to mention one slight difference between the two. Depletion of the red cell mass does not affect the levels of cycling of BFU-E. Why are CFU-M in cycle after platelet depletion?

Burstein: The cycling of CFU-M is not suppressed in the thrombocytotic model, so we have been trying to find alternative explanations. Perhaps some immunologic reaction results in the *in vivo* production of factors which will initiate these cells into the cycle, but which have really nothing to do with either demand for platelets or megakaryocytes. Thrombocytopenia may be regulated differently than thrombocytosis.

McDonald: The increase in CFU-M that occurs after antiserum injection takes approximately 65 hr. We see a very rapid increase in the acetylcholinesterase-positive cells in mice after antiplatelet antisera treatment, peaking at about 8 to 10 hr. Would you care to speculate on what kinds of cells are present in the marrow of mice that are responsible for the CFU-M increase 65 hr after administration of the antiserum?

Burstein: The increase at 65 hr occurs in the spleen, not in the marrow. CFU-M increases in the marrow at 90 and 120 hr. Now, what happens in the spleen? I am not convinced that the increase that is occurring in the spleen or even in the marrow at a later time, has anything to do with megakaryotytic regulation or thrombocytopenia, at all. CFU-M can be triggered into cell cycle within a couple of hours *in vivo*. It may be nonspecific. We have looked at that and with suppression, there is no decrease in the proportion of colonies with low cell numbers. Similarly, with thrombocytopenia, there is no increase in small colonies, either.

Mizoguchi: Do you count two megakaryocytes as a colony?

Burstein: Yes. We find that the two-cell colonies are clonal with respect to plating response.

Chervenik: I would like to ask about the appearance in the increase of megakaryocytes in the spleen before you see them in a marrow. Do you think this is because you have more CFU-M in the spleen or is this perhaps related to migration from the marrow to the spleen, which might have been induced by the antiplatelet antiserum? Have you looked at the CFU-M in the blood of these animals after they receive the antiplatelet serum?

Burstein: We have not looked at CFU-M in the blood. I don't know why the CFU-M in the spleen go up. It may be a specific response to platelet demand and it may also be due to factors induced by the immunologic reaction produced of antiplatelet serum with platelets. We are trying to separate those out now and maybe both nonspecific, as well as some specific responses are involved.

Published 1981 by Elsevier North Holland, Inc.
Evatt, Levine, and Williams, Editors
MEGAKARYOCYTE BIOLOGY AND PRECURSORS:
IN VITRO CLONING AND CELLULAR PROPERTIES

Induction of Human Megakaryocyte Colonies *In Vitro* and Ultrastructural Aspects of Their Maturation

W. Vainchenker, J. Breton-Gorius, and J. Chapman

Unité INSERM U.91. Hôpital Henri Mondor, 94010 Creteil, France and Unité INSERM U.35. Hôpital Henri Mondor, 94010 Creteil, France

Clonal assays for hemopoietic progenitors have permitted a better understanding of the regulation of murine and human granulopoiesis and erythropoiesis (Metcalf, 1977). Megakaryocyte colony assays were first available in the mouse (Burstein et al., 1979; McLeod et al., 1976; Metcalf et al., 1975; Nakeff and Daniels-McQueen, 1976; Williams et al., 1978) and more recently in man (Vainchenker et al., 1979; Weil et al., 1979). Little is known of the stimulators which promote growth *in vitro* of human megakaryoctye progenitors (CFU-M), of the role of serum in megakaryocyte colony formation and of events occurring during maturation which lead to platelet formation *in vitro*.

In the present study, we have compared the stimulating activity of conditioned medium from leukocytes stimulated by PHA (PHA-LCM) to that of erythropoietin (Epo); we have also investigated the serum requirement for megakaryocyte colony formation and the maturation of the different organelles of megakaryocytes in the presence or the absence of stimulating factors.

Methods and Materials

Samples

Samples were collected from the blood of normal volunteers. In one case, the blood sample was obtained from a patient with a chronic myeloid leukemia before treatment. Bone marrow was aspirated from either the sternum of patients exhibiting no hematological disorder or from the posterior iliac crest of normal donors for bone marrow transplantation.

Culture Method

The plasma clot technique (McLeod et al., 1976, 1978) was used as previously described (Vainchenker et al., 1979). The human AB serum was a pool of four different sera from normal blood donors, and was inactivated at 56°C for 30 min.

Light density cells (LDC) were separated by Ficoll-metrizoate centrifugation at 400 × g for 40 min at room temperature (Lymphoprep, Nyegaard, Oslo, Norway, density 1.077 g/ml). LDC were collected and subsequently washed 3 times in cool alpha medium. Adherent cells were removed by incubating 5×10^5 to 1×10^6 LDC/ml overnight in a Petri dish (Iscove et al., 1971). Nonadherent cells were washed once before plating. The stimulating factor was either Epo (Connaught step III) or PHA-LCM. Cultures were observed at 40× magnification with an inverted microscope. At day 11 or 12, cultures were dehydrated *in situ*, and then fixed with 5% glutaraldehyde. Harris hematoxylin counterstaining was performed, and megakaryocytes identified by cytologic examination.

For cultures incubated in the absence of serum, the following ingredients were added, as previously described (Guilbert, 1976): human transferrin (300 μg/ml) (Sigma), $FeCl_3 \cdot 6\ H_2O$ (1.6×10^{-6} M) (Merck), $Na_2SeO_3 \cdot 5.\ H_2O$ (0.173 g/1) (Merck).

In preliminary experiments, an emulsion of egg lecithin and cholesterol was added, but was latter omitted since it was not an obligatory requirement.

1% bovine serum albumin (Sigma) was routinely included in all experiments.

In one experiment, serotonin (Sigma) and calcium were added at concentrations of 1.4×10^{-8} M and 5×10^{-4} M respectively.

In order to test the effect of lipoprotein fractions, four major lipoprotein classes, i.e., very low density lipoprotein (VLDL), low-density lipoprotein (LDL) and high density lipoprotein subclasses HDL_2 and HDL_3 of human serum were isolated by a density gradient ultracentrifugation procedure (Chapman et al., 1979). The hydrated densities of these fractions were <1.017, 1.024–1.050, 1.067–1.100 and 1.100-1.130 g/ml, respectively. All fractions were extensively dialyzed against Hank's solution in the absence of phenol red. Lipoproteins were free of contamination by serum proteins as judged by immunological techniques. The final preparations were diluted in alpha medium to concentrations equivalent to those in the original whole serum.

Preparation of PHA-LCM

PHA-LCM was prepared as previously described (Wu, 1979) with 1% human serum. One batch was a generous gift of Dr. Wu. Some PHA-LCM were recovered at 72 hr of culture, others at day 7. The medium was centrifuged and subsequently filtered through a 0.22 μ filter. The medium was then frozen at -30°C in 1 ml aliquots.

Electron Microscopic Procedure

The fibrin network of cultures which had been grown in the absence or in the presence of a stimulating factor was lyzed with a solution of Pronase (0.1%) (Calbiochem). The free cells from cultures at day 7 until day 12 were first washed in Hank's medium containing fetal calf serum and then twice in Hank's medium alone. The cell pellet was examined by several cytochemical techniques. Platelet peroxidase (PPO) was revealed as previously described (Breton-Gorius et al., 1978); catalase which is present in granules distinct from alpha granules and dense bodies in bone marrow megakaryocytes (Breton-Gorius and Guichard, 1975) was revealed by incubation in alkaline diaminobenzidine medium (Novikoff et al., 1972). Dense bodies which represent serotonin storage organelles in platelets and which could also be detected in mature marrow megakaryocytes (White, 1971) were visualized by two different methods (Richards and DaPrada, 1977; White, 1971). The same methods were applied to bone marrow megakaryocytes.

Results and Discussion

Megakaryocyte Colonies

In the dishes, in the presence or in the absence of PHA-LCM, the first colonies appeared at day 7 and their maximum number was found between day 10 and day 12, as previously reported with Epo (Vainchenker et al., 1979). Scoring of colonies was usually performed at day 11. In all experiments spontaneous megakaryocyte colonies were observed, but the presence of Epo or PHA-LCM increased the number of megakaryocyte colonies (Table 1). However, these colonies were more numerous (1.5 ×) with PHA-LCM than with Epo. Furthermore, the size of the colonies was larger; colonies composed of 40 to 100 cells were more frequently observed. Under all conditions of stimulation, single megakaryocytes were rare. The other advantage

Table 1.

	NUMBER OF MK COLONIES / 10^6 CELLS.		
	SPONTANEOUS MK COLONIES	PHA - LCM 2.5 %	Epo 1.5 u/ml STEP III
I •	11	45	32
II	40	207	150
III	50	203	112
IV •	10	93	76
V	35	120	104
AVERAGE VALUE	29	134	95

COMPARISON OF THE NUMBER OF MK COLONIES WITHOUT STIMULATING FACTOR, WITH PHA - LCM AND WITH Epo.

• SAMPLES FROM DONORS FOR BONE MARROW TRANSPLANTATION.

(6.10^5 light density non adherent bone marrow cells plated.)

of PHA-LCM over Epo was to facilitate scoring of megakaryocyte colonies since cultures were not overcrowded by colonies other than megakaryocytes. In particular, no erythroid colonies were present and at the concentration of 2.5% PHA-LCM, granulocyte colonies were rare in the absence of adherent cells.

Relationship Between the Concentration of PHA-LCM and the Number of Megakaryocyte Colonies

The optimum concentration of PHA-LCM varied from one preparation to another. However, in all preparations, an inhibitory effect was observed when PHA-LCM was added at a concentration in excess of 5%. This inhibitory effect was usually higher for PHA-LCM recovered at day 7 than at day 3 (Fig. 1).

Role of Human Serum in Megakaryocyte Colony Formation

No megakaryocyte colony formation was observed in the absence of serum or lipoproteins, even in the presence of a stimulating factor (Epo) (Table 2). At a higher cellular concentration (6×10^5/ml), a similar number of megakaryocyte

Fig. 1. PHA-LCM recovered at 72 hr of culture has a similar megakaryocyte colony stimulating activity than the PHA-LCM recovered at 7 days of culture. However its inhibitory effect is less evident. This experiment was performed on a bone marrow plated at 2×10^5 cells/ml. Each point is the average value of two Petri dishes.

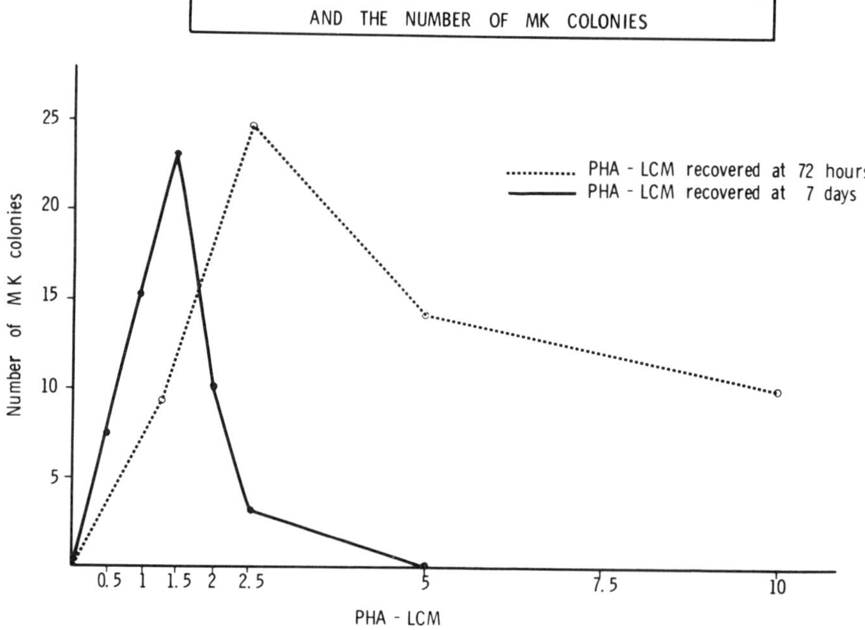

colonies was observed in the presence of 20% serum or the equivalent in LDL. At a lower cellular concentration (4 × 10^5 cells/ml), a slightly higher number of megakaryocyte colonies were grown in presence of LDL than with the whole serum. This result could suggest the presence of an inhibitor in serum and was confirmed by other experiments.

The relationships between the number of cells plated and the number of colonies as a function of serum concentration were studied using PHA-LCM as the stimulating factor (Fig. 2). At the usual concentration of 20% serum, the relationship was not linear. No megakaryocyte colony was observed at 1 × 10^5 plated cells. At 2.5% human serum, the relationship was linear with a regression that passed through the origin. For a concentration of 10%, the relationship was linear down to 2 × 10^5 cells/ml but with an inhibitory effect at 1 × 10^5 plated cells.

At a high cellular concentration (6 × 10^5 cells/ml), a similar number of megakaryocyte colonies was observed from 2.5% to 20% serum (Fig. 3). At a lower cellular concentration (2 × 10^5 cells/ml), an inhibitory effect on megakaryocyte colony formation was observed for serum concentrations greater than 10% (Fig. 3).

Thus, it appeared that low human serum concentrations (2.5–5%) are optimal for growth of megakaryocyte colonies.

In another experiment, the relationship between the number of megakaryocyte colonies and the number of plated cells was studied replacing the serum by LDL (Fig. 4). Under these conditions, a linear relationship with a regression passing

Table 2.

NUMBER OF MLDNA CELLS	NUMBER OF SPONTANEOUS MEGAKARYOCYTE COLONIES					
	20 % SERUM	LDL	VLDL	HDL_2	HDL_3	CONTROL WITHOUT SERUM
4. 10^5 CELLS	27	35	0	5	0	0
6. 10^5 CELLS	36	38	ND	ND	ND	0
6. 10^5 CELLS	26	27	ND	ND	ND	0

INFLUENCE OF SERUM (20 %) AND LOW DENSITY LIPOPROTEINS ON THE NUMBER OF SPONTANEOUS MK COLONIES.

MLDNA CELLS : Marrow light density non adherent cells.
LDL : Low density lipoproteins
VLDL : Very light density lipoproteins
HDL : High density lipoproteins
ND : Not determined

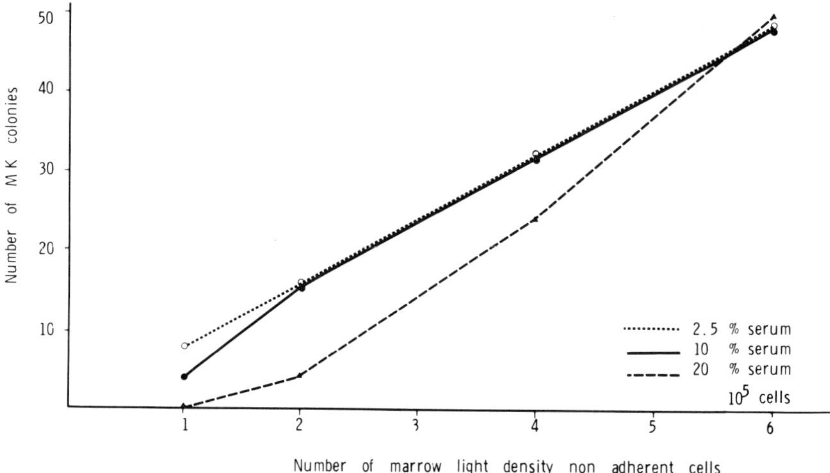

Fig. 2. At 20% serum concentration, the relationship between the number of megakaryocyte colonies and the number of plated cells is nonlinear. The linearity with a regression passing through the origin is obtained at 2.5% serum. Each point is the average value of two Petri dishes.

through the origin was observed when LDL were added to a concentration equivalent to 20% of the whole serum. This linear relationship was observed only with 2.5% serum.

Ultrastructural Studies

The *in vitro* maturation of megakaryocytes started at day 7 of culture and continued until day 12. Similar results were obtained on megakaryocytes cultured in the presence or absence of Epo. At day 7 and day 8, cells with large multilobulated nuclei were identified as megakaryocytes by the presence of platelet peroxidase which was specifically localized in the perinuclear space and endoplasmic reticulum (ER) (Fig. 5). Generally, no other organelle specific for megakaryocytes could be detected. Although a synchronism in the development of demarcation membranes and alpha granules was seen in the majority of megakaryocytes, a minority of megakaryocytes remained very immature at day 12 and resembled those found at day 8 of culture (Fig. 6). As seen *in vivo*, the first demarcation membranes appeared in clusters in a zone lacking ribosomes (Fig. 7). Then, demarcation membranes divided the cytoplasm into platelet territories, more or less regularly (Fig. 8).

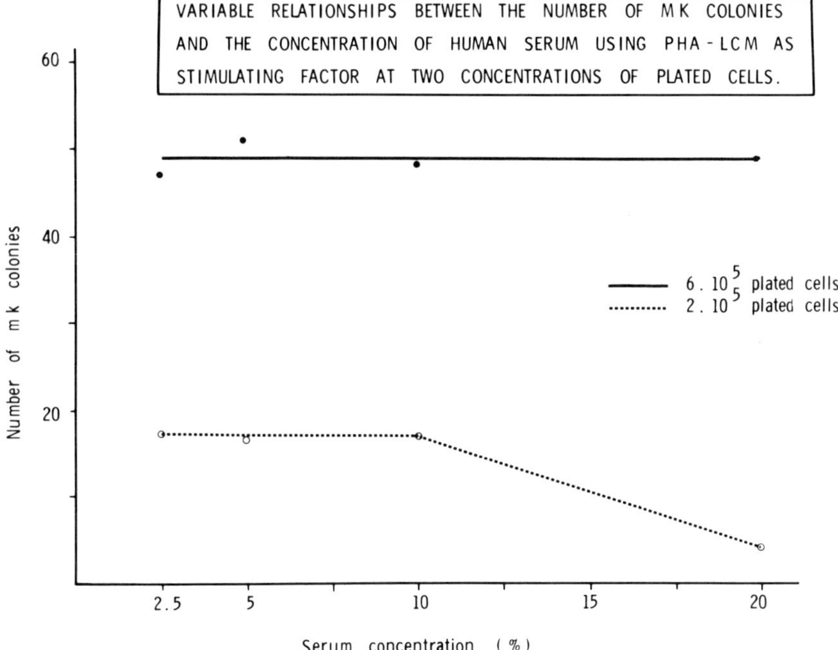

Fig. 3. The results are from the same experiment as in Fig. 2. At 6 × 10^5 plated cells the number of megakaryocyte colonies is identical from 2.5% to 20% serum. At 2 × 10^5 plated cells, the number of megakaryocyte colonies does not vary from 2.5% to 10% serum, but is much lower at 20% serum.

Glycogen particles were observed in clusters either adjacent to the nucleus or in cytoplasmic blebs which were devoid of organelles (Fig. 8). Alpha granules with a typical nucleoid were found in the Golgi zone (Fig. 9). Microfilaments were present at the cell periphery; in addition, this zone may contain smooth endoplasmic reticulum, glycogen particles and a few microtubules but no ribosomes, mitochondria, granules or membranes (Fig. 10). Numerous microtubules could be identified throughout the cytoplasm (Fig. 11). In marrow megakaryocytes, the use of the cytochemical technique for detection of the peroxidatic activity of catalase revealed this enzyme to be present in small granules distinct from alpha granules (Fig. 12). These granules, which resemble microperoxisomes found in many other tissues (Novikoff et al., 1972), were also produced in cultured megakaryocytes (Fig. 13). Another distinct class of granules, dense bodies, could also be demonstrated in mature marrow megakaryocytes, when White's saline (White, 1971) was used (Fig. 14). However, this type of granule was absent from cultured megakaryocytes (Fig. 15). The method based on detection of adenine nucleotide which permitted selectively staining the membrane surrounding dense bodies in megakaryocytes (Richards

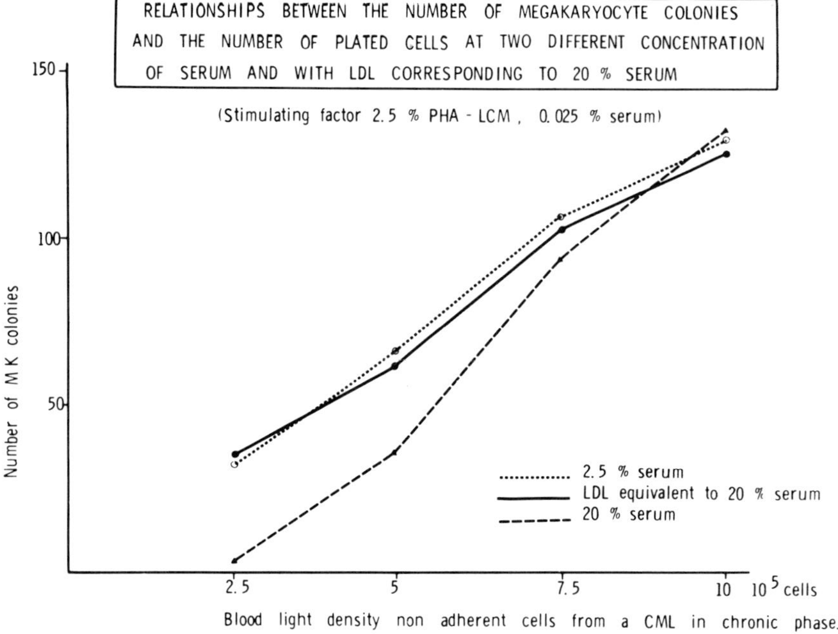

Fig. 4. A linear relationship between the number of megakaryocyte colonies and the number of plated cells with a regression passing through the origin is obtained with LDL corresponding to 20% serum. LDL has the megakaryocyte growth supporting activity of the serum, but not its inhibitory effect.

and DaPrada, 1977) also failed to reveal a reaction in megakaryocytes growing *in vitro*. The addition of serotonin and calcium to the culture medium gave similar results.

It was previously shown that human megakaryocyte colonies (Vainchenker et al., 1979) as well as murine megakaryocyte colonies (McLeod et al., 1976) could be grown by the plasma clot technique using Epo as the stimulating factor. However, in man, spontaneous megakaryocyte colony formation could be observed, but in decreased numbers, thus suggesting that factors other than Epo could stimulate megakaryocyte colony formation (Vainchenker et al., 1979). In the mouse, conditioned media from spleen cells stimulated by a mitogen are able to sustain megakaryocyte colony formation (Burstein et al., 1979; McLeod et al., 1978; Metcalf et al., 1975; Nakeff and Daniels-McQueen, 1976). In a similar way, PHA-LCM is able to sustain human megakaryocyte colony formation, underlining the similarities between human and murine stimulators of megakaryocytopoiesis *in vitro*. It has been previously demonstrated that PHA-LCM contained GM-CSA with two different activities, one necessary for granulocyte colony formation (Cline and Golde, 1974), the second one enhancing the proliferation of granulocyte progenitors

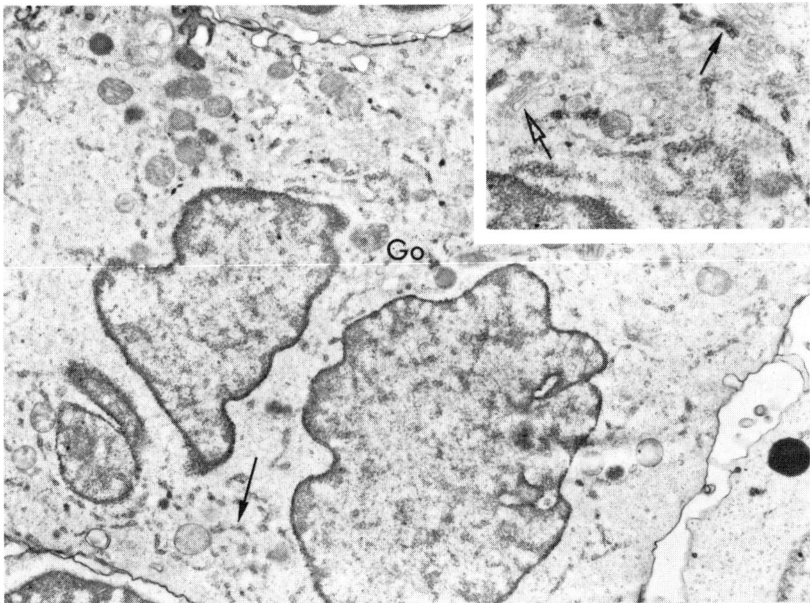

Fig. 5. Ultrastructural aspects of a megakaryocyte from peripheral blood CFU-M, at day 8 of culture, in the presence of Epo. Reaction for platelet peroxidase (PPO). Several nuclear lobes are apparent. A weak PPO reaction is seen in the perinuclear space and cisternae of endoplasmic reticulum (arrow). The cytoplasm does not contain other signs of megakaryocyte maturation. Numerous mitochondria are seen. The Golgi apparatus (Go) is enlarged in inset. 100,000 X. *Inset:* The Golgi cisternae are unreactive (white arrow) while PPO reaction product is visible in ER (black arrow). A similar reaction is observed in megakaryoblasts *in vivo.* 18,900 X.

in suspension culture (Pre-CFU-GM), a burst promoting activity (Fauser and Messner, 1978) and a T lymphocyte growth stimulatory activity (Wu et al., 1977). Fauser and Messner (1978) have also reported that PHA-LCM in association with Epo could stimulate the growth of mixed colonies containing megakaryocytes. Weil et al. (1979), have recently reported that PHA-LCM could stimulate human megakaryocyte colony formation in agar but with a low plating efficiency.

In this study, the stimulating capacity of PHA-LCM for megakaryocyte colony formation was found to be higher than that of Epo since the number and the size of the colonies were larger. This activity is probably different from Epo since no erythroid colonies were observed with PHA-LCM. In fact, the higher cloning efficiency observed with PHA-LCM may underestimate the real number of CFU-M since an inhibitory effect was observed for concentrations of medium over 5%. Purification of these media will be necessary in order to obtain maximum colony growth.

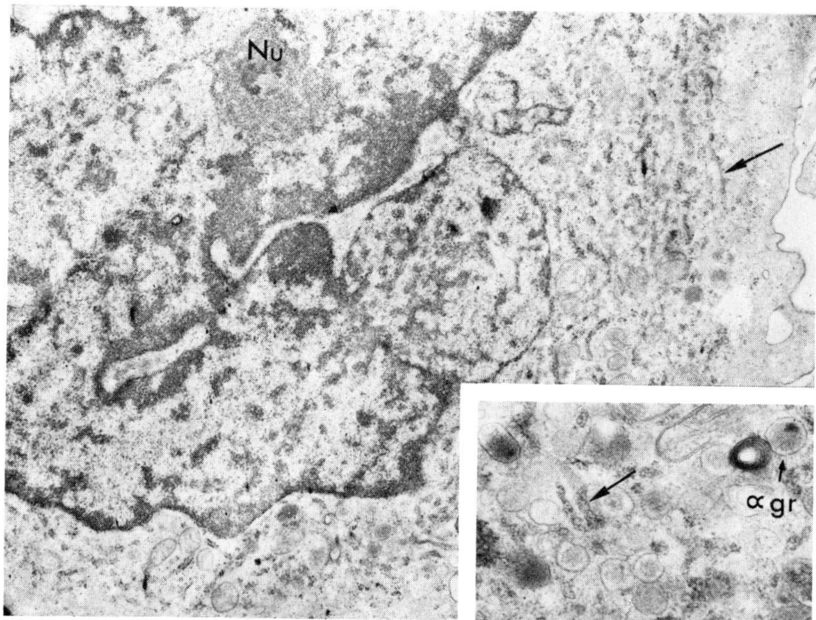

Fig. 6. Ultrastructural aspects of a megakaryocyte from peripheral blood CFU-M, at day 12 of culture, in the absence of Epo. Reaction for peroxidase. The polylobulated nucleus possesses a nucleolus (Nu). The cytoplasm which is very immature contains numerous reactive ER cisternae (arrow). 11,400 X. *Inset:* A higher-power view, however reveals that the synthesis of alpha granules (gr) has started. Note that the megakaryocyte illustrated in Fig. 8, which is very mature, comes from the same culture. 26,000 X.

Using PHA-LCM, a non-linear relationship between the number of megakaryocyte colonies and the number of plated cells was observed using a 20% concentration of serum as previously described for Epo (Vainchenker et al., 1979). This result led us to investigate the exact role of serum in megakaryocyte colony formation. The influence of serum or plasma on megakaryocyte colony formation remains a controversial topic. Metcalf et al. (1977, 1980) have shown that some human sera have a potentiating activity for murine megakaryocyte colony formation and for the growth of multipotential colonies. In the mouse, Nakeff (1977) has shown that horse serum leads to a better maturation than fetal calf serum. Recently, Weil et al. (1979) have also used fetal calf serum and horse serum in preference to human serum for the growth of human megakaryocyte colonies. In man, human sera are able to support the growth of erythroid and granulocytic colonies. Aye et al. (1979) have demonstrated that this ability was linked to LDL. In this study, LDL was also found to correspond to the growth supporting activity of megakar-

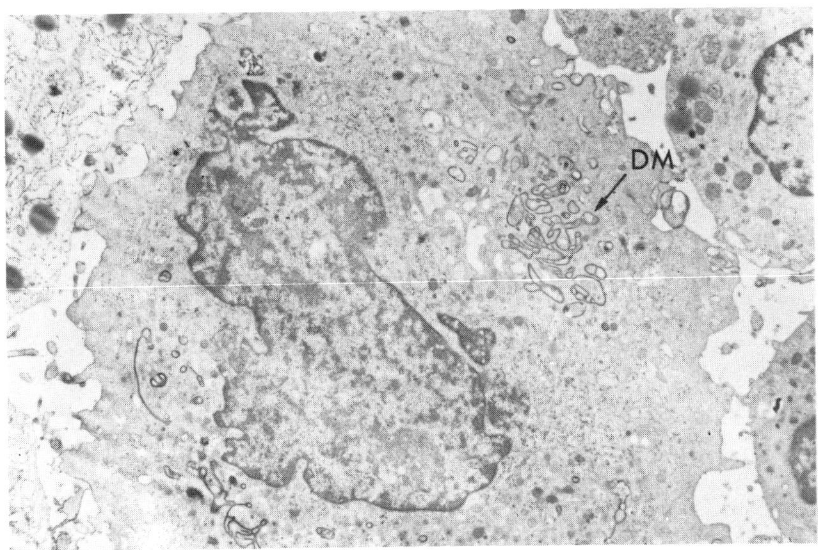

Fig. 7. Megakaryocyte from bone marrow CFU-M at day 12 of culture, in the presence of Epo. This large megakaryocyte is surrounded by macrophages. The nucleus is immature and the cytoplasm exhibits numerous ER and free ribosomes, with the exception of the zone in which a cluster of demarcation membranes (DM) appears. 5,500 ×

yocyte colony formation of human serum. However, as for erythroid and granulocytic colony formation, the exact role (nutritional, regulatory) of LDL remains to be investigated. In addition, the replacement of total serum by LDL has permitted demonstration of the presence of an inhibitor of megakaryocyte colony formation in the whole serum. In fact, using LDL, a linear relationship was observed between the number of plated cells and the number of megakaryocyte colonies. The inhibitory effect of serum could be found only at low cellular concentration. However, when the serum concentration was lowered to 2.5%, this inhibitory effect was suppressed. Thus, for optimal growth, megakaryocyte colony formation requires a low serum concentration.

Several inhibitors have been previously described in human sera. In particular, LDL are known to be able to inhibit the response of human lymphocytes to mitogens (Curtiss and Edgington, 1976). Chan et al. (1971) have noted that normal human sera could inhibit granulocyte colony formation when added to fetal calf serum probably by the intermediary of a lipoprotein. Baker and Galbraith (1978) have shown that human sera contain inhibitory and stimulatory factors which regulate granulocyte colony formation. The inhibitor was thermolabile and was partially destroyed by heat inactivating the sera for 30 min at 56°C. Recently, an inhibitor of cell proliferation has also been described in normal sera (fetal calf or human) which selectively inhibit the multiplication of some cell lines. This inhibitor could

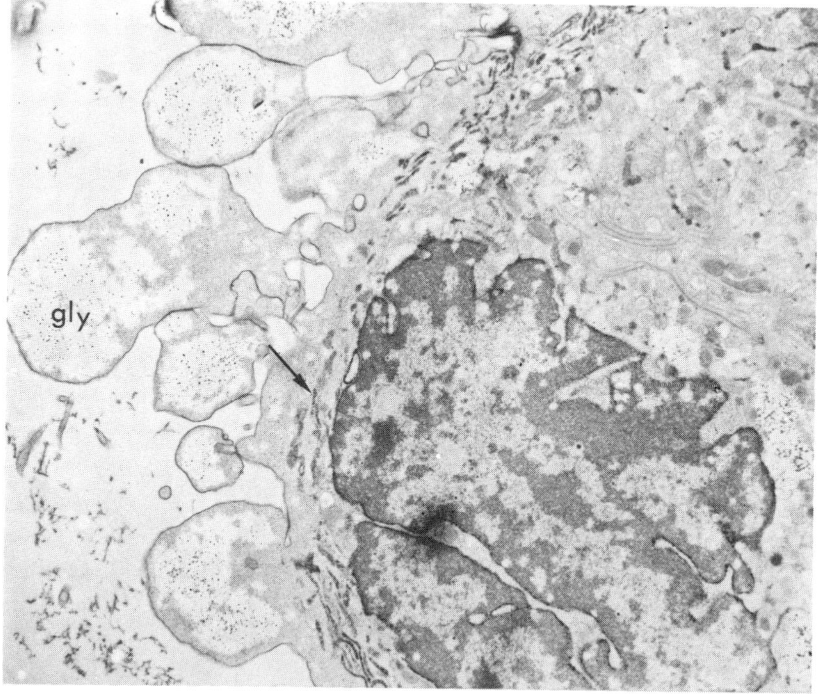

Fig. 8. Megakaryocyte from peripheral blood CFU-M, at day 12 of culture, in the absence of Epo. Reaction for peroxidase. The nucleus with a condensed chromatin exhibits foldings. A strong reaction of PPO is observed (arrow). Demarcation membranes and granules are well developed. Clusters of glycogen (Gly) which are partially extracted are observed, near the nucleus and in cytoplasmic blebs. 10,000 ×

be identified as a protein which migrates as an α globulin (Harrington and Godman, 1980). Until now, no inhibitor of erythroid colonies has been described in human serum. For the megakaryoctye lineage, Burstein et al., (1979) have shown that spleen cell-conditioned media prepared with human plasma stimulates megakaryocyte colony formation to a lesser extent than those prepared in serum-free conditions. However, they did not publish direct evidence for an inhibitor of megakaryocyte colony formation in human plasma. Further studies are now necessary to investigate whether the inhibitor of megakaryocyte colony formation in human serum can be related to those previously described or whether it represents a specific factor of regulation for human megakaryopoiesis.

Cytochemical detection of platelet peroxidase in megakaryocytes growing *in vitro* has shown that this enzyme appears before the specific organelles of megakaryocyte lineage as is the case *in vivo* (Breton-Gorius et al., 1978). Thus, platelet peroxidase constitutes a good marker for identification of these precursors. All the organelles

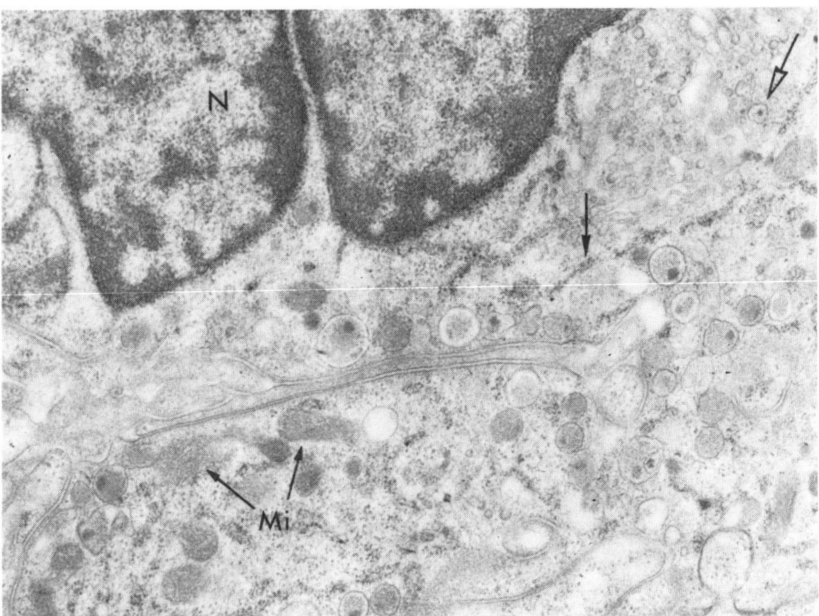

Fig. 9. Megakaryocyte from bone marrow CFU-M, at day 12 of culture, in the presence of Epo. As in the preceeding figures, peroxidase reaction product is seen in perinuclear cisterna and ER (black arrow) but the Golgi cisternae are unreactive. A small immature alpha granule (white arrow) is located at the periphery of the Golgi cisternae. Note the presence of typical alpha granules with dense nucleoids and mitrochondria (Mi) which are surrounded by long sheets of demarcation membranes. 20,300 ×

produced *in vivo*, i.e., alpha-granules and demarcation membranes (Behnke and Perdersen, 1974; Breton-Gorius et al., 1978; Zucker-Franklin, 1979), catalase-containing granules (Breton-Gorius and Guichard, 1975), microtubules (Behnke and Perdersen, 1974), microfilaments (Zucker-Franklin, 1979) and glycogen particles (Breton-Gorius and Reyes, 1976) are also synthesized in megakaryocytes maturing *in vitro*. As described for murine megakaryocyte colony formation (McLeod et al., 1976; Williams et al., 1978), production and shedding of platelets from human megakaryocytes are sometimes observed (Vainchenker et al., 1979; Vainchenker and Breton-Gorius, 1979). However, the dense bodies which normally appear in marrow mature megakaryocytes (White, 1971) are not produced *in vitro*, in spite of the addition of calcium and serotonin. This lack could be due to a defect in the terminal phase of maturation. We are now investigating whether an improvement of the maturation may lead to the production of such dense bodies in megakaryocytes stimulated by PHA-LCM. It has been confirmed that megakaryocyte maturation is identical in culture in the presence or absence of Epo (Vainchenker and Breton-Gorius, 1979). Since the source of Epo from the serum could be elim-

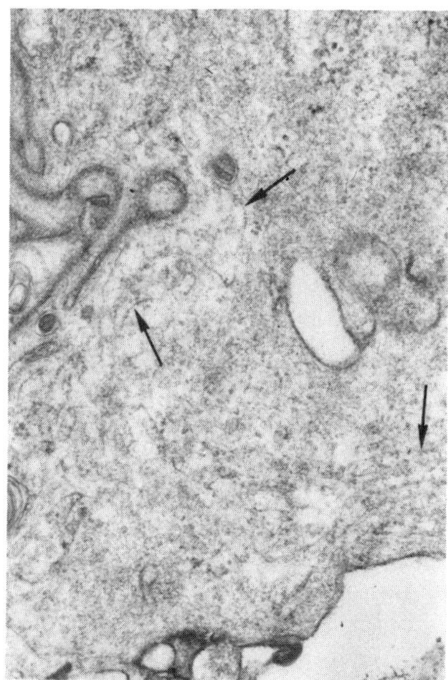

Fig. 10. High-power view of the periphery of a megakaryocyte, from marrow CFU-M at day 12. Demarcation membranes do not penetrate in this region in which numerous microfilaments could be identified (arrows). 32,000 ×

Fig. 11. Portion of a megakaryocyte from blood CFU-M at day 10. Perinuclear cisterna contain platelet peroxidase. Numerous transverse sections of microtubules (arrows) could be seen. 43,500 ×

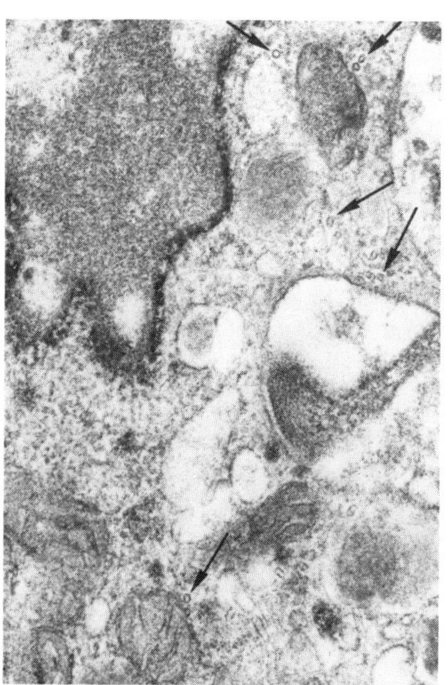

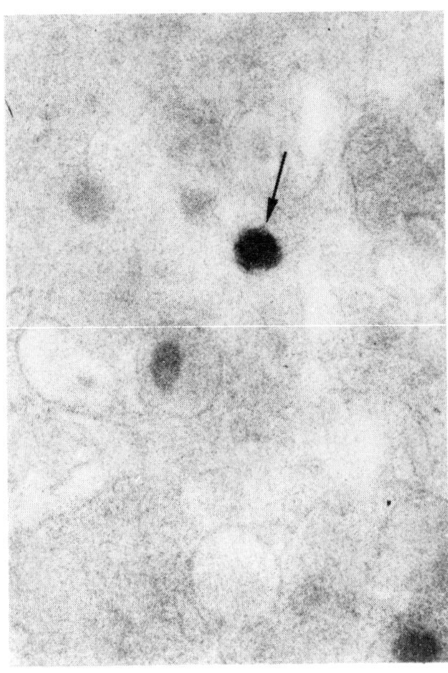

Fig. 12. Bone marrow megakaryocyte incubated in alkaline diaminobenzidine medium. A catalase-containing granule is identified (arrow) while alpha granules are unreactive. 43,600×

inated by its replacement with LDL, it appears unlikely that Epo itself acts on the maturation of megakaryocytes.

Ultrastructural observations reveal that at day 12, a few megakaryocytes remain very immature. It is unknown whether these cells are derived from a more primitive CFU-M analogous to primitive BFU-E and which matures much later than mature BFU-E (Gregory and Eaves, 1977), or whether this delay in maturation represents a defect due to culture conditions.

This study has permitted us to establish conditions permitting use of the CFU-M assay quantitatively even in pathological conditions. We are now developing an immunological technique to permit identification of the abnormal megakaryocytes which may grow in malignant diseases. We anticipate that this culture method used in conjunction with ultrastructural analysis and immunological techniques will permit study of the mechanisms involved in acquired and abnormal megakaryopoiesis.

Summary

Growth of human megakaryocyte colonies in culture has been previously performed using erythropoietin (Epo) as stimulating factor. In the present study, we report that PHA-leukocyte-conditioned medium also induces megakaryocyte colony formation with a slightly higher plating efficiency than Epo, the number of megakaryocyte per colony being increased. An inhibitory effect of human serum

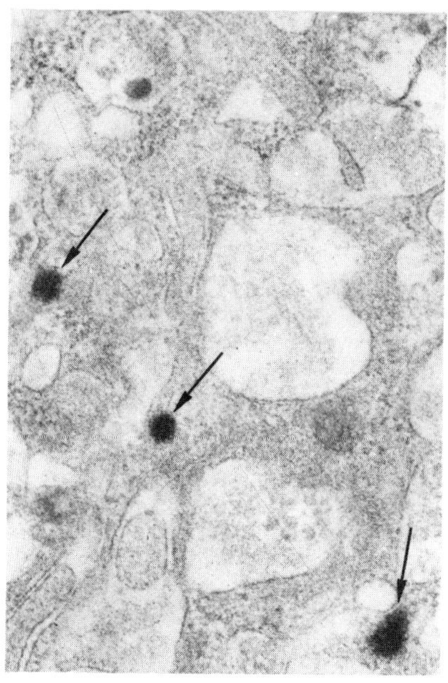

Fig. 13. Megakaryocyte from day 13 culture of blood CFU-M after treatment as in Fig. 12. Catalase-containing granules appear similar to those of marrow megakaryocytes. An unreactive alpha granule is located at the top. 43,600×

Fig. 14. Bone marrow mature megakaryocyte treated by the White's saline. Three dense bodies (arrow) could be very clearly distinguished from alpha granules by the high density of their center. 35,600×

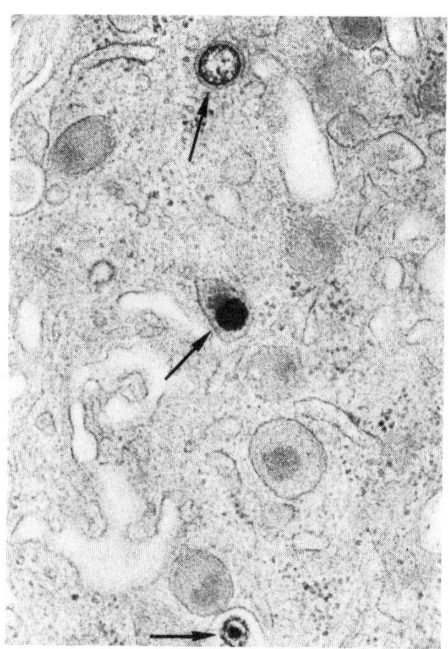

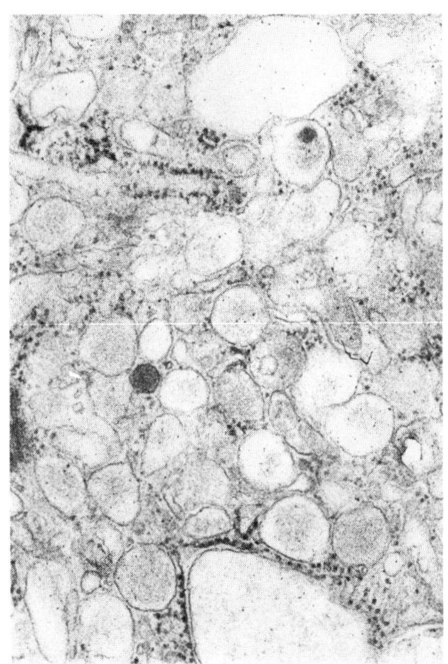

Fig. 15. Megakaryocyte from day 12 culture of blood CFU-M after treatment as in Fig. 14. Numerous granules are produced but dense bodies are absent. 35,600×

on the growth of megakaryocyte colonies was observed at low concentration of cells. This effect could be suppressed by substituting the heat-inactivated-serum by low density lipoprotein, which retained the growth supporting activity of the serum. In order to assess whether this serum inhibitor is specific for megakaryocyte now requires further investigation. Ultrastructural analysis of megakaryocyte maturation showed that platelet peroxidase, the demarcation membrane system, microtubules, glycogen, alpha-granules and catalase-containing granules developed *in vitro*. However, under the present conditions of culture, the dense granules which represent the amine-storing organelles are not produced in cultured megakaryocyte. We anticipate that this culture method used in conjunction with ultrastructural studies will permit investigation of the mechanism involved in some defects of megakaryocytopoiesis.

ACKNOWLEDGMENTS
We thank Dr. Wu for the gift of a batch of PHA-LCM. We are grateful to J. Bouguet and J. Guichard for excellent technical assistance. Supported by INSERM CRL 79 5 011, 78 5 320 and ATP 82 79 114.

References

Aye, M. T., Y. Niho, J. E. Till, and E. A. McCulloch. 1974. Studies of leukemic cell populations in culture. *Blood* 44:205–219.

Aye, M. T., J. A. Seguin, and J. P. McBurney. 1979. Erythroid and granulocytic colony growth in cultures supplemented with human serum lipoproteins. *J. Cell. Physiol.* 99:233–246.

Baker, F. L. and P. R. Galbraith. 1978. Nutritional and regulatory roles of human serum in cultures of human granulopoietic cells. *Blood* 52:241–253.

Behnke, O. and N. T. Perdersen. 1974. Ultrastructural aspect of megakaryoctye maturation and platelet release. In *Platelets: Production, Function, Transfusion and Storage*. Baldini, M. G. and Ebbe, S., eds. New York: Grune and Stratton. pp. 21–31.

Breton-Gorius, J. and J. Guichard. 1975. Two different types of granules in megakaryocytes and platelets as revealed by the diaminobenzidine reaction. *J. Microscop. Biol. Cell.* 23:197–202.

Breton-Gorius, J. and F. Reyes. 1976. Ultrastructure of human bone marrow cell maturation. *Int. Rev. Cytol.* 46:251–321.

Breton-Gorius, J., F. Reyes, G. Duhamel, A. Najman, and N. C. Gorin. 1978. Megakaryoblastic acute leukemia: identification by the ultrastructural demonstration of platelet peroxidase. *Blood* 51:45–60.

Burstein, S. A., J. W. Adamson, D. Thorning, and L. A. Harker. 1979. Characteristics of murine megakaryocytic colonies *in vitro*. *Blood* 54:169–179.

Chan, S. H., D. Metcalf, and E. R. Stanley. 1971. Stimulation and inhibition by normal human serum of colony formation *in vitro* by bone marrow cells. *Br. J. Haematol.* 20:329–341.

Chapman, J., M. C. Taggart, and S. Goldstein. 1979. Density distribution characterization and comparative aspects of the major serum lipoproteins in the common marmoset (*Callithrix Jacchus*), a new world primate with potential use in lipoprotein research. *Biochemistry* 18:5096–5108.

Cline, M. J. and D. W. Golde. 1974. Production of colony stimulating activity by human lymphocytes. *Nature* 248:703–704.

Curtiss, L. K. and T. S. Edgington. 1976. Regulatory serum lipoproteins: Regulation of lymphocyte stimulation by a species of low density lipoprotein. *J. Immunol.* 116:1452–1458.

Fauser, A. A. and H. A. Messner. 1978. Granulo-erythropoietic colonies in human bone marrow, peripheral blood and cord blood. *Blood* 52:1243–1248.

Fauser, A. A. and H. A. Messner. 1979. Identification of megakaryocytes, macrophages and eosinophils in colonies of human bone marrow containing neutrophilic granulocytes and erythroblasts. *Blood* 53:1023–1027.

Gregory, C. J. and A. C. Eaves. 1977. Human marrow cells capable of erythropoietic differentiation *in vitro*. Definition of three erythroid colony responses. *Blood* 49:855–864.

Guilbert, L. J. and N. N. Iscove. 1976. Partial replacement of serum by selenite, transferrin, albumin and lecithin in haemopoietic cell cultures. *Nature* 263:594–595.

Harrington, W. N. and G. C. Godman. 1980. A selective inhibitor of cell proliferation from normal serum. *Proc. Nat. Acad. Sci. USA* 77:423–427.

Iscove, N. N., J. S. Senn, J. E. Till, and E. A. McCulloch. 1971. Colony formation by normal and leukemic marrow cells in culture. Effect of conditioned medium from human leukocytes. *Blood* 37:1–5.

McLeod, D. L., M. M. Shreeve, and A. A. Axelrad. 1978. Culture systems *in vitro* for the assay of erythrocytic and megakaryocytic progenitors. In *In Vitro Aspects of Erythropoiesis*. Murphy, M. J., Jr., ed. New York: Springer-Verlag. pp. 31–36.

McLeod, D. L., M. M. Shreeve, and A. A. Axelrad. 1976. Induction of megakaryoctye colonies with platelet formation *in vitro*. *Nature* 261:492–494.

Metcalf, D. 1977. Megakaryocyte colony formation. In *Hemopoietic Colonies. In vitro Cloning of Normal and Leukemic Cells*. Berlin: Springer-Verlag. pp. 150–159.

Metcalf, D., G. R. Johnson, and A. W. Burgess. 1980. Direct stimulation by purified GM-CSF of the proliferation of multipotential and erythroid precursor cells. *Blood* 55:138–147.

Metcalf, D., H. R. McDonald, N. Odartchenko, and B. Sordat. 1975. Growth of mouse megakaryoctye colonies *in vitro*. *Proc. Nat. Acad. Sci. USA* 72:1744–1748.

Nakeff, A. 1977. Colony forming unit-megakaryocyte (CFU-M): its use in elucidating the kinetics and humoral control of the megakaryocytic committed progenitor cell compartment. In *Experimental Hematology Today*. Baum, S. J. and Ledney, G. D., eds. New York: Springer-Verlag. pp. 111–123.

Nakeff, A. and S. Daniels-McQueen. 1976. *In vitro* colony assay for a new class of megakaryoctye precursor: Colony-forming unit megakaryoctye (CFU-M). *Proc. Soc. Exp. Biol. Med.* 151:587–590.

Novikoff, A. B., P. M. Novikoff, C. Davis, and N. Quintana. 1972. Studies on microperoxisomes. II. A cytochemical method for light and electron microscopy. *J. Histochem. Cytochem.* 20:1006–1023.

Richards, J. G. and M. DaPrada. 1977. Uranaffin reaction: a new cytochemical technique for the localization of adenine nucleotides in organelles storing biogenic amnes. *J. Histochem. Cytochem.* 25:1322–1336.

Vainchenker, W. and J. Breton-Gorius. 1979. Differentiation and maturation *in vitro* of human megakaryocytes from blood and bone marrow precursors. In *Cell Lineage, Stem Cells and Cell Determination*. INSERM Symposium no. 10. Douarin, N. Le, ed. Elsevier North Holland Biochemical Press. pp. 215–226.

Vainchenker, W., J. Guichard, and J. Breton-Gorius. 1979. Growth of human megakaryocyte colonies in culture from fetal, neonatal and adult peripheral blood cells. Ultrastructural analysis. *Blood Cells* 5:25–42.

Vainchenker, W., J. Bouguet, J. Guichard, and J. Breton-Gorius. 1979. Megakaryocyte colony formation from human bone marrow precursors. *Blood* 54:940–945.

Weil, S. C., N. Williams, H. Jackson, T. P. McDonald, E. M. Rabellino, and M.A.S. Moore. 1979. The effect of erythropoietin, thrombopoietin and PHA-leukocyte conditioned medium on the *in vitro* growth of human megakaryocytic precursors. *Blood* 54 (Suppl. 1):167.

White, J.G. 1971. Serotonin storage organelles in human megakaryocytes. *Am. J. Pathol.* 63:403–408.

Williams, N., H. Jackson, A. P. C. Sheridan, M. J. Murphy, Jr., A. Elste, and M.A.S. Moore. 1978. Regulation of megakaryopoiesis in long-term bone marrow cultures. *Blood* 51:245–255.

Wu, A. M. 1979. Properties and separation of T lymphocyte growth stimulatory activity (TL-CSA) and of granulocyte-macrophage colony stimulatory activity (GM-CSA) produced separately from two human T lymphocyte subpopulations. *J. Cell. Physiol.* 101:237–250.

Wu, A. M., R. Ruscetti, and R. C. Gallo. 1977. Cell-factor interaction in normal human and leukemia cell populations. In *Experimental Hematology Today*. Baum, S. J., and Ledney, G. D., eds. New York: Springer-Verlag. pp. 165–176.

Zucker-Franklin, D. 1979. The ultrastructure of megakaryocytes and platelets. In *Regulation of Hematopoiesis* (Vol. II). Gordon, A. S., ed. New York: Appleton Century Crofts. pp. 1553–1586.

Discussion

Groopman: Is there a difference between the number of mixed colonies that were mixed megakaryocyte-erythroid versus pure megakaryocytic with the lymphocyte-conditioned medium versus erythropoietin?

Breton-Gorius: We use a very low concentration of leukocyte-conditioned medium. We observed a minimum number of granulocytic colonies, since monocytes in the cultures are the only source of CSF. With this medium, no erythroid colonies occur.

Williams: So you have only pure megakaryocytic colonies.

Breton-Gorius: Yes. It is an advantage to use this medium to easily quantitate the number of megakaryocyte colonies, since they constitute the majority of colonies.

Groopman: Have you tried serum from hyperlipemic patients, who have high circulating LDL?

Breton-Gorius: No, but it is our intention.

Warheit: I didn't see, in your cultured megakaryocytes, the platelet-peroxidase reaction along the nuclear envelope that you have shown so beautifully in *in vivo* megakaryocytes.

Breton-Gorius: There is a reaction, but probably you cannot see it very easily because to detect all the organelles we have to counterstain the section with lead and thus, since the chromatin is very heavily stained, it is very difficult to see the reaction in the perinuclear system. I must add that the majority of cells possess the platelet peroxidase reaction, but a few megakaryocytes are totally devoid of it. I don't know if it is the equivalent feature to acetylcholinesterase in the mouse, but it seems that some enzymes typical of megakaryocytes are not present in cultured megakaryocytes.

Long: Have you seen in culture the small uninuclear platelet peroxidase-positive cells that you have reported before, and that are found in bone marrow? Also, are there any colonies composed of these cells in culture?

Breton-Gorius: Yes. You see these cells more frequently early, at day 7 of culture, and generally, in the adult, there are only a few small cells at day 12. But it is not the same phenomena in the fetal liver or cultures derived from fetal blood. In these circumstances, many colonies are constituted of very small megakaryocytes, very difficult to recognize by morphology alone, and in these cases the use of electron microscopy and fluorescent antiplatelet antiserum helps to detect the megakaryocytic nature of these cells.

Levine: With the use of the antiplatelet serum, how many more megakaryocytes or megakaryocyte colonies could you detect than with other methods?

Breton-Gorius: There may be a few colonies with just one large megakaryocyte, but several small ones that probably would not have been detected without the fluorescent reaction. We have not counted the difference at this time.

Levine: I have a comment concerning the dense bodies in platelets and megakaryocytes. They are often very difficult to see by routine electron microscopy and, as you know, it's been reported that preincubation with serotonin helps their detection. Additionally, a short culture with ethidium bromide has led to the observation of large numbers of dense bodies even in the youngest megakaryocytes. Such a procedure might be useful in future studies.

Zuckerman: I noticed that you use low density, nonadherent cells for culturing. Was there a specific reason for that? Have you found that the adherent cells, presumably mostly monocytes, influenced the incidence of megakaryocyte colonies?

Breton-Gorius: The major reason was to remove fibroblasts which are extremely numerous in human bone marrow cultures.

Zuckerman: Does removal of monocytes from peripheral blood change the number of megakaryocyte colonies grown?

Breton-Gorius: We are now investigating this question. Since spontaneous megakaryocyte colonies are obtained without adding stimulating factors, probably mononuclear cells produce specific factors for megakaryocyte colony growth.

Williams: In our hands, removal of T cells from human peripheral blood has not influenced the numbers of megakaryocyte colonies grown.

Hoffmann: I have worked with plasma clot cultures for a number of years and have always been impressed by the distortion of morphology of mononuclear cells in these cultures. I have found it difficult to quantitate numerically, megakaryocyte colonies, just using light microscopy. What are your morphological criteria for defining the megakaryocyte in such a system?

Breton-Gorius: First, the density and appearance of the nucleus. The nucleus is multilobulated and very dense compared with the round nucleus of the macrophage. The appearance of the periphery of the cell which shows cytoplasmic blebs which are characteristics of megakaryocytes, but not macrophages. These characteristics are, however, not sufficient. Platelet-peroxidase is useful only for electron microscopy and it is absolutely necessary to develop a marker analogous to acetylcholinesterase.

Hoffmann: The conclusions that you are drawing from your ultrastructural studies are difficult for me to understand. You are not plucking individual colonies. You are lysing a clot and getting a cell pellet. The megakaryocytes may be those that survive in culture for 10-12 days and are not necessarily the progeny of CFU-M.

Williams: That is a concern. Many single large megakaryocytes are found in the cultures, and these must also be present in your preparation.

Breton-Gorius: It is possible to examine individual colonies. Since megakaryocytes are so spread, it is so much work to cut one colony and to have many sections and observe only 2 or 3 cells. We have preferred to take all megakaryocytes growing in colonies in culture to examine the megakaryocyte morphology using the electron microscope.

Walz: At what time do you first observe the structure that you identify as the alpha granule? Does that structure only appear at a particular time?

Breton-Gorius: In general, they occur between day 8 and 9 of culture.

Walz: Can you quantify these structures per megakaryocyte?

Breton-Gorius: We have not quantitated the alpha granule production. Sometimes they are irregularly distributed.

Williams: Have you repeated the experiments of Fauser and Messner, of adding PHA-leukocyte-conditioned medium and erythropoietin several days apart, using your better culture conditions?

Breton-Gorius: We observed mixed colonies with the two factors, but the number of pure megakaryocyte colonies is not increased by the addition of both erythropoietin and leukocyte-conditioned medium.

Published 1981 by Elsevier North Holland, Inc.
Evatt, Levine, and Williams, Editors
MEGAKARYOCYTE BIOLOGY AND PRECURSORS:
IN VITRO CLONING AND CELLULAR PROPERTIES

Megakaryocytes in Human Mixed Hemopoietic Colonies

H. A. Messner and A. A. Fauser

The Ontario Cancer Institute, Department of Medicine, University of Toronto and Institute of Medical Science, University of Toronto, Ontario, Canada.

Since the initial observation by Metcalf et al. (1975) and Nakeff and Daniels-McQueen (1976), a number of investigators have reported the clonal growth of megakaryocytes in cultures of murine hemopoietic cells. They can either be detected as pure megakaryocytic colonies (Metcalf and MacDonald, 1976; Nakeff and Daniels-McQueen, 1976; Freedman et al., 1977; Williams et al., 1978; Penington, 1979; Burstein et al., 1979) or alternatively in mixed hemopoietic colonies (McLeod et al., 1976; Johnson and Metcalf, 1977; Hara and Ogawa, 1978; Humphries et al., 1979) associated with cells of other cell lineages. Particularly, the association of megakaryocytes and erythropoietic cells in culture was repeatedly reported (McLeod et al., 1976, Hara and Ogawa, 1978; Humphries et al., 1979). These observations suggest that it might be feasible to define progenitors of megakaryocytopoiesis in culture at various levels of maturation. Furthermore, information is available that different regulator molecules are required for early and late stages of megakaryocyte development (Williams et al., 1978; Burstein et al., 1979).

Recently, conditions were described that support the growth of human megakaryocytes in culture. Vainchenker et al. (1979) observed in cultures of human fetal and adult blood pure megakaryocytic colonies as well as megkaryocytes associated with erythroid bursts. We have demonstrated megakaryocytes within mixed colonies that contain, besides erythroblasts, granulocytes and macrophages and were derived from human pluripotent hemopoietic progenitors (Fauser and Messner, 1979).

It is the purpose of this communication to examine the pattern of nuclear and cytoplasmic maturation in megakaryocytes that developed in human mixed hemopoietic colonies as progeny of pluripotent precursors.

Methods and Materials

Patients

Bone marrow samples were obtained from normal bone marrow transplant donors and centrifuged in LSM (Litton, Bionetics, Kensington, Md.) to yield a mononuclear cell suspension of density less than 1.077 g/ml.

Culture Conditions

Mixed hemopoietic colonies were grown as previously described (Fauser and Messner, 1978, 1979). Briefly, mononuclear cells were mixed with alpha-medium (Flow Laboratories) or modified Dulbecco's MEM (Aye et al., 1979), with 30% fetal calf serum (FCS), 5% medium conditioned by leukocytes in the presence of phytohemagglutinin (PHA-LCM) (Aye et al., 1974), and 0.9% methylcellulose as viscous support. Aliquots of 0.9 ml were placed in 35 mm Petri dishes and 1 unit of erythropoietin was added on day 4 of culture. The cultures were incubated at 37°C in a humidified atmosphere, supplemented with 5% CO_2. Mixed colonies were usually evaluated 12–14 days after initial plating.

Identification of Megakaryocytes

Megakaryocytes in culture can be either observed within mixed hemopoietic colonies or in loose aggregates of 3–10 cells without the presence of additional hemopoietic cell lineages. They can be identified directly in culture by their large size and well demarcated, translucent, hyaline cytoplasm (Fig. 1), using an inverted microscope of high optical quality (Zeiss, IM 35). These typical properties facilitate their distinction from more yellowish granulated macrophages. The nature of these cells can be verified by cytological analysis. For this purpose, colonies were removed from the culture dishes by micropipette. Slides were prepared either by cytocentrifugation (Fauser and Messner, 1978) (Shandon Southern Instruments Ltd.) or by blowing the colony containing microdroplets directly onto glass slides. These preparations were examined routinely by Wright stain. Some specimens were stained for acid phosphatase using 10% formalin in 0.1 m acetate buffer as fixative and alpha-naphthol AS-MX phosphate in combination with fast blue RR diazonium salt as substrate (Markovic and Shulman, 1977).

Results and Discussion

Frequency of Megakaryocytes in Mixed Hemopoietic Colonies

Approximately 20% of mixed hemopoietic colonies contain in addition to erythroblasts, granulocytes and macrophages, also megakaryocytes as identified by direct observation (Fauser and Messner, 1979) and cytological analysis by Wright stain. In addition, these cells stained positive for acid phosphatase. The frequency of megakaryocytes within individual colonies ranged from 2–50. Replating of primary colonies yielded occasionally secondary mixed colonies that contained megakaryocytes. No predominant association between megakaryocytes and erythroid colonies or bursts was observed under the employed culture conditions.

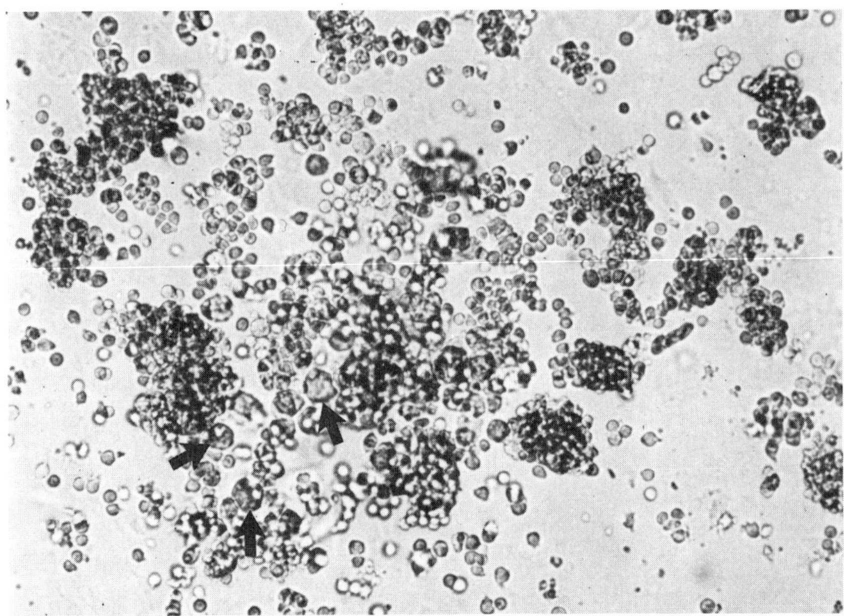

Fig. 1. Mixed hemopoietic colony with large, translucent cells (see arrows) that were subsequently verified as megakaryocytes by cytological analysis.

In addition, within the same plates, loose aggregates of 3–10 megakaryocytes were observed independent of mixed colonies or erythroid bursts.

Maturation of Megakaryocytes in Mixed Colonies

A total of 163 megakaryocytes identified in cytological preparations of mixed colonies and stained by Wright stain were assessed for their degree of polyploidy and cytoplasmic maturation. The number of nuclear segments was counted for each megakaryocyte and the cytoplasm evaluated on the basis of the following staining properties (Fig. 2): (1) homogeneous dark blue cytoplasm, (2) homogeneous light blue or grayish cytoplasm, (3) light blue cytoplasm with fine granulation, (4) light blue cytoplasm with coarse granulation.

The majority of megakaryocytes (Fig. 3) contained 2–5 nuclear segments (75%). Only 25% demonstrated a higher degree of polyploidy, with as many as 12 or 16 segments. Cells with all stages of cytoplasmic maturation were observed. Simultaneous assessment of polyploidy and cytoplasmic maturation (Fig. 4) indicated that cells with a small number of nuclear segments were heterogeneous with respect to their cytoplasmic appearance; that is, cells of all 4 levels of maturation were observed. Megakaryocytes with a large number of nuclear segments usually demonstrated mature cytoplasmic features.

Megakaryocytes from human bone marrow can be grown under culture conditions that support the formation of mixed hemopoietic colonies derived from pluripotent

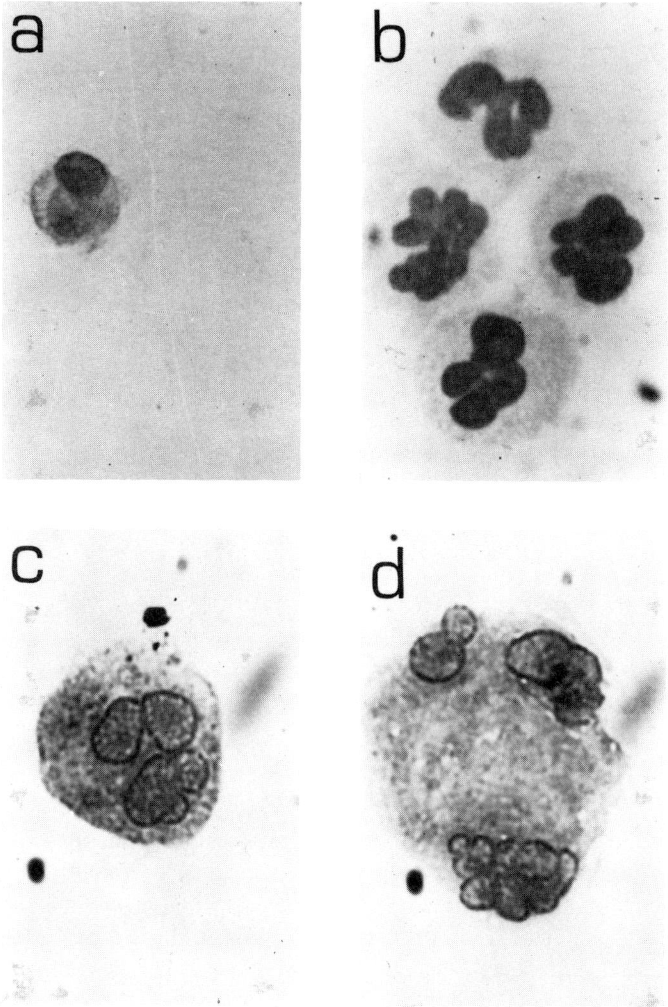

Fig. 2. Representation of 4 different stages of megakaryocyte maturation observed in culture (Wright stain): (a) cells with one or two nuclear segments and dark blue, homogeneous cytoplasm; (b) cells with light blue homogeneous cytoplasm; (c) cells with granulated cytoplasm; (d) cells with multilobed nuclei and coarse cytoplasmic granulation.

hemopoietic progenitors. Their identification is feasible either by direct observation using an inverted microscope or by cytological analysis. Megakaryocytes can be detected within mixed colonies. Occasionally they form loose aggregates of indi-

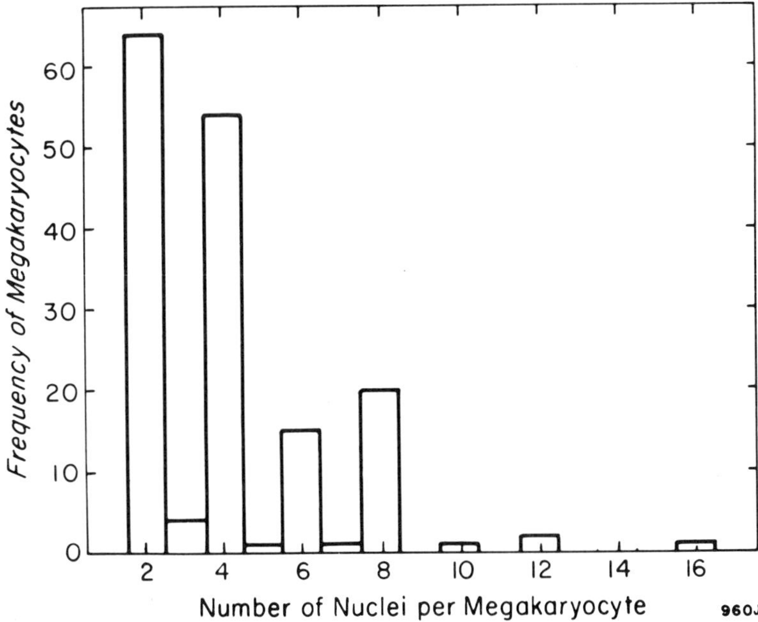

Fig. 3. Assessment of the degree of polyploidy in 163 megakaryocytes grown within mixed hemopoietic colonies. Each bar represents the number of megakaryocytes with the indicated number of nuclear segments.

vidual megakaryocytes without association of other hemopoietic cell lineages. Morphologically, the maturation pattern of megakaryocytes in culture was found to resemble that observed in direct preparations of bone marrow; however, more commonly cells with 2–4 nuclear segments were identified rather than cells with a high degree of polyploidy and mature coarse granulated cytoplasm. The coarse cytoplasmic granulation of multinucleated megakaryocytes is suggestive of platelet production, but so far no proof has been obtained for their formation in culture.

We conclude that megakaryocytes can be identified in cultures of human bone marrow as progeny of pluripotent hemopoietic progenitors. This observation is further supported by the demonstration of megakaryocytes in secondary mixed hemopoietic colonies. In contrast to data reported for murine megakaryocyte cultures (McLeod et al., 1976; Hara and Ogawa, 1978; Humphries et al., 1979), a consistent association between megakaryocytes and erythroid components was not established. Since maturation can be observed in cultures, it is conceivable that the system will serve as a useful tool to examine mechanisms that regulate commitment of human pluripotent progenitors towards megakaryocytopoiesis and promote megakaryocyte maturation in culture. The observation of molecules regulating early and late events in murine megakaryocytopoiesis (Williams et al., 1978; Humphries et al., 1979) may lead to some understanding of certain human megakaryocyte disorders.

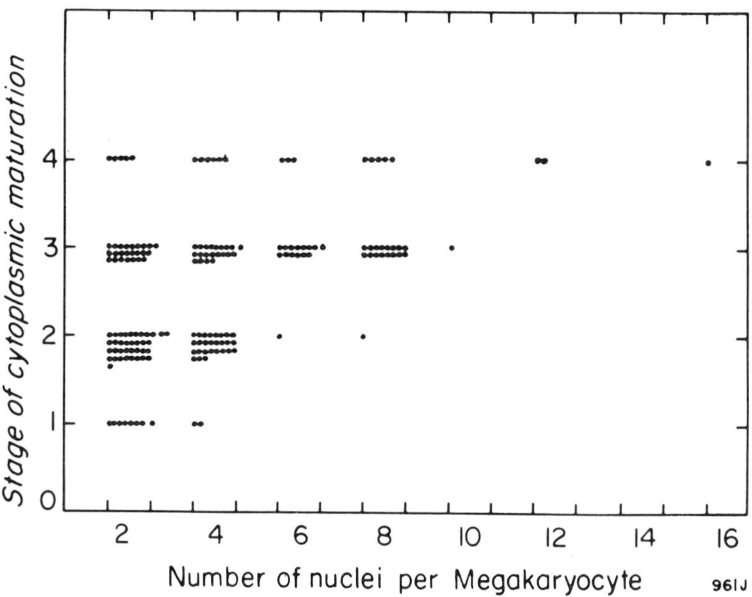

Fig. 4. Simultaneous assessment of polyploidy and cytoplasmic maturation as observed by different staining properties (see text and Fig. 2).

Summary

Pluripotent hemopoietic progenitors form mixed colonies when cultured with medium conditioned by leukocytes in the presence of PHA (PHA-LCM) and erythropoietin. Approximately 20% of these mixed colonies contain megakaryocytes in addition to granulocytes and erythroblasts. Using an inverted microscope they can be identified directly within some colonies by their size and their hyaline translucent appearance. These properties facilitate their distinction from macrophages that usually display a more granular yellowish cytoplasm. Analysis of individual mixed colonies by Wright stain revealed that the number of megakaryocytes within a single colony may vary between two and fifty. Cells with morphological criteria of megakaryocytes vary in their degree of maturation within the same colony from immature megakaryocytes with one or two nuclear segments and dark bluish cytoplasmic stain to cells with as many as twelve or sixteen nuclear segments and the typical granular light blue cytoplasm. The majority of megakaryocytes was found to be of the mature type with two to eight nuclear segments. These cells stain positive for acid phosphatase using 10% formalin in 0.1 molar acetate buffer as fixative and alpha-naphthol AS-MX phosphate in combination with fast blue RR diazonium salt as substrate.

In conclusion, megakaryocytes can be identified in mixed hemopoietic colonies indicating the pluripotent nature of their ancestors. The assays provide a unique system to examine regulatory mechanisms involved in the commitment towards megakaryocytopoiesis.

ACKNOWLEDGMENTS
Supported by MRC, University of Toronto, A. A. Fauser is a Fellow of Medical Research Council of Canada.

References

Aye, M. T., Y. Niho, J. E. Till, and E. A. McCulloch. 1974. Studies of leukemic cell populations in culture. *Blood* 44:205–219.

Aye, M. T., J. A. Seguin, and J. P. McBurney. 1979. Erythroid and granulocytic colony growth in cultures supplemented with human serum lipoproteins. *J. Cell. Physiol.* 99:233–246.

Burstein, S. A., J. W. Adamson, S. K. Erb, and L. A. Harker. 1979. Murine megakaryocytopoiesis: Evidence for early and late regulatory control. *Blood* 54 (Suppl. 1):165a.

Fauser, A. A. and H. A. Messner. 1978. Granulo-erythropoietic colonies in human bone marrow, peripheral blood, and cord blood. *Blood* 52:1243–1347.

Fauser, A. A. and H. A. Messner. 1979. Identification of megakaryocytes, macrophages, and eosinophils in colonies of human bone marrow containing neutrophilic granulocytes and erythroblasts. *Blood* 53:1023–1026.

Freedman, M. H., T. P. McDonald, and E. F. Saunders. 1977. The megakaryocyte progenitor (CFU-M): *In vitro* proliferative characteristics. *Blood* 50 (Suppl. 1):146.

Hara, J. and M. Ogawa. 1978. Murine hemopoietic colonies in culture containing normoblasts, macrophages and megakaryocytes. *Am. J. Hematol.* 4:23–24.

Humphries, R. K., P. B. Jacky, F. J. Dill, A. C. Eaves, and C. J. Eaves. 1979. CFU-S in individual erythroid colonies derived *in vitro* from adult mouse marrow. *Nature* 297:718–720.

Johnson, G. R. and D. Metcalf. 1977. Pure and mixed erythroid colony formation *in vitro* stimulated by spleen conditioned medium with no detectable erythropoietin. *Proc. Nat. Acad. Sci. USA.* 74:3879–3882.

McLeod, D. L., M. M. Shreeve, and A. A. Axelrad. 1976. Induction of megakaryocytic colonies with platelet formation *in vitro*. *Nature* 261:492–494.

Markovic, O. S. and N. R. Shulman. 1977. Megakaryocyte maturation indicated by methanol inhibition of acid phosphatase shared by megakaryocytes and platelets. *Blood* 50:905–914.

Metcalf, D., H. R. MacDonald, N. Odarthchenko, and B. Sordat. 1975. Growth of mouse megakaryocyte colonies *in vitro*. *Proc. Nat. Acad. Sci. USA* 72:1744–1748.

Nakeff, A. and S. Daniels-McQueen. 1976. *In vitro* colony assay for a new class of megakaryocyte precursor: Colony-forming unit megakaryocyte (CFU-M). *Proc. Soc. Exp. Biol. Med.* 151:587–590.

Penington, D. G. 1979. Megakaryocyte colony culture using a liver cell conditioned medium. *Blood Cells* 5:13–23.

Vainchenker, W., J. Guichard, and J. Breton-Gorius. 1979. Growth of human megakaryocyte colonies in culture from fetal, neonatal, and adult peripheral blood cells: Ultrastructural analysis. *Blood Cells* 5:25–42.

Williams, N., H. Jackson, A. C. P. Sheridan, M. J. Murphy, A. Elste, and M. A. S. Moore. 1978. Regulation of megakaryocytopoiesis in long-term murine bone marrow culture. *Blood* 51:245–255.

Discussion

Williams: The ratio of mixed colonies to megakaryocyte colonies seems to be quite different from what Dr. Breton-Gorius reported before. Have you discussed your differences with her? She sees many pure megakaryocyte colonies and few mixed colonies and you see a lot of mixed colonies and few pure megakaryocyte colonies. Do you have any explanation for that?

Messner: I have not focused on pure megakaryocytic colonies. Also, our culture systems are different. We are using the methyl cellulose system which probably influences the cellular composition. Arthur Axelrad would suggest that the majority of erythroid bursts in plasma clot cultures contain megakaryocytes, while the most common association in our methyl cellulose cultures is granulopoiesis and erythropoiesis. The culture conditions that we employ are all different. I have not made the attempt to use the serum source of human material that might give us an advantage to stimulate megakaryocyte production. The percentage of PHA-leukocyte-conditioned medium we have used is 5% and allows the maturation of granulocytes, as well; which is again slightly different from that given by Dr. Breton-Gorius.

Williams: This is a terribly important issue. If a number of laboratories want to use the same methods to look at mixed colonies, particularly in disease states, it is going to be terribly difficult to make comparisons.

Messner: I agree with you.

Williams: You must have started to look at these mixed colonies in leukemic situations. Do you see a change in the proportion of mixed colonies which contain megakaryocytes in any syndrome?

Messner: I can't answer that question. I have looked at mixed colonies in patients with leukemia, chronic as well as acute. One can observe these mixed colonies in these patients, but I have not performed any specific examination of megakaryocytes (*Blood Cells* [in press]).

Inoue: You mentioned that the plating efficiency was improved by using 2-mercaptoethanol. Does that refer only to the mixed colonies, or all types of colonies?

Messner: Three types of colonies were compared. The mixed colonies show a significant increase up to about 4- to 5-fold. The erythroid colonies derived from the BFU-E show a 40 to 50 percent increase; however, the frequence of CFU-C usually doesn't change significantly.

Inoue: What type of cells do you think that it is affecting?

Messner: I don't know. First of all, cells live longer; which means you can observe them quite healthy after 18, 20, and 21 days, so it could very well be that the sulfhydryl groups sustain cell membranes, and the effect is rather non-specific. Since the erythroid components are the most vulnerable cells in our culture conditions, 2-mercaptoethanol may just be an improvement to bring out the erythroid components.

Levine: First, I would like to say that I agree completely that those are very probably all megakaryocytes; I agree with your criteria. Second, it is misleading to speak about megakaryocyte nuclei. The megakaryocyte was named 90 years ago to distinguish the cell with one nucleus from the cell with many nuclei, the osteoclast, and there have been a number of further observations: Paulus showed that each nuclear lobe is not an equal integer. I'm skeptical that you can make an accurate count of the number of nuclear lobes. Some of the youngest megakaryocytes definitely have more than 2–4 lobes. I think that lobe counts are always questionable.

Messner: I think that that criticism is very well taken. It was an attempt to try and do nothing more than identify cells of various types and see whether one can, in fact, observe a maturation pattern in the cultures.

Nakeff: How did you attempt to quantitate maturation in your system? Did you place the cells into four maturation categories? How did you determine these?

Messner: By simple cytoplasmic criteria: Very dark bluish stains of the most primitive form; a light bluish stain with a fairly coarse granulation, for the more mature forms of megakaryocytes.

Shulman: Did you use a nuclear/cytoplasmic ratio as one criterion? Do you think that is possible?

Messner: We did not look at that parameter.

Levine: I think so, based on information to be published (*Brit. J. Haemat.* 45:487, 1980).

Published 1981 by Elsevier North Holland, Inc.
Evatt, Levine, and Williams, Editors
MEGAKARYOCYTE BIOLOGY AND PRECURSORS:
IN VITRO CLONING AND CELLULAR PROPERTIES

Ploidization of Megakaryocyte Progenitors *In Vitro*

J. M. Paulus, M. Prenant, J. Maigne, M. Henry-Amar, and J. F. Deschamps

Institut de Pathologie Cellulaire, INSERM U.48, l'Association Claude Bernard et l'Ecole Pratique des Hautes Etudes, Hôpital de Bicêtre, Le-Kremlin Bicêtre, France, l'Unité de Recherches Statistiques, INSERM U. 21, Villejuif, France and the Departement de Clinique et Pathologie Médicales, Institut de Médecine, Université de Liége, Belgium

In normal rodents and man platelets are released from megakaryocytes which have arrested DNA synthesis at the 8c, 16c, 32c or 64c ploidy level, that is after 2–5 endoreduplications (Ebbe, 1976; Paulus, 1970). Little is known about the factors which determine the final ploidy level reached by the endoreduplicating megakaryoblasts. Experimental or pathological alterations of the platelet count induce changes in megakaryocyte ploidy, cytoplasmic size and rate of DNA synthesis which tend to restore platelet count to normal levels (Harker and Finch, 1969; Odell et al., 1976; Paulus, 1970). Potentiators have been described which increase the percentage of megakaryocytes having high ploidy values (Williams et al., 1981). In the present experiments *in vitro* plasma clot cultures were used to study the relationship between ploidization and the mitotic amplification of megakaryocyte progenitors. It was found that the number of endoreduplications (NbE) expressed by colony megakaryocytes was inversely related to the number of cells produced per colony, in terms of the number of doublings (NbD) undergone by the colony progenitor. The results might suggest that a deficient culture medium causes a block in ploidization for colonies originating from the most ancestral or proliferative progenitors or that the differentiation of megakaryocyte progenitors may involve a mechanism insuring a balance between mitotic and endoreduplicative amplifications.

Methods and Materials

Bone marrow from C57B1 mice was cultured in the plasma clot medium supplemented with 3 U. erythropoietin (NIH, Bethesda) per ml (McLeod et al., 1978). After 5, 7 or 9 days incubation the preparations were fixed with glutaraldehyde and sequentially stained for acetylcholinesterase (AchE) and DNA. Ultrastructural examination of untreated and cultured mouse bone marrow showed that AchE can be first demonstrated in the Golgi apparatus of megakaryoblasts and also is present in the endoplasmic reticulum in more mature megakaryocytes (Fig. 1). Maps recording the x–y coordinates of each megakaryocyte were systematically drawn. In those preparations plated at the concentration of 3×10^4 nucleated cells per ml, colonies were generally well separated and the NbD undergone in the megakaryocyte pathway by the colony progenitor was calculated as 1.44 times the napaian logarithm of the number of megakaryocytes. Megakaryocyte ploidy was measured *in situ* by scanning cytophotometry (Paulus et al., unpublished data) and NbE was calculated as (1.44 log ploidy)-1. Ultrastructural demonstration of AchE was done according to Tranum–Jensen and Behnke (1977).

Results and Discussion

Ploidy determinations were made on two sets of cultures, plated at the concentrations of 2×10^5 cells/ml (one of the six identical dishes examined on day 7) or 3×10^4 cells/ml (two groups of nine dishes, sampled equally on days 5, 7 and 9). In the first set, the ploidy distribution was as follows: 2c 8.1%; 4c 21.8%; 8c 30.5%; 16c 22.8%; 32c 15.3% and 64c 1.5%. Separate mapping of 2c–4c and 32c–64c megakaryocytes clearly showed that the former tend to be concentrated in aggregates while the latter were generally isolated cells (Fig. 2a). Because many aggregates were confluent, their allocation to individual colonies often proved impossible. For this reason a coefficient of closeness was calculated for each megakaryocyte as $\Sigma\ (1/d^2)f(d)$ where f(d) was the number of megakaryocyte neighbors at distance d (with $0 < d < 150$ mμ). The factor $1/d^2$ served to give proportionately more weight to close neighbors than to distant ones. Using this mathematical formulation it was shown that greater proximity was associated with low ploidy (Fig. 2b), indicating that the ploidy of a given megakaryocyte tends to decrease as a function of the number and proximity of its megakaryocyte neighbors.

Since in the above experiment ploidy could not be correlated with colony size because of confluence, experiments were conducted at the lower plating concentration of 3×10^4 cells/ml. Two types of colonies were found, some purely megakaryocytic (M-colonies) or others mixed with erythroblasts and sometimes granulocytes (ME-colonies). Both types of colonies could contain in addition dispersed macrophages. Compared to M-col, ME-col were less numerous (33.0 ± 5.7 vs $96.0 \pm 12.7/10^5$ cells), peaked later (7–9 days vs 3–5) and underwent a greater number of doublings (3.9 ± 1.2 vs 1.1 ± 1.3), as will be detailed elsewhere (Paulus et al., unpublished data). The two types of colonies also differed in the mean NbE undergone by their megakaryocytes (Fig. 3), which was 1.3 ± 0.7 for ME–col vs 2.2 ± 0.8 in M–col ($p < 0.001$).

For each M- or ME-col, the mean NbE was plotted as a function of the number of doublings completed by the colony progenitor (Fig. 4). A significant negative correlation was found between these two variables at days 5 ($r = -0.39$; $p < 0.01$), 7 ($r = -0.05$; $p < 0.01$) and 9 ($r = -0.44$; $p < 0.01$). In ME–col, the mean NbE tended to be uniformly low so that no significant correlation was found between mean NbE and NbD.

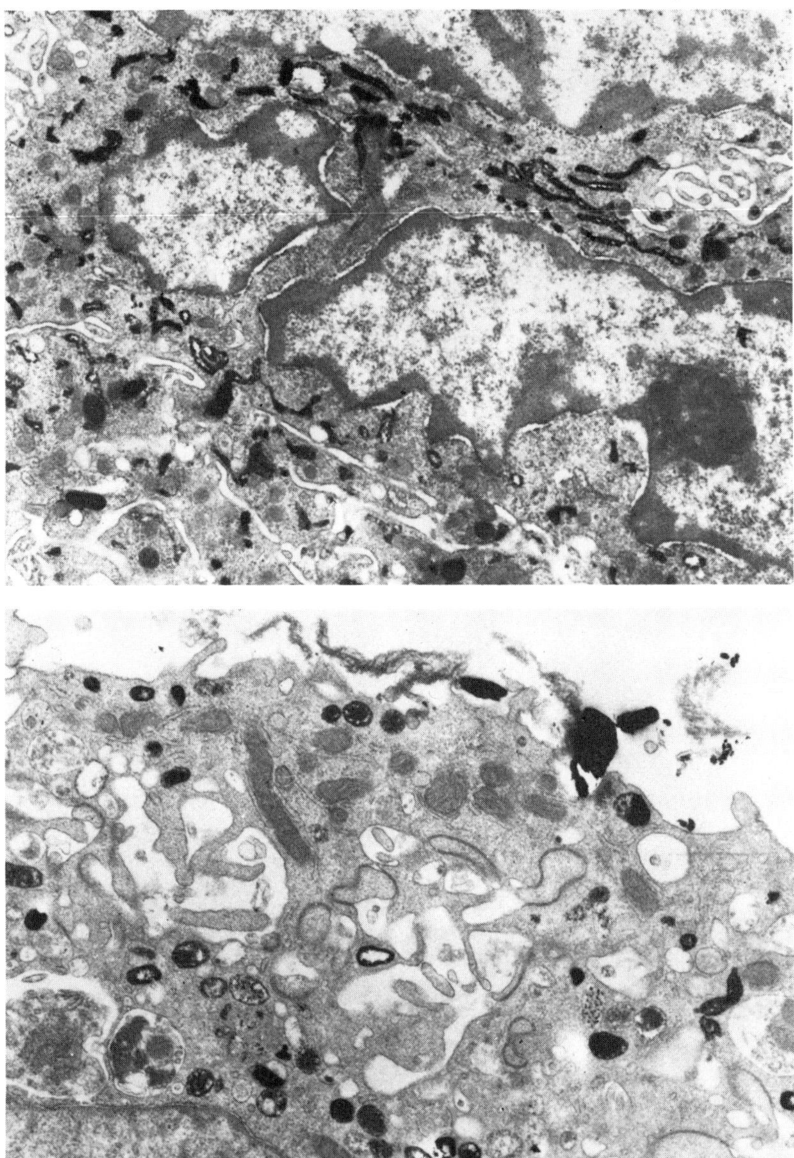

Fig. 1. Demonstration of AchE in megakaryocytes. Above: Normal mouse bone marrow. Note the staining of endoplasmic reticulum and not of demarcation membranes. Below: A 7-day culture of mouse bone marrow. The enzymatic activity is localized in dilated tubules of the endoplasmic reticulum.

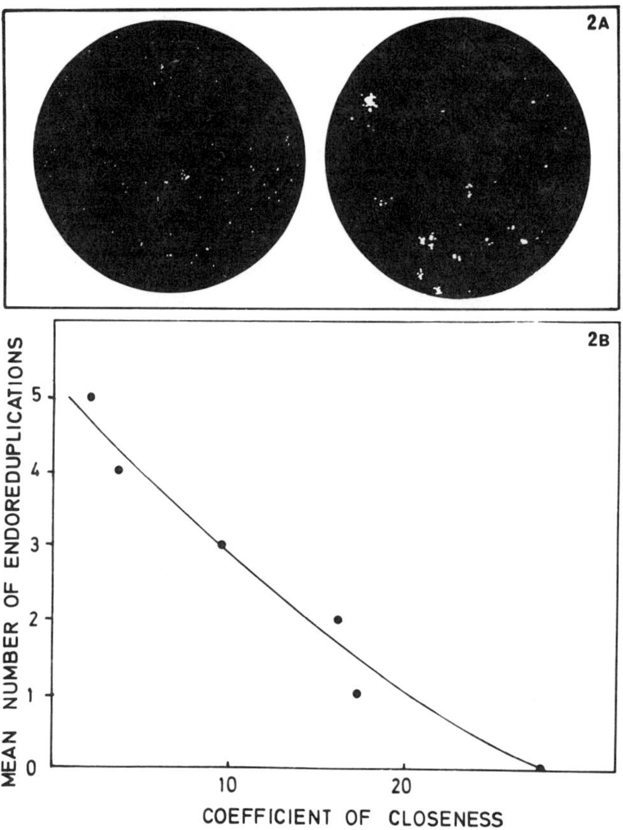

Fig. 2a. Computer maps of megakaryocyte growth in one Petri dish inoculated with 2 × 10^5 cells per ml. Each dot represents a megakaryocyte whose ploidy was measured after combined staining for acetylcholinesterase and DNA. At left 32c–64c cells are mostly isolated whereas at right 2c–4c megakaryocytes tend to belong to large aggregates

2b. Inverse relationship between mean NbE and coefficient of closeness, showing that megakaryocytes having many close megakaryocyte neighbors undergo fewer endoreduplications than isolated ones.

The apparent balance between the number of mitotic and endoreduplicative cycles could have meant that, by day 7 for instance, megakaryocytes in larger colonies did not have sufficient time to reach high ploidy values. If this explanation were true, then colonies of a given size would show an increase in mean number of endoreduplications from day 5 to 7 and 9, i.e., the regression line between the latter variable and the number of doublings would shift with time. However, the ordinate at origin and slope of the regression lines for days 5, 7 and 9 were not significantly different (Fig. 4). In the meanwhile the cultures continued to grow, as judged from the increase in mean number of doublings or in number of colonies containing more than 16 megakaryocytes (Paulus et al., unpublished data). Thus, despite continuing

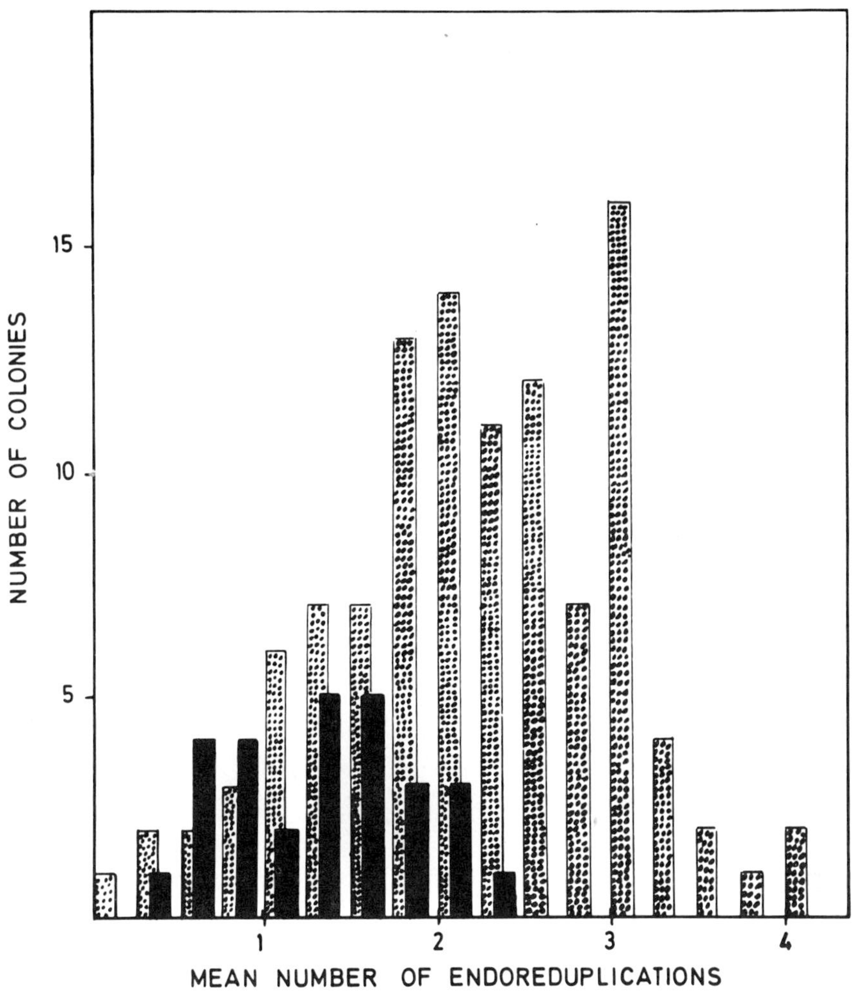

Fig. 3. Frequency distribution of mean NbE in pure megakaryocyte colonies (dotted columns) and in mixed egakaryocyte-erythroblastic colonies (black columns).

mitotic proliferation in megakaryocyte progenitors, a stationary ploidization pattern was established between days 5–9 wherein the number of endoreduplications undergone by each megakaryocyte was related to the mitotic history of the cell.

The observation that the ploidy level reached by megakaryocytes is related to their past proliferative history is compatible with two recently proposed models of colony development, derived from the study of cumulative doubling distributions of progenitors (Paulus et al., unpublished data). To illustrate the difference between these two models it is best to compare the pattern of maturation they imply for the progenitors revealed in 7-day cultures.

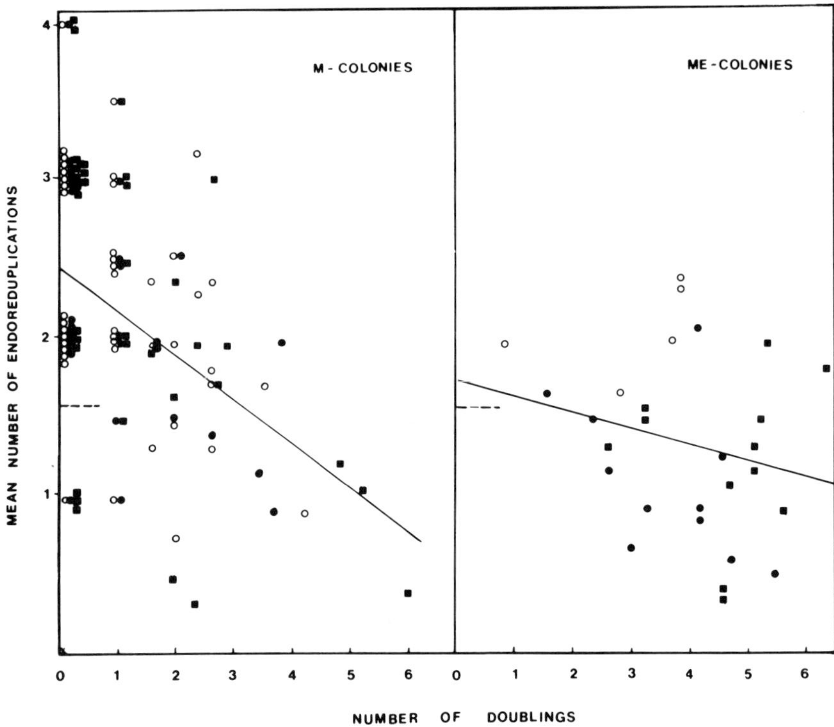

Fig. 4. Relationship between mean NbE and NbD in M and ME-colonies. Cultures were measured at 5(0), 7(●) or 9(■) days. Since no significant difference was found between cultures sampled at different times the regression lines were calculated on the pooled data for each type of colony.

In a first model, 7-day CFU–M and 7-day CFU-ME constitute two separate heterogeneous cell populations organized according to their generation age. A deficiency in the culture medium causes a block in ploidization for colonies originating from the most ancestral progenitors, which have not been stimulated *in vivo* by a hypothetical ploidization factor. Late progenitors would escape that block if they have been sampled after receiving the adequate stimulation *in vivo*.

In a second model, the 7-day progenitors of pure and mixed colonies compose two distinct homogeneous compartments of relatively quiescent cells. Cultivation and stimulation of these cells produces a wave of mitoses whose number appears to be distributed according to two or three exponential functions. The switch to ploidization is then a random event whose probability increases at successive stages of progenitor maturation and the final ploidy level is determined by the stage at which this switch occurs. Further experiments are in progress to discriminate between these two models.

Summary

The ploidization pattern of megakaryocytes in pure megakaryocyte colonies (M-col) and megakaryocyte-erythroblast colonies (ME-col) was studied in plasma clot cultures of mouse marrow incubated at 2×10^5 or 3×10^4 cells/ml with 3 U. of erythropoietin. The cultures were double-stained by the reactions of Karnovsky and Roots (1964) for acetylcholinesterase (AchE) and Feulgen for DNA. Electron microscopic demonstration of AchE indicated that the enzyme is a reliable marker of the megakaryocyte line. Each colony was characterized by the number of doublings (NbD) completed by its progenitor and the mean number of endoreduplications (NbE) per megakaryocyte. Compared to M-col, ME-col has a significantly greater NbD and smaller mean NbE. In M-col, there was a linearly decreasing relationship between mean NbE and NbD and the regression line was invariant with time despite continuing progenitor proliferation, showing reduced ploidization in the larger colonies. The data are compatible with two recently proposed models of megakaryocyte progenitor maturation: (a) there is an age structure in these progenitors together with a block in ploidization for colonies originating from the most ancestral progenitors; or (b) the switch to ploidization is a random event whose probability increases at successive stages of CFU-M/ME maturation and the final ploidy level is determined by the stage at which this switch occurs.

ACKNOWLEDGMENTS

This work was supported by grants from the Institut National de la Santé et de la Recherche Médicale, l'Ecole Pratique des Hautes Etudes and l'Association Claude Bernard (Paris, France). Grateful thanks are extended to Dr. A. Ball for the gift of human erythropoietin. J. M. Paulus is Maître de Recherches at the Fonds National de la Recherche Scientifique, Bruxelles, Belgium. The authors express their gratitude to Prof. M. Bessis for his generous support and advice. The help of Prof. J. Serra in analyzing the data of Figure 2 is also acknowledged.

References

Ebbe, S. 1976. Biology of megakaryocytes. In *Progress in Hemostasis and Thrombosis*, (Vol. 3). Spaet, J. H., ed. New York: Grune and Stratton. pp. 211–229.

Harker, L. A. and C. A. Finch. 1969. Thrombokinetics in man. *J. Clin. Invest.*, 48:963–974.

Karnovsky, M. J. and L. Roots. 1964. A "direct-coloring" thiocholine method for cholinesterases. *J. Histochem. Cytochem.* 12:219–221.

McLeod, D., M. M. Shreeve, and A. Axelrad. 1978. Culture systems *in vitro* for the assay of erythrocytic and megakaryocytic progenitors. In *In Vitro Aspects of Erythropoiesis*. Murphy, M. J., Jr., ed. New York: Springer-Verlag. pp. 31–36.

Odell, T. T., Jr., J. R. Murphy, and C. W. Jackson. Stimulation of megakaryocytopoiesis by acute thrombocytopenia in rats. *Blood* 48:765–775.

Paulus, J. M. 1970. DNA metabolism and development of organelles in guinea-pig megakaryocytes: A combined ultrastructural, autoradiographic and cytophotometric study. *Blood* 35:298–311.

Tranum-Jensen, J. and O. Behnke. 1977. Electron microscopical identification of the committed precursor cell of the megakaryocyte compartment of rat bone marrow. *Cell Biol. Internat. Reports* 1:445–452.

Williams, N., H. M. Jackson, R. R. Eger, and M. Long. 1980. The separate roles of factors in murine megakaryocyte colony formation.*Megakaryocytes In Vitro*.

Evatt, Levine, and Williams, Editors
MEGAKARYOCYTE BIOLOGY AND PRECURSORS:
IN VITRO CLONING AND CELLULAR PROPERTIES

The Effects of Acute Thrombocytopenia Upon Megakaryocyte-CFC and Granulocyte-Macrophage-CFC in the Bone Marrow and Spleen of Mice: with Studies of Ploidy Distribution in Megakaryocyte Colonies

Jack Levin, Francine C. Levin, Donald Metcalf, and David G. Penington

The Johns Hopkins University School of Medicine and Hospital, Baltimore, Maryland and The Walter and Eliza Hall Institute of Medical Research and the University of Melbourne, Melbourne, Australia

The administration of heterologous platelet antiserum produces acute thrombocytopenia, which results in alterations in megakaryocytopoiesis during the 60-hr period following acute thrombocytopenia (Ebbe, 1968; Penington and Olsen, 1970). Recently, techniques have become available for the culture *in vitro* of cells which produce megakaryocyte colonies (Metcalf et al., 1975; Williams and Jackson, 1978). However, few data are available concerning the effects of perturbation of thrombopoiesis in the donors on the characteristics of their subsequently cultured hematopoietic cells (Nakeff and Bryan, 1978). Similarly, although acute thrombocytopenia causes an increase in ploidy (DNA content) in bone marrow megakaryocytes (Penington and Olsen, 1970), no information is available concerning the effects of acute thrombocytopenia on the progeny of megakaryocyte colony-forming cells (Meg-CFC). Therefore, we have studied the effects of thrombocytopenia on Meg-CFC obtained from mice in which acute thrombocytopenia had been produced with platelet antiserum and on the ploidy distribution in individual megakaryocyte colonies derived from the bone marrow and spleen of normal and thrombocytopenic mice.

Methods and Materials

C57BL/6J mice, approximately 8 weeks of age, were used. After cervical dislocation, bone marrow and spleen cells were obtained, as previously described (Levin et al., 1980). All cultures were performed in an agar medium (final concentration of agar was 0.3% in Dulbecco's modified Eagle's medium) (Levin et al., 1980). Final concentration of fetal calf serum was 20%. One ml of a cell suspension in agar medium was cultured with 0.2 ml of pokeweed mitogen stimulated spleen cell conditioned medium, and incubated for 7 days in a humidified atmosphere of 10% CO_2 in air.

Cultures were scored for colony formation after 7 days of incubation, the time at which megakaryocyte colonies reached their maximum frequency. All colonies considered to be possibly megakaryocytic were removed from two of the three culture dishes, which constituted each triplicate set, with a finely drawn Pasteur pipette. Culture dishes were deliberately "overpicked" to assure that all megakaryocyte colonies were removed and identified histochemically, following staining for acetylcholinesterase for 1.5 hr.

Acute thrombocytopenia (platelet counts $<5\%$ of normal) was produced by the administration of platelet antiserum (PAS), prepared in rabbits. Both PAS and normal serum (NS) were adsorbed three times each with washed, platelet-free thymocytes, macrophages, granulocytes, and red blood cells from C57BL/6J mice. Neither PAS or NS produced detectable alterations in hematocrit levels or total white blood cell counts.

Microdensitometric measurement of DNA in the individual cells of megakaryocyte colonies was carried out using an M-85 scanning microdensitometer (Vickers Instruments, England), after intact colonies were placed on glass slides and stained with the Feulgen-Schiff reaction.

Results and Discussion

Effects of Platelet Antiserum or Normal Serum on Granulocyte, Macrophage, and Megakaryocyte Colony-Forming Cells

At time periods from 1 hr to 14 days after administration of platelet antiserum (PAS), normal serum (NS), or phosphate buffered saline, different groups of animals were sacrificed and their bone marrow and splenic cells cultured. At no time studied were there any alterations in the number of macrophage, granulocyte, or megakaryocyte colonies detected when femoral bone marrow was cultured at 5 or 10 $\times$ 10^4 cells/ml. In marked contrast, from 3 to 7 days following the administration of PAS and from 4 to 7 days following the administration of NS, there was an increase in granulocyte and macrophage colony-forming cells (GM-CFC) in the spleen (Fig. 1). A similar response was noted in the number of Meg-CFC in the spleen (Fig. 1). A maximum 6-fold increase in Meg-CFC occurred 4 days after administration of PAS. At no time was the frequency of Meg-CFC significantly increased above control levels in animals that had received NS.

Total nucleated cells in the spleen increased 24 hr after injection of PAS or NS and remained relatively elevated for 9 days. Similarly, although spleen weights did not significantly increase above normal, there was a significant difference between spleen weights during the 24-hr period following PAS or NS and the subsequent

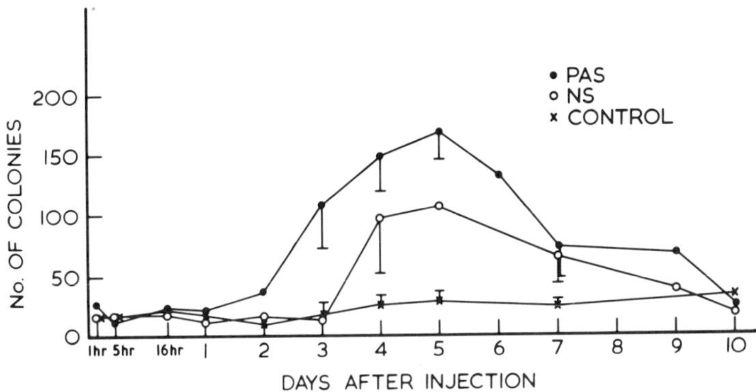

Fig. 1. Effect of platelet antiserum (PAS) or normal serum (NS) upon granulocyte-macrophage-CFC (GM-CFC) in spleen. Mean total number of GM-CFC at various times after administration of 0.05 ml PAS or NS is shown. Normal controls received phosphate-buffered saline. 1×10^6 nucleated spleen cells were cultured. Four to six experiments were performed on each of days 1, 2, 3, 4, 5, and 7; two to three experiments were performed at each of the other times shown (only 1 experiment at 16 hr). 1 S.E. is indicated only where statistically significant differences occurred. The number of granulocyte and macrophage colonies was increased from three to seven days following PAS ($p < 0.05$). (From Levin et al., *Blood* 56, 1980; with permission of Grune and Stratton).

13 days, when spleen weights were greater. As a result of increased numbers of nucleated cells in the spleen in association with the increase in frequency of splenic CFC, total splenic CFC increased 5-fold, 4 to 5 days after NS and 8–10 fold, 3 to 5 days after PAS. At the time of maximum increase, total GM-CFC/spleen was approximately 15,000 and 27,000, following NS and PAS, respectively. Total splenic Meg-CFC increased 3-fold, 4 days following NS and 6–7 fold, 4 to 6 days following PAS. At the time of maximum increase, total Meg-CFC/spleen was approximately 1,600 and 3,600, respectively.

Types of Megakaryocyte Colonies

Two types of megakaryocyte colonies were detected in soft agar cultures, confirming previous observations (Metcalf et al., 1975; Williams and Jackson, 1978). One is composed of essentially big, mature appearing, acetylcholinesterase positive cells (Fig. 3, top); and the other of larger numbers of a more heterogeneous population, many of which were acetylcholinesterase negative (Fig. 3, bottom).

Measurement of DNA

DNA analyses revealed that the morphological differences between these two types of colonies, designated "big cell" and "heterogeneous," respectively, were associated with distinctly different ploidy patterns. The normal ploidy distributions of the two types of colonies are shown in Figure 4 and Figure 5. Big cell colonies

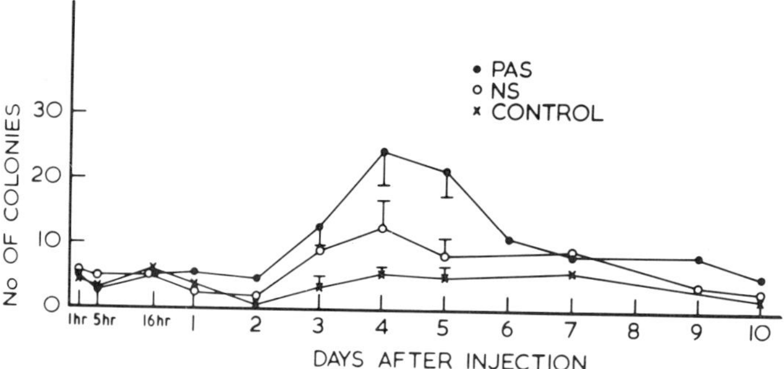

Fig. 2. Effect of platelet antiserum (PAS) or normal serum (NS) upon megakaryocyte-CFC (Meg-CFC) in spleen. Mean total number of Meg-CFC at various times after administration of 0.05 ml of PAS or NS is shown. Normal controls received phosphate-buffered saline. 1 $\times$ 10^6 nucleated spleen cells were cultured. Four to six experiments were performed on each of days 1, 2, 3, 4, 5, and 7; two to three experiments were performed at each of the other times shown (only 1 experiment at 16 hr). 1 S.E. is indicated only where statistically significant differences occurred. The number of megakaryocyte colonies was increased from three to five days following PAS ($p < 0.05$). (From Levin et al., *Blood* 56, 1980; with permission of Grune and Stratton).

demonstrated the relative ploidy frequencies characteristic of megakaryocytes detectable in bone marrow (Fig. 4). In contrast, heterogeneous colonies contained a marked preponderance of 2N and 4N cells (Fig. 5). The big cell type of megakaryocyte colony normally reached maximum ploidy levels after 4 days in soft agar culture. In contrast, heterogeneous colonies did not achieve maximum ploidy levels until 7 days in culture (Levin et al., in press).

Effects of Thrombocytopenia on Ploidy Distribution

Ploidy levels in both types of megakaryocyte colonies were measured in 7-day cultures, derived from cells obtained at various times after production of acute thrombocytopenia. Figure 6 demonstrates that big cell colonies, obtained from bone marrow and spleen 4 days after acute thrombocytopenia had been produced in the donors, showed an increase in ploidy levels. There was a marked increase in the frequency of both 32N and 64N cells, the latter never being detected in colonies from normal animals. The number of cells/colony did not increase. Ploidy distribution was similarly altered 5 days after production of thrombocytopenia in the donors, but was normal 2 and 7 days after thrombocytopenia. Surprisingly, heterogeneous colonies did not demonstrate a shift in ploidy profiles after induction of acute thrombocytopenia.

An increase in Meg-CFC was demonstrated in cultures of hematopoietic cells obtained from mice in which acute thrombocytopenia had been produced. However, an increase in Meg-CFC was detected only in the spleen, not the bone marrow, during a period from 1 hr to 14 days following induction of acute thrombocytopenia;

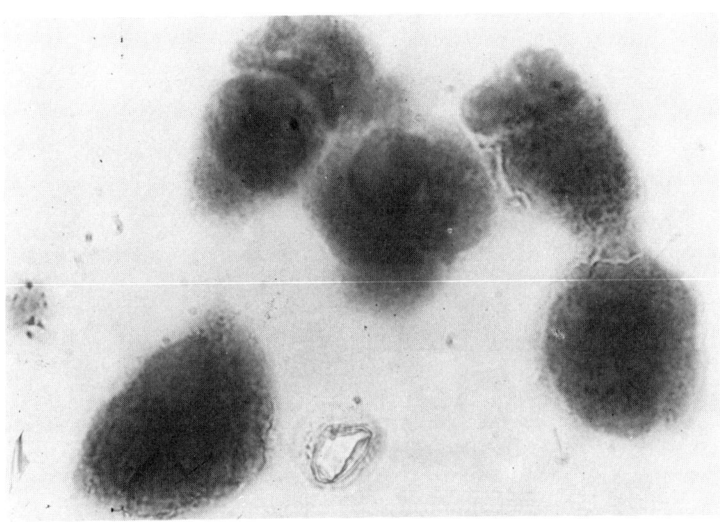

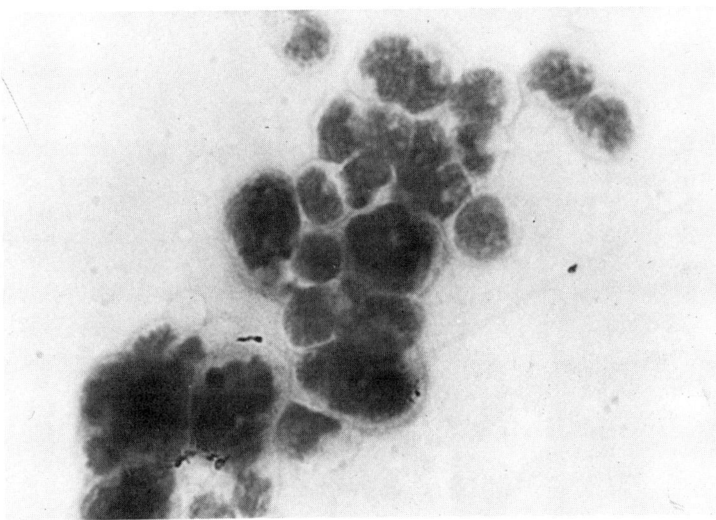

Fig. 3. Big cell colonies (top) contained large, mature appearing acetylcholinesterase (AChE) positive cells. Heterogeneous colonies (bottom) contained large, intermediate, and small cells. Many of the small cells were AChE negative, as indicated by their clear cytoplasm. The presence of AChE is indicated by the dark, granular color in the cytoplasm. The nucleus was counterstained with methyl green (original magnification × 1,000).

and the increase in frequency of splenic Meg-CFC occurred only after a lag of 2 days and did not reach a maximum until 4 to 5 days following acute thrombocytopenia. This indicates that Meg-CFC are not the cells immediately responsible for changes in bone marrow, including an increase in recognizable megakaryocytes,

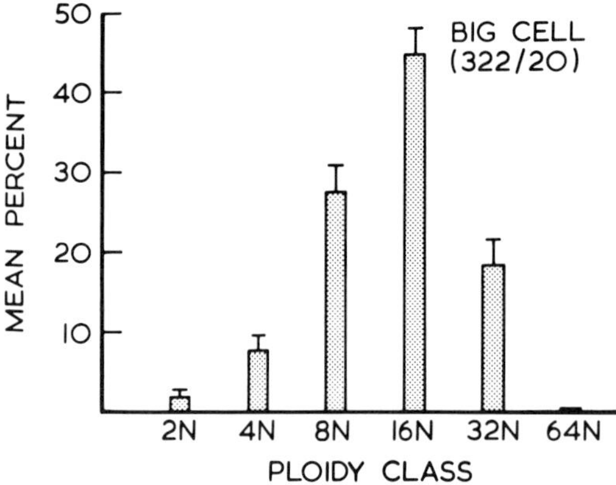

Fig. 4. Normal ploidy distribution of the big cell type of megakaryocyte colony. Mean percent ± 1 S.E. of total cells in each ploidy class is shown. The numbers in parentheses indicate total number of megakaryocytes in which DNA content was measured/total number of colonies. After 7 days of culture, big cell colonies contained 16 ± 2.3 acetylcholinesterase positive cells and the ploidy distribution was characteristic of recognizable megakaryocytes in bone marrow. The mean ploidy level was 16.8 ± 0.8/cell.

during the 60-hr period after induction of acute thrombocytopenia (Odell et al., 1976; Penington and Olsen, 1979; Odell et al., 1969). Our results and those of Burstein et al. (1979) differ from those of Nakeff and Bryan (1978), who reported a 2.5-fold increase in bone marrow Meg-CFC within 2 hr of induction of acute thrombocytopenia with a rapid return to normal after 6 hr.

Therefore, Meg-CFC detected in soft agar cultures may represent cells more immature or primitive than those immediately responsive to acute thrombocytopenia. Their increased number, 4 to 5 days after acute thrombocytopenia, at a time when the donor animals had normal or slightly increased levels of circulating platelets, indicates that the feedback mechanism, usually considered to operate during perturbation of platelet counts or mass, does not prevent an increase in Meg-CFC once the initial stimulus of thrombocytopenia has occurred. Goldberg et al. (1977) also concluded that only committed megakaryocyte precursors were responsive to this physiologic feedback mechanism, whereas pluripotential stem cells (and presumably Meg-CFC) were not. The time of appearance of increased Meg-CFC in our mice supports this hypothesis as do the data of Williams et al. (1979) that Meg-CFC in bone marrow were not suppressed by hyper-transfusion of platelets.

Measurement of the DNA content of individual megakaryocytes revealed two patterns of ploidy distribution in megakaryocyte colonies. The "big cell" type of colony demonstrated a ploidy pattern that closely resembles the ploidy distribution of recognizable megakaryocytes in the bone marrow of normal rodents. The consistent presence of a relatively few 2N and 4N cells in this type of colony suggests that their proportion in these colonies resembles their relative frequencies in bone

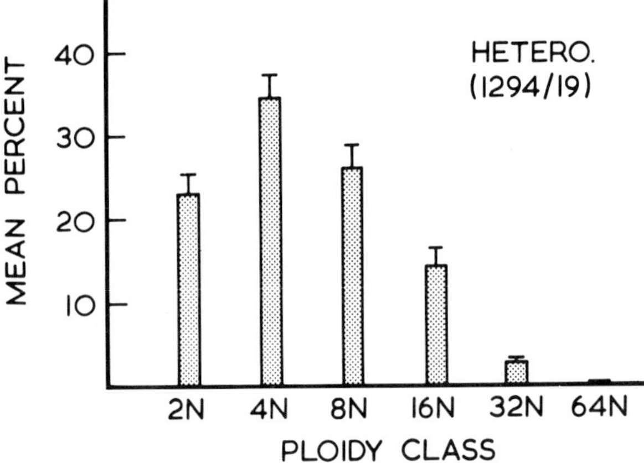

Fig. 5. Normal ploidy distribution of the heterogeneous type of megakaryocyte colony. Mean percent ± 1 S.E. of total cells in each ploidy class is shown. The numbers in parentheses indicate total number of megakaryocytes in which DNA content was measured/total number of colonies. After 7 days of culture, heterogeneous colonies contained 44 ± 9.6 cells/colony (some of which were acetylcholinesterase negative). The mean ploidy level was 6.8 ± 0.7 cell.

marrow, where they are usually not identified.

The population contained in "heterogeneous" colonies included small, acetylcholinesterase (AChE) negative cells. Although the smallest AChE negative cells could not be morphologically identified as megakaryocytes, their lack of resemblance to any of the other cell types present after seven days in this culture system, their negativity with benzidine, and the presence of 4N cells in this sub-population make it most likely that they represent megakaryocyte precursors or immature megakaryocytes. A more detailed discussion of these small cells is presented in another report (Levin et al., in press). Importantly, the ploidy distribution in heterogeneous colonies was almost exactly opposite to that observed in big cell colonies, i.e., 2N and 4N cells accounted for 57% of the cells and only 17% were 16N and 32N. Heterogeneous colonies were not simply big cell colonies coincidently associated with an additional group of immature cells, because the relative frequencies of 8N, 16N, and 32N cells in the two types of colonies were different and no colonies which contained only 2N, 4N, and 8N cells were ever observed.

Following induction of acute thrombocytopenia, there is an increase in the ploidy of megakaryocytes in the bone marrow of rodents. Mean ploidy levels reach a peak within 48 hr and after 4 to 5 days, ploidy distribution returns to normal (Odell et al., 1976; Penington and Olsen, 1970). Similar increases in ploidy were observed in big cell colonies after induction of acute thrombocytopenia in the donors of hematopoietic cells for culture. Big cell colonies demonstrated a marked increase in the proportion of 32N and 64N cells. However, increased ploidy was detected

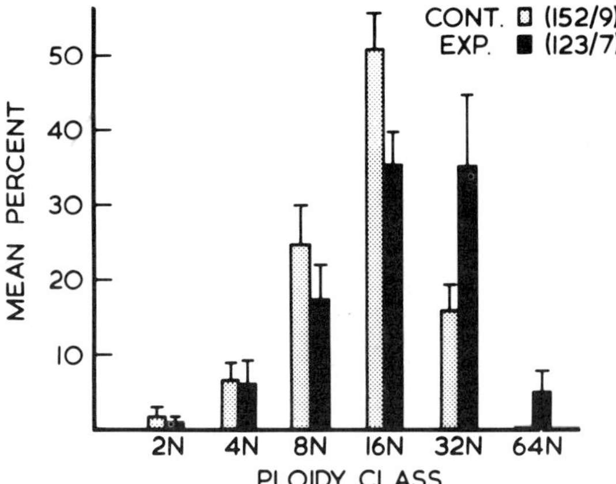

Fig. 6. Effect of acute thrombocytopenia upon ploidy distribution in big cell type megakaryocyte colonies. CFC were obtained 4 days following induction of acute thrombocytopenia in the donor mice (blank bars). Control (Cont.) animals (stippled bars) had normal platelet counts. Mean percent ± 1 S.E. of total cells in each ploidy class is shown. The numbers in parentheses indicate total number of megakaryocytes in which DNA content was measured/total number of colonies. Colonies were removed for DNA analysis after 7 days of culture. p values [Control vs. Experimental (4 days after acute thrombocytopenia)] for mean ploidy/cell, <0.005; % 32N, <0.01; % 64N, <0.05.

only in colonies derived from cells obtained 4 to 5 days after the production of acute thrombocytopenia, the same time at which the frequency of these colonies also was maximally increased (see Fig. 2). By this time, both megakaryocytes and platelet levels in the donors had already returned to normal. Therefore, Meg-CFC are not responsible for the acute changes in ploidy distribution in bone marrow. Altered Meg-CFC represent a direct but delayed effect of acute thrombocytopenia or a secondary effect of the bone marrow alterations, perhaps as a consequence of increased transit of immediate precursors of megakaryocytes into the megakaryocyte compartment. Importantly, heterogeneous colonies did not alter following production of thrombocytopenia in the donor animals, as was observed in big cell colonies.

Therefore, our data indicate the existence of two types of Meg-CFC that produce megakaryocyte colonies which are discrete and distinguishable from each other by number of cells per colony, presence or absence of marked heterogeneity of cell size, DNA content and ploidy distribution, and responsiveness to the stimulus of acute (or chronic) thrombocytopenia. The predominance of 2N and 4N cells in heterogeneous colonies, in contrast to the big cell type colony, is associated with the presence of many AChE negative cells, and suggests that a more primitive colony-forming cell produces this type of colony. Importantly, the pattern of ploidy in megakaryocytes appears to be a property of individual stem cells, each of which gives rise to progeny which are heterogeneous with respect to ploidy. No colony was homogeneous for a single ploidy class.

Meg-CFC produced big cell colonies with increased ploidy levels, although they were obtained from the bone marrow or spleen of animals that had already apparently recovered from thrombocytopenia, and in which megakaryocyte number and ploidy had returned to normal. Therefore, culture conditions allowed expression of this potential of "stimulated" Meg-CFC, indicating they had been programmed by previous stimuli *in vivo*, although feedback mechanisms present in the physiologic environment prevent Meg-CFC from manifesting this potential. Accordingly, *in vivo* feedback mechanisms which ultimately regulate megakaryocyte ploidy must operate before CFC can produce megakaryocytes with increased ploidy, i.e., upon unrecognized and as yet undefined precursors. Concomitantly, CFC or their progeny *in vivo*, are prevented from maturing into increased numbers of platelet-producing megakaryocytes, because a second, delayed wave of rebound thrombopoiesis does not occur following acute thrombocytopenia.

The significant increase in granulocyte-macrophage-CFC (GM-CFC) in the spleen, from 3 to 7 days after acute thrombocytopenia and to a lesser extent in mice that received normal rabbit serum, was unanticipated. Ebbe et al. (1971) obtained increased numbers of CFU-S from spleens of mice 4 to 6 days after they had received platelet antiserum (PAS). Also comparable to our data were lack of similar increases of CFU-S in bone marrow and concomitant proportional increases of both hematopoietic colonies and number of megakaryocytes in the spleens of irradiated recipients of spleen cells from PAS-treated donors (Ebbe et al., 1971). We have concluded that the presence of antigen-antibody complexes, particulate material, or heterologous antigens in the rabbit sera produced increased levels of CSF, which subsequently resulted in increased GM-CFC, as reported previously (Levin et al., 1980; McNeill, 1970; Apte and Pluznik, 1976). It appears that as a result of our experimental manipulation, stimulation of the reticulo-endothelial system occurred, which resulted in an increase in GM-CFC. Massive platelet destruction following administration of PAS, with subsequent phagocytosis, presumably played a major role in producing this response (Apte and Pluznik, 1976).

Our data indicate the importance of evaluating all colonies in culture to determine if a presumably specific stimulus has actually perturbed more than one type of hematopoietic cell and its precursors. The necessity to continue observations of cells from bone marrow and spleen, after peripheral blood counts have returned to normal, is also demonstrated.

Summary

The effects of acute thrombocytopenia, produced by platelet antiserum (PAS), on both megakaryocyte colony-forming cells (Meg-CFC) and granulocyte-macrophage colony-forming cells (GM-CFC) were studied. During the 1-hr to 14-day period following acute thrombocytopenia (platelet counts $<5\%$ of normal), bone marrow and splenic cells of C57BL mice were obtained and cultured for 7 days in 0.3% agar. Megakaryocyte colonies were identified by positive stain for acetylcholinesterase after removal from culture. At no times were alterations in frequency of GM-CFC or Meg-CFC detected in femoral bone marrow. In contrast, the frequency of GM-CFC in spleen was increased from 3 to 7 days after PAS (maximum 8-fold increase on day 5), and from 4 to 7 days after normal serum (NS) (maximum 5-fold increase on day 5). Maximum 6-fold increase in Meg-CFC in spleen occurred 4 days after PAS with lesser, not significant increase after NS. DNA content of individual megakaryocytes was determined. One type of megakaryocyte colony had a mean ploidy level of 16.8 ± 0.8/cell and a ploidy distribution characteristic of recognizable megakaryocytes in bone marrow. Four to five days after acute thrombocytopenia, these colonies demonstrated a marked increase in DNA content; frequency of 32N cells increased from 17% to 30% and 64N cells from 0% to 6%. The delayed increase in Meg-CFC indicates that they are unlikely to be responsible for the altered megakaryocytopoiesis previously reported in bone marrow after acute thrombocytopenia. Delayed appearance of altered ploidy suggests that the pool of stem cells, from which committed megakaryocyte precursors are derived, responds perhaps indirectly to the stimulus of platelet depletion. Increase in GM-CFC may reflect stimulation of the reticulo-endothelial system by heterologous proteins.

ACKNOWLEDGMENTS

Supported in part by Research Grant HL 01601 from the National Heart, Lung, and Blood Institute, Research Grants CA 22556 and CA 25972 from the National Cancer Institute, National Institutes of Health, Bethesda, Maryland, and by the National Health and Medical Research Council of Australia. J. Levin was a Faculty Scholar of the Josiah Macy, Jr. Foundation during the performance of these studies. The Australia-United States Program of the National Science Foundation provided travel support for J. Levin.

References

Apte, R. N. and D. H. Pluznik. 1976. Control mechanisms of endotoxin and particulate material stimulation of hemopoietic colony-forming cell differentiation. *Exp. Hematol.* 4:10–18.

Burstein, S. A., J. W. Adamson, S. K. Erb, and L. A. Harker. 1979. Murine megakaryocytopoiesis: Evidence for early and late regulatory control. *Blood* 54 (Suppl. 1):165a.

Ebbe, S. 1968. Megakaryocytopoiesis and platelet turnover. *Ser. Haematol.* 1:65–98.

Ebbe, S., E. Phalen, J. Overcash, D. Howard, and F. Stohlman, Jr. 1971. Stem cell response to thrombocytopenia. *J. Lab. Clin. Med.* 78:872–881.

Goldberg, J., E. Phalen, D. Howard, S. Ebbe, and F. Stohlman, Jr. 1977. Thrombocytotic suppression of megakaryocyte production from stem cells. *Blood* 49:59–69.

Harker, L. A. 1968. Kinetics of thrombopoiesis. *J. Clin Invest.* 47:458–465.

Levin, J., F. C. Levin, and D. Metcalf. 1980. The effects of acute thrombocytopenia on megakaryocyte-CFC and granulocyte-macrophage-CFC in mice: Studies of bone marrow and spleen. *Blood* 56:274–283.

Levin, J., F. C. Levin, D. G. Penington, and D. Metcalf. Measurement of ploidy distribution in megakaryocyte colonies obtained from culture: With studies of the effects of thrombocytopenia. *Blood*, in press.

McNeill, T. A. 1970. Antigenic stimulation of bone marrow colony-forming cells. III. Effect *in vivo*. *Immunol.* 18:61–72.

Metcalf, D., H. R. MacDonald, N. Odartchenko, and B. Sordat. 1975. Growth of mouse megakaryocyte colonies *in vitro*. *Proc. Nat. Acad. Sci. USA* 72:1744–1748.

Nakeff, A. and J. E. Bryan. 1978. Megakaryocyte proliferation and its regulation as revealed by CFU-M analysis. In *Hematopoietic Cell Differentiation*. Golde, D. W., Cline, M. J., Metcalf, D., and Fox, C. F., eds. New York: Academic Press. pp. 241–259.

Penington, D. G. and T. E. Olsen. 1970. Megakaryocytes in states of altered platelet production: Cell numbers, size and DNA content. *Br. J. Haematol.* 18:447–463.
Odell, T. T., J. R. Murphy, and C. W. Jackson. 1976. Stimulation of megakaryocytopoiesis by acute thrombocytopenia in rats. *Blood* 48:765–775.
Odell, T. T., Jr., C. W. Jackson, T. J. Friday, and D. E. Charsha. 1969. Effects of thrombocytopenia on megakaryocytopoiesis. *Br. J. Haematol.* 17:91–101.
Williams, N. and H. Jackson. 1978. Regulation of the proliferation of murine megakaryocyte progenitor cells by cell cycle. *Blood* 52:163–170.
Williams, N., T. P. McDonald, and E. M. Rabellino. 1979. Maturation and regulation of megakaryocytopoiesis. *Blood Cells* 5:43–55.

Discussion

Paulus: In the colonies in which you see an increase in the number of megakaryocytes from day 4 to day 7, can you exclude that there are a few small precursors that by themselves would have the capability of proliferation, while the big cells don't have it?

Levin: That is an important question. We have had no difficulty in recognizing small cells in large megakaryocyte colonies. I believe that we would see at least some small cells in this type of colony of 12–18 cells, if they were there. If present, they are very difficult to see, and very rare.

Burstein: I have also seen cells like the ones you showed in culture, very large megakaryocytes that appear to be dividing. Half the nuclear material seemed to be on one side of one cell flowing, essentially like a dumbbell shape, to the other cell. I have seen this numerous times, now, and really didn't know what it meant. I still am not sure whether it isn't an *in vitro* artifact.

Levin: Yes, of course, we have to be careful about extrapolating from the culture to real life, but I agree with you. We have just finished looking at some three-day plates, where we can now see some of our colonies, and we have seen that we call doublets: big cells that look like they are in the process of dividing into two cells.

Levine: I have occasionally seen the same thing in cultures of differentiated megakaryocytes and with Normarski Optics found that the constrictions are merely twists in spread megakaryocytes. Even with prolonged observations on warm stage, nuclear or cell division was never observed. I have a high magnification photomicrograph of this phenomenon.

Bessman: Did the increment of 64N megakaryocytes come from a deficit in the 16N compartment?

Levin: These data are only proportions of the total population and do not allow an inference as to the definite pathway. Since the number of megakaryocytes doubled in the colonies between days 4 and 7, but the mean ploidy stayed the same, one could speculate that each cell doubles, with each daughter cell having the same ploidy as its mother cell. Because so few 2N and 4N positive cells are present in the colonies, it is likely that the 64N cells come not from them, but rather from the intermediate stages.

Levine: You are asking us to believe two somewhat radical ideas. The first I can accept, but the second does not automatically follow. If there are indeed, two classes of megakaryocyte colony-forming units, why couldn't it be that one of them matures quickly and the cells from that achieve their final ploidies based on some predetermined influence, and then later the second class takes longer to develop, so that you are seeing a second group come up. I'm skeptical that differentiated megakaryocytes can actually double their numbers.

Levin: Your skepticism is reasonable and well-founded. I can only report our numbers. Both types of colonies are always present together and one type of colony is neither added to or replaced by the other type. One final point about how megakaryocytes mature: We examined 290 individual colonies at different stages and found none in which all the megakaryocytes had the same ploidy value.

Evatt, Levine, and Williams, Editors
MEGAKARYOCYTE BIOLOGY AND PRECURSORS:
IN VITRO CLONING AND CELLULAR PROPERTIES

Murine Megakaryocytopoiesis: The Effect of Thrombocytopenia

Sigurdur R. Petursson, Paul A. Chervenick, and T. P. McDonald

Department of Medicine, University of Pittsburgh, Pittsburgh, Pennsylvania and University of Tennessee Memorial Research Center, Knoxville, Tennessee

Previous studies by a number of investigators have shown that platelet production by megakaryocytes (MK) is influenced by the level of circulating blood platelets (McDonald and Clift, 1979). Ebbe et al. (1968) reported an increase in the size of megakaryocytes and an increased rate of platelet production in rats made thrombocytopenic by transfusions. Injection of mice with anti-platelet antiserum resulted in thrombocytopenia which was followed by a thrombocytotic recovery phase (McDonald and Clift, 1979). A humoral substance (thrombopoietin) is thought to stimulate the production of platelets (McDonald and Clift, 1979). Whether this stimulating substance affects megakaryocytes and MK progenitor cells (CFU-M) is not clear. The present studies were done in order to assess the effects of acute and chronic thrombocytopenia on the production of megakaryocytes and megakaryocyte progenitor cells (CFU-M).

Methods and Materials

Mice obtained from Jackson Laboratories (Bar Harbor, Maine) were injected intraperitoneally with 0.1 ml rabbit anti-mouse latelet serum (RAMPS) either as a single injection for acute thrombocytopenia or at three-day intervals for inducing chronic thrombocytopenia. Noninjected mice were used as controls. Blood platelet concentration, total marrow nucleated cells and megakaryocytes, and megakaryocyte progenitor cells (CFU-M) per humerus were determined at various intervals following RAMPS. Venous blood was obtained by retroorbital sinus puncture and platelets were counted in a hemocytometer after a 1:10 dilution in 1% ammonium oxalate under phase contrast microscopy at 400×. Bone marrow megakaryocyte concentrations were estimated by the technique reported by Ebbe and Phalen (1978). A suspension of marrow cells was stained with 0.5% new methylene blue and megakaryocytes were counted in a hemocytometer at 40 × (Ebbe and Phalen, 1978). CFU-M were assayed by their ability to form megakaryocyte colonies in the *in vitro* soft gel (methylcellulose) culture system. Stimulation of megakaryocyte colony formation was obtained with pokeweed-stimulated spleen cell conditioned medium (Nakeff and Daniels-McQueen, 1976). Marrow cells in a concentration of 7.5×10^4 cells/ml were incubated in the soft gel culture system and scored for MK colonies after 9 days of incubation at 10% CO_2 and 37°C. Cell cycle characteristics of mouse CFU-M were measured *in vitro* by determining the degree to which colony forming capacity of the cultured cells was lost after tritiated thymidine (^{3}H-TdR) suicide (Joyce and Chervenick, 1977).

Results and Discussion

After a single injection of RAMPS, circulating platelets were reduced to 0 within two hr (Fig. 1). Rapid recovery of blood platelets occurred between days 2 and 4, and was followed by a rebound thrombocytosis to 175% of control on day 6. Platelets then returned to normal. Marrow megakaryocytes were increased slightly at 24 hr and increased rapidly between 2 and 3 days following RAMPS reaching a value of 295% of control by day 3. During the next week, platelets gradually declined to normal. Total CFU-M/humerus increased sharply between day 1 and 2 after RAMPS reaching 250% of control on day 2. Thereafter, CFU-M declined to normal by day 5.

The effects of a single injection of RAMPS on the percent of CFU-M in S phase of DNA synthesis were also studied. ^{3}H-TdR suicide studies revealed a slight increase over controls in the number of CFU-M in S phase on the second day following RAMPS (135% of control; p = NS). The number of progenitor cells in DNA synthesis increased to a level twice that of controls by day 3 (223% of control; $p < 0.05$) after which there was a rapid decline to normal by day 5.

Repeated administration of RAMPS at three day intervals (Fig. 2) resulted in a rapid reduction of peripheral platelets to 0, with platelets remaining at less than 50% of controls throughout the experiment. Megakaryocytes increased by day 2, and remained elevated at 220–295% of controls throughout the experiment. CF%-M increased after 24 hr and remained elevated at 166–244% of controls for the duration of the experiment.

During the chronic administration of RAMPS, CFU-M in the S phase increased slightly at 2 days, and reached values of 313% of control by day 4. This increase persisted at two to three times control levels for the remainder of the experiment.

Observations by several investigators have shown that megakaryocyte and platelet production respond to changes in the blood platelet levels (McDonald and Clift, 1979; Ebbe et al., 1968). An increase in the rate of platelet production in response to peripheral blood thrombocytopenia in rats was demonstrated by Ebbe et al. (1968) who also demonstrated that peripheral platelet levels markedly influence megakaryocyte size. Odell et al. (1969) showed an increase in megakaryocyte concentration in the marrow of rats following administration of antiplatelet serum. The

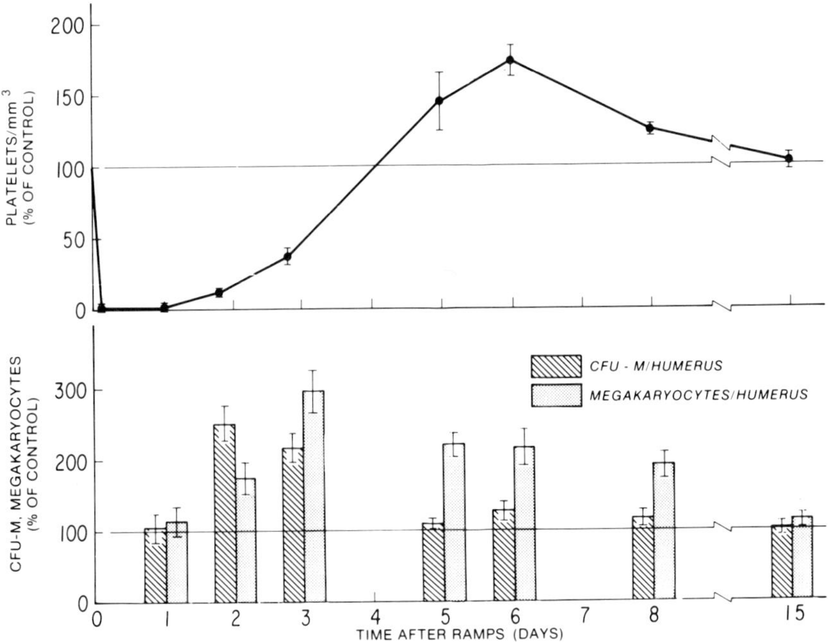

Fig. 1. The response of blood platelets to a single injection of antiplatelet serum (RAMPS) is illustrated in the top panel. The responses of the marrow megakaryocytes (MK) and MK precursor cells (CFU-M) are illustrated in the bottom panel. Control values for platelets: 1,056,000 ± 34,000/mm^3; MK: 5198 ± 409/humerus; CFU-M; 1296 ± 89/humerus.

results of the present studies indicate that there is a marked increase in the marrow megakaryocytes and CFU-M in response to thrombocytopenia. Marrow megakaryocytes were increased within 24 hr after a single antiplatelet serum injection and reached peak values of 3 times that of controls. The CFU-M were also increased within 48 hr, reaching peak values of 2 1/2 times control. In animals made chronically thrombocytopenia with repeated injections of antiplatelet serum, megakaryocytes and CFU-M remained elevated at 2–3 times the control values. These results suggest that blood platelet and marrow megakaryocyte production are at least in part regulated by the blood platelet level through its effect on the CFU-M compartment. Our results are at slight variance with those of Burstein et al. (1979) who did not observe an increase in marrow CFU-M until 90 hr following the induction of thrombocytopenia by antiplatelet serum. Whether these differences in results are due to differences in culturing techniques, antiserum, or other factors is unclear at the present time.

The *in vivo* pulse labeling studies of megakaryocytes with tritiated thymidine done by Feinendegen et al. (1962) and Ebbe and Stohlman (1965) allowed some inferences to be made with respect to DNA synthesis in the morphologically uni-

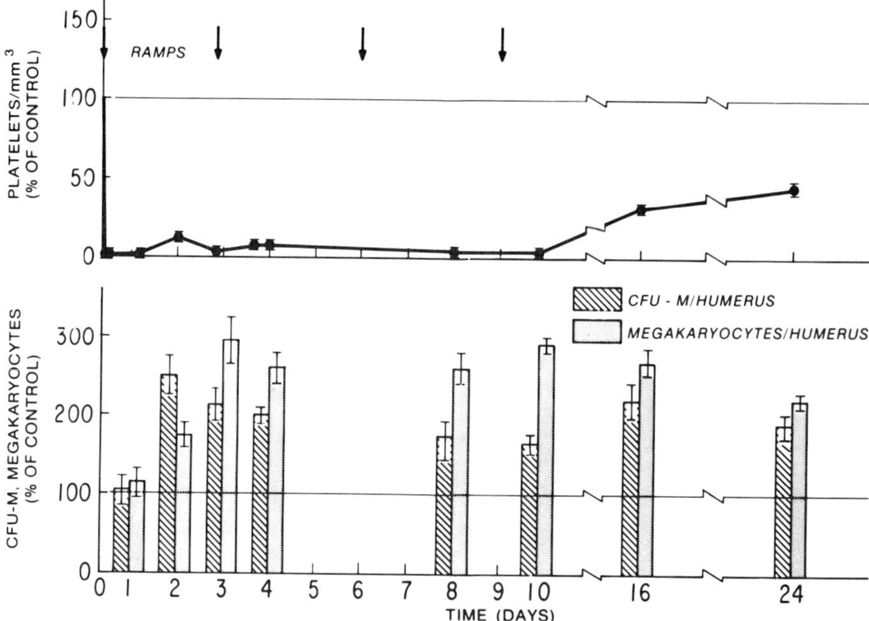

Fig. 2. The response of blood platelets to repeated injections of RAMPS is illustrated in the top panel. The responses of the marrow MK and CFU-M are illustrated in the bottom panel. Control values are the same as in Figure 1.

dentifiable precursor compartment. The studies indicated that these precursor cells were essentially all in DNA synthesis during the two to three-day period before their progression into the megakaryocyte compartment. The process of DNA synthesis appeared to be nearly continuous prior to the entry of the precursor cell into the recognizable megakaryocyte compartment. Our *in vitro* tritiated thymidine suicide studies are in agreement with these earlier observations. A high proportion of CFU-M were in the DNA synthesis phase (27.8 ± 1.8%; mean and SE) during the 20 min of exposure to tritiated thymidine. Furthermore, the fraction of cells in DNA synthesis was influenced by peripheral platelet levels. The proportion of CFU-M in S phase increased two-fold following acute thrombocytopenia. This increase in DNA synthesis was increased further and sustained during the 3 1/2 weeks of chronic RAMPS administration. These changes in DNA synthesis of the CFU-M in response to changes in blood platelet levels again suggest that a major point of regulation of megakaryocyte and platelet production is at the MK progenitor cell (CFU-M).

Chronic administration of RAMPS resulted in chronic thrombocytopenia and a persistent increase in megakaryocytopoiesis. A previous study of chronic thrombocytopenia in rats by Odell et al. (1969) showed a persistent increase in marrow megakaryocyte concentration during administration of antiplatelet serum. This increase was maintained at about 250% of controls and was associated with an increase in endomitosis as well. The present studies are in agreement with those reported by Odell et al. (1969) and show that megakaryocyte production in mice can increase at least 295%, CFU-M production 250%, and CFU-M S phase activity 300% following thrombocytopenia. These figures indicate the capacity of the marrow for megakaryocyte and platelet production.

Summary

The effect of thrombocytopenia on megakaryocytopoiesis was studied in mice made thrombocytopenic with rabbit anti-mouse platelet serum (RAMPS). Mice were injected with 0.1 ml RAMPS either as a single injection or at 3-day intervals for several weeks. At various times thereafter, platelets, total number of marrow megakaryocytes and megakaryocyte progenitor cells (CFU-M) were determined. CFU-M were assayed by their ability to form megakaryocyte colonies in the soft gel *in vitro* culture system. Stimulation of megakaryocyte colony formation was by pokeweed stimulated spleen cell conditioned medium. Following a single injection of RAMPS, platelets decreased to 1.2% of control values by 2 hr. By 2 1/2 days platelets began to increase, reached normal values by day 4, continued to increase to peak values of 215% of control by day 6, and decreased to normal by day 8. By 2 1/2 days megakaryocytes had increased 2-fold (13,800 ± 1230 vs 7090 ± 1190) and returned to near normal values by day 6. CFU-M increased markedly beginning at 2 1/2 days, reached a peak value of 250% of control (2400 ± 230 vs 970 ± 400) at 3 1/2 days then gradually declined to normal by day 12. These increases in CFU-M are of a substantially higher magnitude compared to the results reported by Burstein et al. (*Blood* 54:426, 1979). Repeated injections of RAMPS at 3-day intervals resulted in a chronic suppression of platelets to less than 50% of control value. Megakaryocyte numbers increased by 130% by 40 hr, and reached peak values of 250–300% of controls over a 3 1/2 week period. CFU-M increased to 166% of control by 40 hr and gradually rose to 270% of control values over 3 1/2 weeks. This study indicates that megakaryocytes and CFU-M increase in response to acute thrombocytopenia and remain elevated during chronic thrombocytopenia suggesting that megakaryocytes and CFU-M production are regulated in part by the level of circulating platelets.

References

Burstein, S. A., J. W. Adamson, S. K. Erb, and L. A. Harker. 1979. Murine megakaryocytopoiesis: Evidence for early and late regulatory control. *Blood* 54;165a.

Ebbe, S. and E. Phalen. 1978. Regulation of megakaryocytes in W/W^v mice. *J. Cell. Physiol.* 96:73–79.

Ebbe, S. and F. Stohlman, Jr. 1965. Megakaryocytopoiesis in the rat. *Blood* 26:20–35.

Ebbe, S., F. Stohlman, Jr., J. Overcash, J. Donovan, and D. Howard. 1968. Megakaryocyte size in thrombocytopenic and normal rats. *Blood* 32:383–392.

Feinendegen, L. E., N. Odartchenko, H. Cottier, and V. P. Bond. 1962. kinetics of megakaryocytic proliferation. *Proc. Soc. Exp. Biol. Med.* 111:177–182.

Joyce, R. A. and P. A. Chervenick. 1977. Corticosteroid effect on granulopoiesis in mice after cyclophosphamide. *J. Clin. Invest.* 60:277–283.

McDonald, T. P. and R. Clift. 1979. Effects of thrombopoietin and erythropoietin on platelet production in rebound-thrombocytotic and normal mice. *Am. J. Haematol.* 6:219–228.

Nakeff, A. and S. Daniels-McQueen. 1976. *In vitro* colony assay for a new class of megakaryocyte precursor: Colony-forming unit megakaryocyte (CFU-M). *Proc. Soc. Exp. Biol. Med.*151:587–590.

Odell, T. T., C. W.Jackson, T. J. Friday, and D. E. Charsa. 1969. Effects of thrombocytopenia on megakaryocytopoiesis. *Br.J. Haematol.* 17:91–101.

Discussion

Shulman: Do you conclude from the chronic thrombocytopenia studies that the compensatory capacity to produce megakaryocytes is only about 2–1/2 times normal?

Petursson: Yes, although we only injected the mice every 3 days, and it may be able to be pushed a little more by injecting every day.

Nakeff: What is the proliferative capacity of the cells forming the colonies after chronic thrombocytopenia?

Petursson: We haven't looked at colony size in this study.

Isolated Megakaryocytes

Evatt, Levine, and Williams, Editors
MEGAKARYOCYTE BIOLOGY AND PRECURSORS:
IN VITRO CLONING AND CELLULAR PROPERTIES

The Biology of Isolated Megakaryocytes

An Introduction

Richard F. Levine

Veterans Administration Medical Center and George Washington University, Washington, D. C.

The first two sections of this book are concerned with how megakaryocytes come to be; this section focuses more on what megakaryocytes are and how they function as cells. This distinction is somewhat arbitrary, but a different orientation will be apparent in the following chapters on the biology of isolated megakaryocytes and on the characteristics of megakaryocytes.

The study of a cell is greatly facilitated by the opportunity to examine and manipulate a large and purified population of that cell. The ready availability of erythrocytes has made possible many great advances in membrane biochemistry over the last 25 years. Until more recently, observations on megakaryocytes were limited by the low frequency of these cells in the marrow. In 1964, Paulus reported the first attempt to isolate megakaryocytes and some progress was made with velocity sedimentation techniques (Nakeff and Maat, 1974; Pretlow and Stinson, 1976). Three more recent advances in megakaryocyte isolation are reviewed here briefly: the harvest of megakaryocytes from marrow tissue, a more effective means of isolation of megakaryocytes, and the achievement of an essentially pure population for biochemical studies.

In a sense megakaryocytes are immature platelets. Megakaryocytes contain platelet components and are capable of undergoing degranulation and topographic changes like those of platelets (Levine and Fedorko, 1976). In the disaggregation of marrow tissue, megakaryocytes may be affected, as platelets would also be, by any adenosine diphosphate or proteases released from damaged erythrocytes and leukocytes. The use of various substances (citrate, adenosine, theophylline, and others) which inhibit platelet aggregation not only prevented degranulation and dilatation of the demarcation channels in the megakaryocytes but also gave increases in the yield of megakaryocytes (Table 1). These substances were evaluated by direct

Table 1. Comparison of Megakaryocyte Content of Paired Femurs with Different Harvesting Media.

Comparison	Composition of medium	Number of observations	Nucleated cell counts (megakaryocytes/10^3 cells) mean ± SD		Statistical significance
(a)	HBSS CMFH	4	1.77 2.88	0.15 0.20	< 0.005
(b)	CMFH CMFH + citrate 0.38% (C)	12	2.48 2.98	0.42 0.46	< 0.005
(c)	CMFH, pH 6.8-7.0 CMFH, pH 7.2–7.4	6	3.14 3.17	0.49 0.50	NS
(d)	CMFH + C CMFH + C + adenosine 10^{-3} M (A)	8	2.96 3.46	0.17 0.47	< 0.01
(e)	CMFH + C CMFH + C + theophylline 2×10^{-3} M (T)	6	2.96 3.85	0.44 0.51	< 0.005
(f)	CMFH + C + A CMFH + C + A + T	8	2.95 3.60	0.78 0.68	< 0.005
(g)	CMFH + C + A CMFH + C + T	8	3.41 3.39	0.63 0.42	NS
(h)	CMFH + C CMFH + C + PGE, 10^{-5} M	8	3.66 3.89	0.79 0.88	< 0.05
(i)	HBSS CMFH + C + A + T	6	2.15 3.59	0.46 0.50	<<0.0005

(Source: Reproduced from J. Cell Biol. 69:165, 1976, by copyright permission of the Rockefeller University Press)

comparisons of the quantitative recovery of megakaryocytes obtained with each agent. With calcium- and magnesium-free Hanks' solution, significantly more megakaryocytes were recovered than with regular Hanks' balanced salt solution. Similarly, the successive additions of citrate, adenosine, and theophylline all produced significant increments in recovery. The effect of each was additive to that of the others, e.g., adenosine and theophylline together were better than either alone. A 67% quantitative improvement was achieved with the use of all of these additions in the CATCH medium of Levine and Fedorko (1976).

One might infer from the characteristic large size of megakaryocytes that they could be separated easily from other marrow cells by velocity sedimentation techniques. According to Stoke's Law, the major determinant of the rates of fall of different cell types through the same medium is the cell size (Fig. 1). In a velocity sedimentation run, the gravitational force, viscosity, and medium density are essentially constant. In many tissues, the cell densities are fairly similar, so that the only variable influencing the velocity is the cell size. However, megakaryocytes have a different, lighter density profile from most other cells (Leif et al., 1975; Levine and Fedorko, 1976; Rabellino et al., 1979) so that megakaryocytes do not fall as fast as expected from consideration of only their sizes. In fact, separation by density gradient centrifugation gave consistently better isolation than did velocity sedimentation. Subjecting this light density fraction of cells to a subsequent velocity sedimentation amplified the purification possible with either technique alone (Levine and Fedorko, 1976). Further purification to 70% has been achieved routinely with the use of a second velocity sedimentation (Nachman et al., 1977); the repetition is effective because it is carried out with a smaller, more optimal cell load (Miller and Phillips, 1969).

The concept of megakaryocyte population purity in terms of the percentage megakaryocytes present is an underestimate of purity in the biochemical sense. In an experiment using six isolated populations, the average megakaryocyte diameter was 36 times that of the nonmegakaryocytes (Table 2). After an overnight culture, in which the nonmegakaryocytes proliferated but the megakaryocytes did not change in number, the cell populations averaged 44.7% megakaryocytes. With the assumption that the average cell sizes were unchanged, the megakaryocytes should have comprised 96.1% of the cell mass. After a similar overnight incubation in the presence of ^{3}H-leucine, radioautography was carried out. The ratio of leucine incorporation by megakaryocytes and nonmegakaryocytes confirmed this calculation of the megakaryocyte proportion of the total cell mass. Thus, because megakaryocytes are larger than the other marrow cells they comprise nearly the entire cell mass of isolated populations.

Fig. 1. Stoke's Law.

$$\text{Velocity of Sedimentation} = \text{radius}^2 \cdot (\text{density}_{\text{cell}} - \text{density}_{\text{medium}}) \cdot \frac{2}{9} \cdot \frac{\text{gravitational force}}{\text{viscosity}}$$

Table 2. Megakaryocytes are Larger than Other Marrow Cells and Comprise Most of the Cell Mass of Isolated Megakaryocyte Populations. Mega, Megakaryocytes; Non, Nonmegakaryocytes.

		Diameter μm	Volume μm^3	Volume ratio	Frequency in pellet %	Proportion of pellet %
Isolated guinea pig megakaryocyte populations (average values, N = 6)	Mega	30.6	15,000	35.7	64.5	98.5
	Non	9.3	420		35.5	
After culture (not measured but assuming no size change)				Mega	44.7	96.1
				Non	55.3	
Grain counts after (^{3}H) – leucine, on 1 μm sections.				Mega	44.7	96.2

(Source: Data from Nachman, Levine, and Jaffee [1977])

Now that we can obtain essentially pure populations of differentiated megakaryocytes (Levine and Fedorko, 1976; Nachman et al., 1977; Rabellino et al., 1979) the following subjects can be better addressed:

- morphologic characteristics and morphometrics;
- biochemical and immunologic markers, and their appearance during maturation;
- cell biology of platelet organelle genesis and the mechanism whereby a megakaryocyte can produce a few thousand approximately equal platelets;
- biochemical differences between platelets and megakaryocytes, the biochemical origins of platelet associated substances, and the development of the capacity for membrane ''stickiness;''
- effects of thrombopoietin and other hypothetical stimulatory substances directly on differentiated megakaryocytes; and
- the effects of various drugs and antibodies on megakaryocytes.

The following chapters constitute an exciting beginning on this ambitious program.

References

Leif, R. C., S. B. Smith, R. L. Warters, L. A. Dunlap, and S. B. Leif. 1975. Buoyant density separation of cells. I. The buoyant density distribution of guinea pig bone marrow cells. *J. Histochem. Cytochem.* 23:378–389.

Levine, R. F. and M. E. Fedorko. 1976. Isolation of intact megakaryocytes from guinea pig femoral marrow. Successful harvest made possible with inhibitors of platelet aggregation; enrichment achieved with a two-step separation technique. *J. Cell Biol.* 69:159–172.

Miller, R. G. and R. A. Phillips. 1969. Separation of cells by velocity sedimentation. *J. Cell. Physiol.* 73:191–202.

Nachman, R., R. Levine, and E. A. Jaffe. 1977. Synthesis of factor VIXI antigen by cultured guinea pig megakaryocytes. *J. Clin. Invest.* 60:914–921.

Nakeff, A. and B. Maat. 1974. Separation of megakaryocytes from mouse bone marrow by velocity sedimentation. *Blood* 43:591–595.

Paulus, J. M. 1964. Isolement du mégakaryocyte. *C. R. Soc. Biol.* 158:1747–1749.

Pretlow, T. G. and A. J. Stinson. 1976. Separation of megakaryocytes from rat bone marrow cells using velocity sedimentation in an isokinetic gradient of ficoll in tissue culture medium. *J. Cell. Physiol.* 88:317–322.

Rabellino, E. M., R. L. Nachman, N. Williams, R. J. Winchester, and G. D. Ross. 1979. Human megakaryocytes. I. Characterization of the membrane and cytoplasmic components of isolated marrow megakaryocytes. *J. Exp. Med.* 149:1273–1287.

Evatt, Levine, and Williams, Editors
MEGAKARYOCYTE BIOLOGY AND PRECURSORS:
IN VITRO CLONING AND CELLULAR PROPERTIES

Criteria for the Identification of Megakaryocytes

Richard F. Levine

VA Medical Center and George Washington University, Washington, D. C.

Consideration of specific characteristics of megakaryocytes may be useful to detect not only the larger, easily recognized megakaryocytes, but also the smaller and younger megakaryocytes. These less mature megakaryocytes can sometimes be confused with young fat cells, macrophages, myeloblasts, and proerythroblasts, or just not recognized. This review will attempt to provide criteria for identification of the entire population of differentiated megakaryocytes and will focus particularly on distinguishing the earliest, least obvious forms.

Size

Size has always been the salient feature of megakaryocytes. Marrow giant cells were first described in 1849 by Robin. In 1890, Howell named the megakaryocytes to distinguish them from the "polykaryocytes," which we now know as the much less frequent osteoclasts. In 1943, Japa published a size range for megakaryocytes as 30–100 μm and this estimate without data has been widely repeated.

The correct range appears to be about 10–50 μm. Figure 1 is a histogram of cell sizes from 1000 megakaryocytes (solid line) and 1000 nonmegakaryocytes (dashed line). These glutaraldehyde-fixed guinea pig bone marrow cells in suspension were individually examined with an optical micrometer by phase contrast microscopy at 1000×. The glutaraldehyde fixation did not affect cell sizes (Levine and Fedorko, 1976). Solely morphologic criteria (see below) were used to decide if each cell were a megakaryocyte or not. The megakaryocytes ranged from 10–50 μm in diameter and the other cells were mostly 5–13 μm. Similar results to these have been obtained with human (Levine, 1980), monkey, and rat marrow cells. The important conclusion to be drawn from this laboriously determined distribution of megakaryocyte and nonmegakaryocyte sizes is that the megakaryocytes comprise

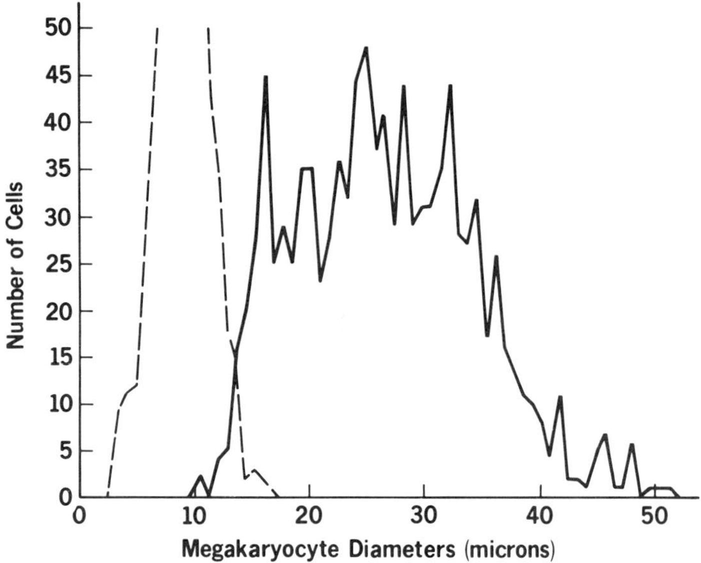

Fig. 1. Diameters of guinea pig megakaryocytes (solid line, N = 1000) and nonmegakaryocytes (dashed line, N = 1000).

a distinct and somewhat separate population within the marrow, most of them distinguishable on the basis of size alone. In the data of Figure 1, a size of 13 μm was generally sufficient to confirm that a given cell was a megakaryocyte. Only 2% of the megakaryocytes were smaller than this threshold and 2% of the nonmegakaryocytes larger. Size greater than 13 μm is probably the simplest objective criterion to detect megakaryocytes. With a little practice this distinction may be made routinely at 160–250 × magnification, without relying on a micrometer.

This criterion is a quantitative refinement of the popular concept that megakaryocytes are larger than other marrow cell types. Although essentially only megakaryocytes are larger than this threshold, some megakaryocytes do occur within the size range of the nonmegakaryocytes; these smaller megakaryocytes are harder to detect by phase contrast microscopy, even at 1000 × . The problem of identifying the very earliest differentiated megakaryocytes is discussed below.

Ploidy

The name "megakaryocyte" refers not to the cell size but to the large nuclear size. The modern concept of increased nuclear material derived from repeated endomitoses was described by Jolly in 1923 and quantitatively confirmed in 1964 by Garcia and by de Leval. In the next chapter, Dr. Paul Bunn presents data on megakaryocyte ploidy values. I should like, first, to consider the notion of a threshold ploidy value as a sufficient criterion for the identification of megakaryocytes.

When over 6000 unselected guinea pig bone marrow cells were examined by Feulgen microdensitometry on a Vickers M-85 Scanning Microdensitometer, the classical distribution of 2N, 4N, and transitional cells was seen (Fig. 2). 2N cells were predominant, with a distinct 4N peak also visible. The 4N peak dropped off sharply virtually to zero at 5.2N. All cells with higher ploidies (not all plotted) had nuclear morphology, nuclear size, and cell size consistent with those of megakaryocytes (Levine et al., 1980). A major benefit of this ploidy threshold was the enhanced identification of formerly questionable Feulgen stained megakaryocyte nuclei; such morphology/ploidy correlations helped refine the morphologic criteria for recognition of immature megakaryocytes.

A similar result was obtained when 206,000 cells were examined by flow cytometric analysis of propidium iodide-stained nuclei. Figure 3 shows the results with an isolated guinea pig population (27% megakaryocytes). The 8N population of megakaryocytes formed a separate peak apart from the tail of the 4N population

Fig. 2. Ploidy values of unselected guinea pig bone marrow cells, determined by Feulgen microdensitometry (data plotted only to 6N; complete data in Levine et al., 1980).

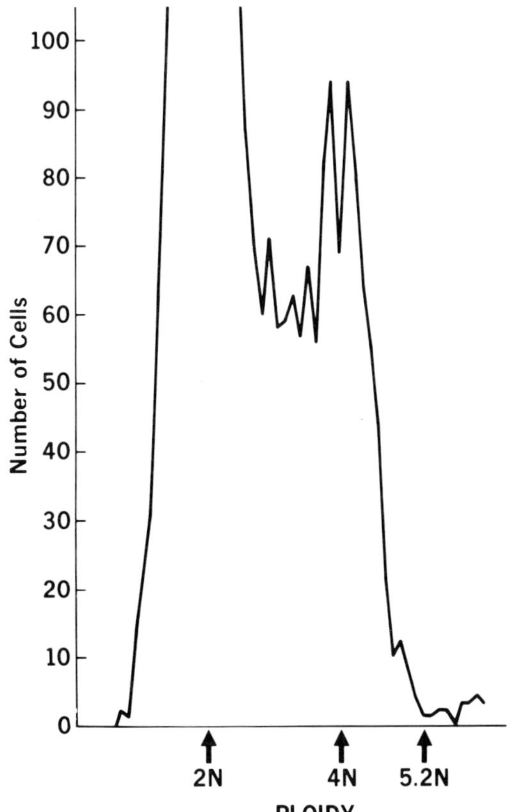

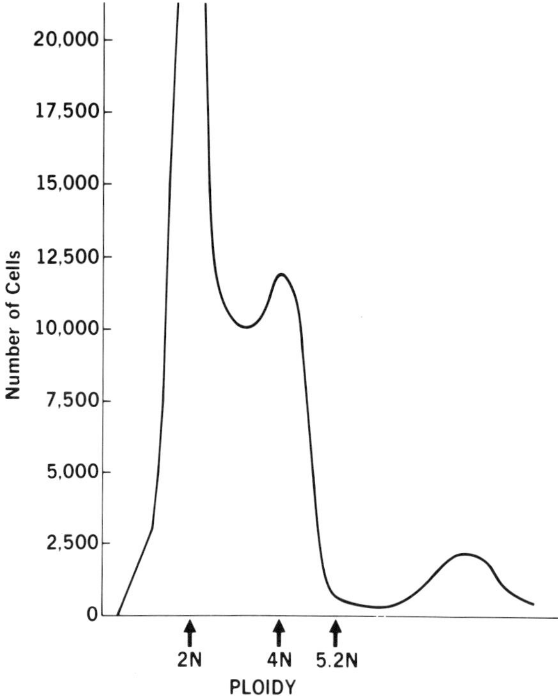

Fig. 3. Tracing of flow cytometric analysis of guinea pig isolated megakaryocyte population (data plotted only to 10N).

which ended at approximately 5.2N. Because of the sharp drop off of that tail and because ploidy values from approximately 6.5 to 9.5N are apparently statistically encompassed in the 8N distribution, those cells with 5.2–6.5N are probably megakaryocytes in transition from 4N to 8N. Direct proof for this idea has not yet been obtained.

Although many dividing cell types achieve 4N ploidy levels in the $g_2 \rightarrow M$ phase, megakaryocytes are the only significant marrow cell to achieve normally states greater than 4N. Osteoclasts might achieve high ploidy levels but are quite rare; I have seen only a dozen osteoclasts in several years of working with marrow cells. The 4N megakaryocytes are difficult to detect microscopically with only the Feulgen stain, and are impossible to sort out by single parameter flow cytometry from other 4N cells. There probably exist differentiated megakaryocytes of 2N as well, but these need other, more complicated approaches to detect routinely. The chapter by Dr. Manfred Mayer presents one answer to this problem.

Morphologic Characteristics

At some point in the enumeration of megakaryocytes, the standard of reference has to be the morphologic characteristics of the megakaryocyte line. The phase contrast,

Wright-Giemsa, and electron microscopic appearances have been examined systematically elsewhere (Levine, 1980). Because the granular and mature megakaryocytes are fairly easy to recognize (they are generally > 20 μm in diameter and have ploidy values > 8N), this chapter will show morphologic criteria for only the immature megakaryocytes.

By phase contrast the most immature megakaryocytes appear to have little cytoplasm but two to four nuclear lobes with prominent nucleoli. The cells in Figure 4 are 9–15.5 μm in diameter, but this appearance can be found in cells from 9 to 22 μm diameter. Although other cell types may occasionally achieve diameters up to 20 μm, their size is generally due to their cytoplasmic mass. Thus, the combination of cell size, high nuclear/cytoplasmic ratio, and lobulated nucleus is sufficient to identify the megakaryoblast. No other marrow cell type has this appearance.

The distinction from immature fat cells (Fig. 5), however, is often subtle. Fat cells are found in rabbit marrow of any age and in mature subjects of any species. Large, obvious fat cells are probably disrupted at the time of marrow disaggregation or are removed during washes, owing to their lipid content and very low specific gravity. Immature fat cells are 10–20 μm and by phase constrast have numerous cytoplasmic inclusions, some of which may be suggestive of nucleoli, although their nuclei are not prominent like those of megakaryocytes. Fat cells can be distinguished by the presence of small (1–2 μm) golden inclusions and an overall yellowish glow, different from the whitish refractility of granular megakaryocytes.

In the second megakaryocyte maturation stage (Fig. 6), the nucleus becomes invaginated and the lobes appear to have a ''U'' shaped configuration if viewed side on. The nuclear/cytoplasmic ratio may be 1:1 to 1:2 and the cell size may be 14–30 μm. Although this stage has a superficial resemblance to metamyelocytes, the latter cells are post-mitotic, and thus have a smaller cell size and nuclear size as well as prominent nuclear chromatin without nucleoli. This megakaryocyte stage, with the beginnings of visible granule formation and increasing amounts of cytoplasm, can probably be called the promegakaryocyte. Following this stage is the

Fig. 4. Phase contrast photomicrographs of rat megakaryoblasts, ranging from 15.5–9 μm. (1940 ×)

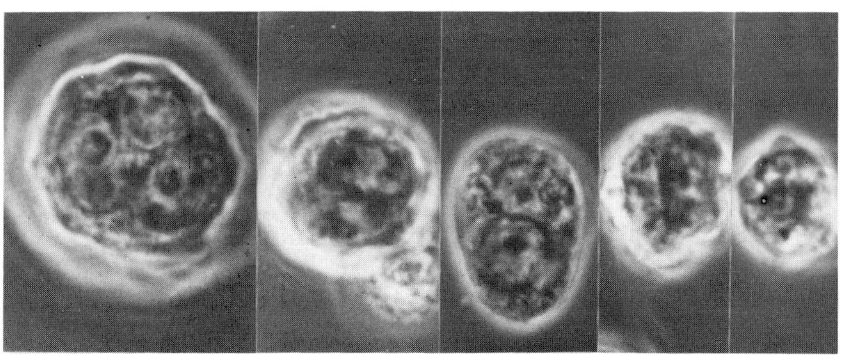

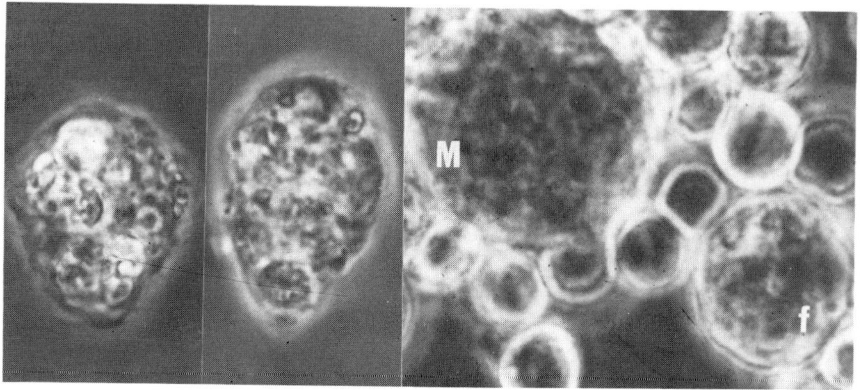

Fig. 5. Left: two rabbit fat cells, both about 15 μm. Right: human megakaryocyte (M, 20 μm diameter) and fat cell (f, 15.6 μm). (1940×)

more familiar granular megakaryocyte with separated nuclear lobes and large amounts of cytoplasm.

In an electron micrographic study published in 1971, MacPherson found the youngest megakaryocytes (10–15 μm) to have bilobed or trilobed nuclei, sparse cytoplasm with polyribosomes, and a few alpha granules and demarcation membranes. These last two findings are, of course, specific components of megakaryocytes. Figure 7 shows a very immature cell with multiple nucleoli, a section diameter of 8.8 μm, some demarcation membranes, and many free ribosomes/polysomes in a volume of cytoplasm less than that of the nucleus. In Figure 8 is a megakaryocyte of the same stage (diameter 7.8 μm), with at least 3 lobes, one of which contains portions of two nucleoli. The presence of an alpha granule further confirms its identify. Again, the cytoplasm is filled with ribosomes. No other cell type would have this nuclear/cytoplasmic ratio and a lobulated nucleus.

By bright-field microscopy of Wright-Giemsa stained marrow cells the heavy ribosome concentration of the youngest megakaryocytes may be seen as a prominent basophilia, dark enough usually to obscure the nucleus (Fig. 9). The Wright stain solution was applied to air dried cells on cover slips for one min, diluted with buffer for one min, washed, and then diluted Giemsa stain was applied for 1 min. The intensity of staining is distinctly greater than that of a similar hue in proerythroblasts and plasma cells. These almost black cells, 8–22 μm in diameter, are easily distinguished from other marrow cell types. It is proposed that this intense basophilia is a simple criterion for the identification of these immature megakaryocytes, which, by analogy with the cytologic features of proerythroblasts and myeloblasts, are the megakaryoblasts.

Specific Chemical Reactions

Although size, ploidy value, and staining reaction are simple objective criteria for

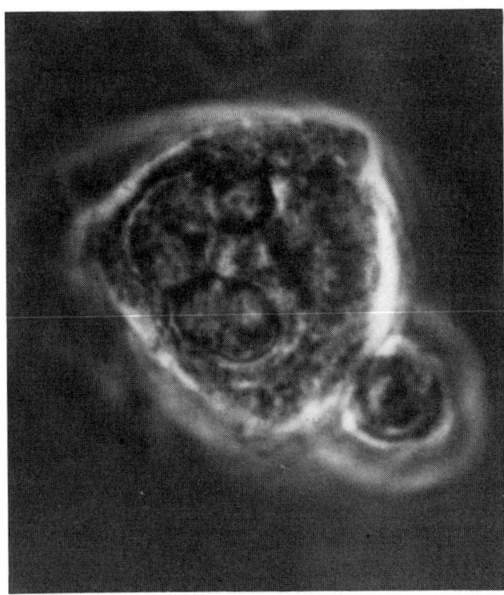

Fig. 6. Rat promegakaryocyte (19 μm diameter). (1940×)

detection of megakaryocytes, these techniques have their limitations. All of the above characteristics are threshold phenomena; they require a certain amount of development for detection. What occurs in these cells from the moment of initiation of differentiation to these thresholds of detectability is as yet unknown. What is needed to explore this gap are more sensitive methods that are equally specific.

Two similar approaches exist, both based on the early appearance in megakaryocyte differentiation of specific platelet substances. A histochemical reaction can be carried out in rat, mouse, or rabbit marrow cells for the enzyme acetylcholinesterase and immunofluorescence studies can be performed in any species with a variety of anti-platelet sera. Several other chapters contain findings with these techniques, but I should like to consider two issues in the literature on acetylcholinesterase that had been unclear to me.

The first is what are the exact sizes of the small acetylcholinesterase positive cells and the second is what is the nature of their relationship with mature megakaryocytes. Figure 10 shows a number of rat marrow cells unequivocally positive for acetylcholinesterase after a 3 hr incubation. They ranged from 5.6 to 36 μm. Figure 11 shows the size distribution of rat acetylcholinesterase positive cells (solid line) air dried on glass cover slips as well as that of rat megakaryocytes in suspension detected by phase contrast microscopy at 1000× (dashed line). The two curves are quite similar, except for the left side of the distributions. The phase contrast criteria, as discussed above, cannot detect megakaryocytes smaller than about 10 μm in diameter. About 10% of the enzyme positive cells were smaller than 10 μm and about 18% were smaller than the 13 μm threshold. The acetylcholinesterase reaction was found in a continuous distribution down to and including cells of 5 μm diameter.

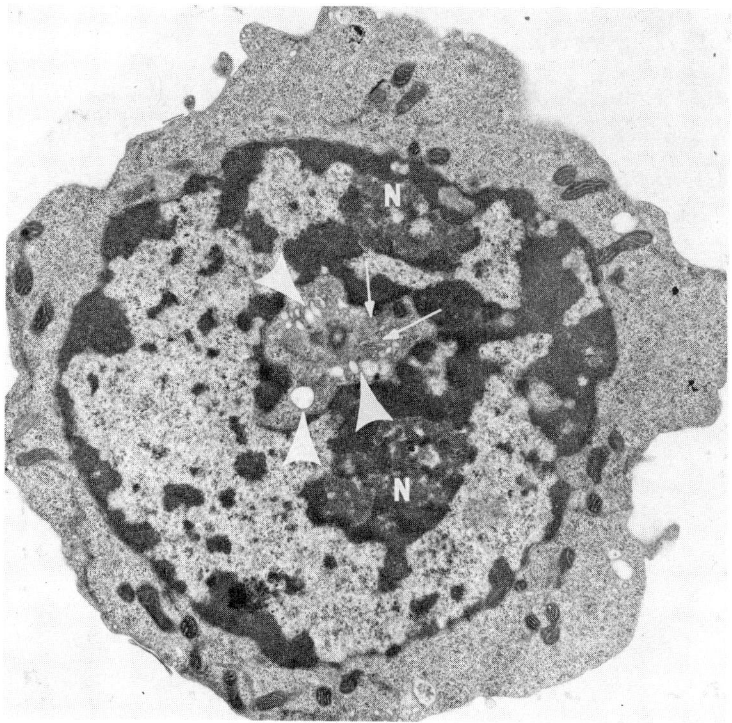

Fig. 7. Guinea pig megakaryoblast. Single irregular nucleus with 2 nucleoli (N). Cytoplasm filled with many ribosomes (tiny black dots). In the cell center, besides Golgi complexes (small arrows) are 2 areas (larger arrows) of slightly dilated demarcation channels. Cell section diameter is 8.8 μm. (12,900×). Harker (1968) estimated that for megakaryocytes in paraffin sections of marrow tissue the average section diameter was 0.87 of the maximum diameter; if the same is true of ultra thin sections, the true equatorial diameter of this cell would be likely to be > 10 μm

This size distribution of cells with the same differentiation marker suggests that these small positive cells are already megakaryocytes, the earliest ones. Dr. Michael Long has convincing evidence (Long: 293) that these small enzyme containing cells are newly differentiated megakaryocytes that mature into the recognizable megakaryocytes. Dr. Mayer (Chapt. 26) and Dr. Mazur (Chapt. 24) suggest that their sensitive immunologic detection methods will confirm that these are the earliest cells of the megakaryocyte lineage. As these very small cells seem to be minimally differentiated, even less developed than the megakaryoblasts, an appropriate designation for them might be "just differentiated megakaryocytes." Ultrastructural studies are underway to confirm that these small cells are not erythroid cells which contain small amounts of this enzyme (Zajicek and Datta, 1953) and to determine

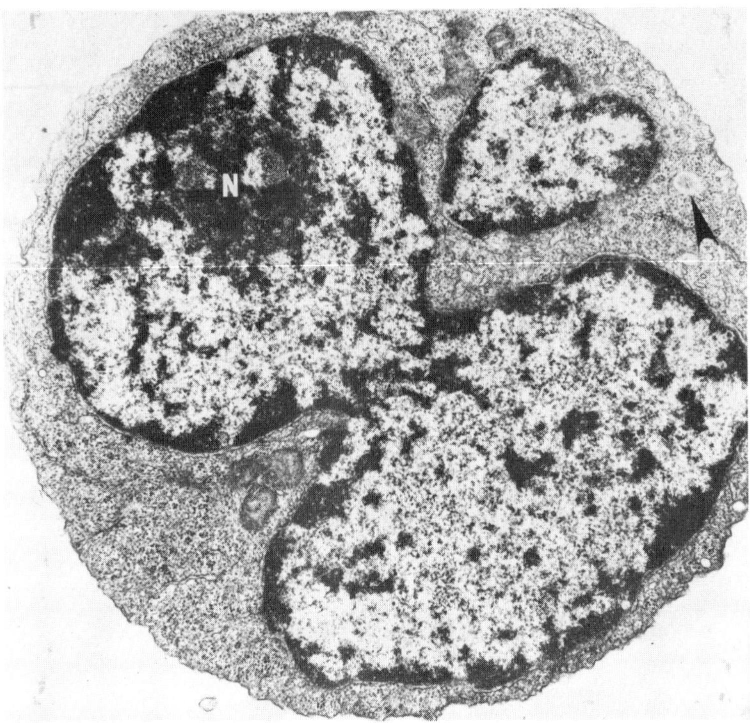

Fig. 8. Human megakaryoblast. Trilobed nucleus with single nucleolus (N). Cytoplasm filled with ribosomes. Single alpha granule (arrow) is present. Cell section diameter is 7.8 μm. (16,000×)

whether they are merely smaller versions of the megakaryoblasts shown above or are indeed morphologically undifferentiated.

Discussion

The appropriate criteria to use for the identification of megakaryocytes depend on the purpose for which the information is sought. These goals can be arbitrarily divided, like the organization of this book, into two different areas, based on interest in the megakaryocytes themselves or in the precursors of megakaryocytes.

There are a number of questions where the focus is on the population of differentiated megakaryocytes: (1) clinical evaluation of marrow megakaryocytes, (2) examination of the quantitative and qualitative responses to experimental manipulation of thrombopoiesis *in vivo*, (3) monitoring megakaryocyte isolation procedures, (4) estimation in isolated populations of the megakaryocyte purity for biochemical studies, etc. For such investigations megakaryocyte size and, for the

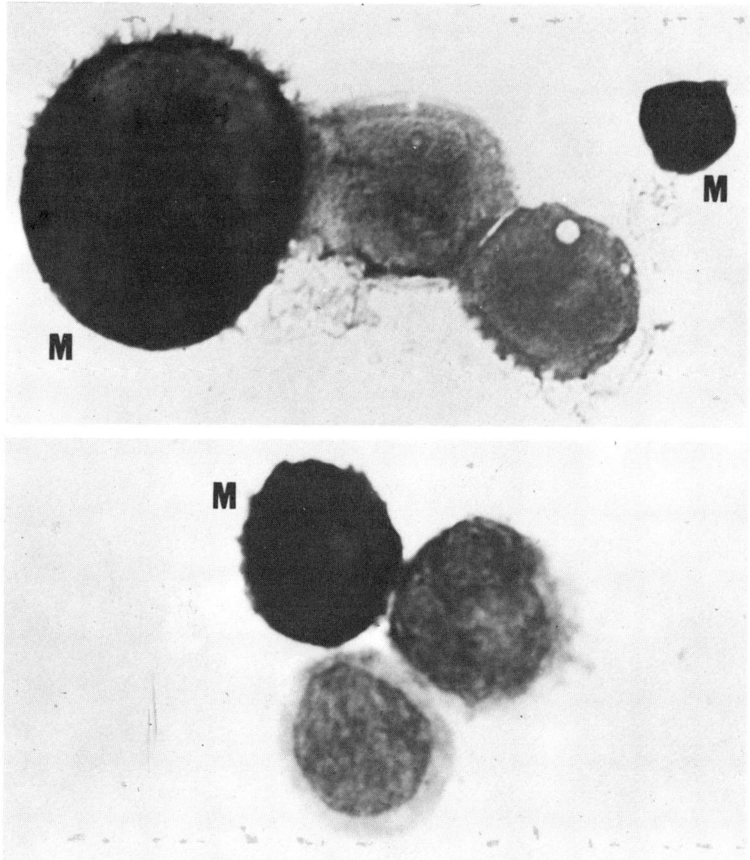

Fig. 9. Guinea pig promegakaryoblasts (M) and nonmegakaryocytes. The megakaryocytes diameters are 7.8–2.5 μm. (1940×)

smaller/younger megakaryocytes, size plus morphologic characteristics, can serve to detect almost the entire megakaryocyte population. This approach is simple, rapid, and consistent.

In a different direction are questions about the steps preceding differentiation into megakaryocytes but for which the assays depend on the enumeration of megakaryocytes: (1) the need to count all of the progeny of megakaryocyte precursors and to determine the complete distribution of cell types and their spatial associations, (2) quantitation of megakaryocytes in semi-solid media in petri dishes which together prevent the use of 160–250× magnification, (3) early detection of the differentiation step, (4) study of the beginning of maturation, (5) timing of all the maturation steps, etc. For these investigations the certainty of the specific chemical reactions seems desirable, although these procedures are done on fixed cells and additional

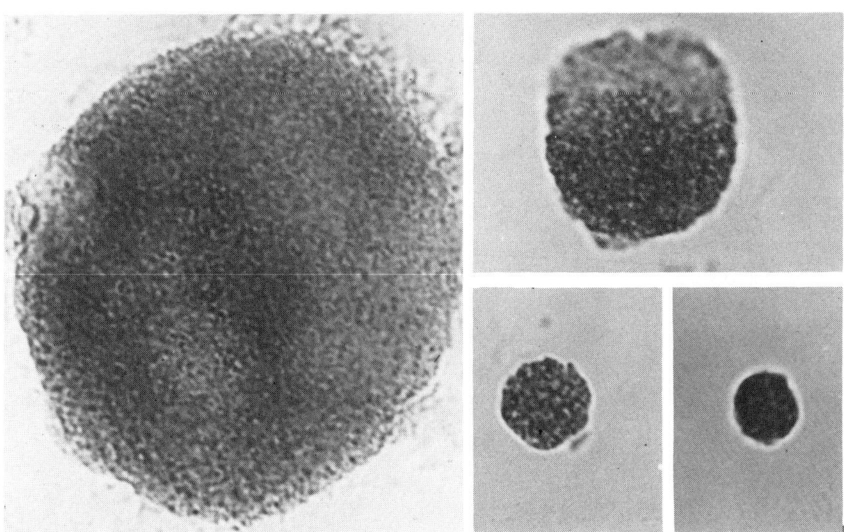

Fig. 10. Acetylcholinesterase positive cells in rat bone marrow. These cells are, respectively, 36, 16, 7.2, and 5.6 μm in diameter. (1940×)

or simultaneous histochemical, autoradiographic, or other morphologic observations may be precluded or difficult.

Summary

Several objective criteria exist for the identification of megakaryocytes. A size of 13 μm or more was the simplest criterion for recognition of megakaryocytes in all

Fig. 11. Diameters of acetylcholinesterase positive cells in rat bone marrow (solid line, N = 428) and of rat megakaryocytes in suspension examined by phase contrast (dashed line, N = 375).

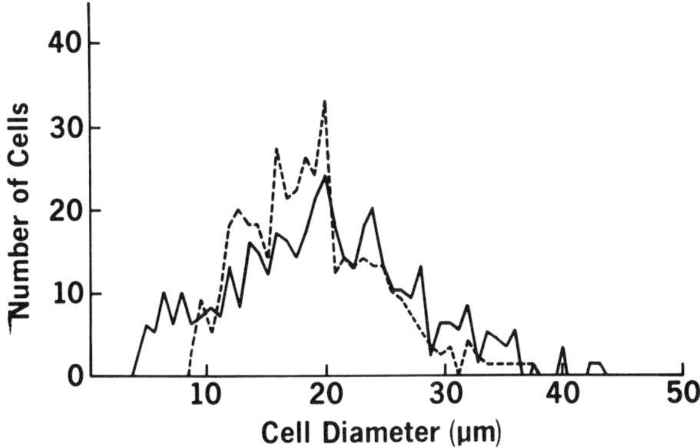

species and without special procedures. Ploidy greater than 5.2N was also found to be a sufficient criterion and was most useful in validating the morphologic criteria for the immature megakaryocytes. Megakaryoblasts were shown to have the following specific morphologic characteristics: by phase constrast microscopy, a lobulated nucleus and a high nuclear/cytoplasmic ratio; by electron microscopy, the same and prominent ribosomes and polysomes, plus the presence of alpha granules or demarcation channels; and by Wright-Giemsa staining, intense basophilia. The just differentiated megakaryocytes, less developed than the megakaryoblasts, could be detected by specific chemical reactions.

ACKNOWLEDGMENT
This work was supported by the United States Veterans Administration.

References

de Leval, M. 1964. DNA levels in normal guinea pig megakaryocytes. *C. R. Soc. Biol.* 158:2198–2201.

Garcia, A. M. 1964. Feulgen — DNA values in megakaryocytes. *J. Cell. Biol.* 20:342-345.

Harker, L. A. 1968. Megakaryocyte quantitation. *J. Clin. Invest.* 47:452–457.

Howell, W. H. 1890. Observations on the occurrence, nature, and function of the giant cells of the bone marrow. *J. Morphol.* 4:117–130.

Japa, J. 1943. A study of the morphology and development of the megakaryocytes. *Brit. J. Exp. Path.* 24:73–80.

Jolly, J. 1923. *Traité Technique d'Hématologie*. Paris:Maloine.

Levine, R. F. 1980. Isolation and characterization of normal human megakaryocytes. *Br. J. Haematol.* 45:487–497.

Levine, R. F., and M. E. Fedorko. 1976. Isolation of intact megakaryocytes from guinea pig femoral marrow. *J. Cell Biol.* 69:159–172.

Levine, R. F., P. A. Bunn, Jr., K. C. Hazzard, and M. L. Schlam. 1980. Flow cytofluorometric analysis of megakaryocyte ploidy. Comparison with Feulgen microdensitometry and discovery that 8N is the predominant ploidy class in guinea pig and monkey marrow. *Blood* 56:210–217.

MacPherson, G. G. 1971. Development of megakaryocytes in bone marrow of the rat: an analysis by electron microscopy and high resolution autoradiography. *Proc. Roy. Soc. B. Biol Sci.* 177:265–274.

Robin, C. 1849. Sur l'existence de deux especes nouveaux d'elements anatomiques qui se trouvent daus le canal medullaire des os. *Gaz. Med.* 4: 992–993.

Zajicek, J., and N. Datta. , 1953. Investigations on the acetylcholinesterase activity of erthrocytes, platelets, and plasma in different animal species. *Acta Haematol.* 9:115–121.

Discussion

Hempling: Did you use diameters of these glutaldehyde fixed cells to calibrate a particle size analyzer? What were your findings when unfixed cells placed in an isotonic culture medium were analyzed by this technique?

Levine: The glutaldehyde effect on cell diameter was monitored in two ways. Red cells, which are sensitive to osmotic effects, were measured before and after glutaldehyde treatment. Also, by making the comparison that you have just suggested; i.e., by looking at the fixed versus fresh megakaryocytes in an electronic size analyzer. I measured diameters by the microscope and the size analyzer measured volumes which can be converted directly into diameters. The identity of the two distributions was striking. That will be published in the future.

P. Schick: You described threshold size criteria of glutaldehyde fixed cells. With megakaryocytes put on glass slides, wouldn't you expect some spreading? Do you need another size threshold for cells on glass?

Levine: When cells are put on glass, the megakaryocytes do spread very slightly. Other cell types spread a little bit more, so that the quantitative threshold does have to be adjusted somewhat; more to the range of about 15 microns. That data will also be published. I think that it was the spreading of other cell types that was responsible for the popular estimate that megakaryocytes start at 30 microns, because some of these cell types do spread on glass up to 20 microns.

Nakeff: What is the present level of recovery of megakaryocytes in your isolation procedure? How confident do you feel that the cells that you are obtaining are representative of the total poulation?

Levine: Our goal is, of course, to isolate a large sample of intact megakaryocytes. These methods give that result, although they do lose some of the younger, smaller megakaryocytes, as shown by Dr. Bunn. A meaningful answer to your first question requires the critical application of rigorous criteria for the identification of megakaryocytes in the marrow cell population. Much published data is incomplete because of lack of attention to this important matter. In fact, I am not sure just how comparable our recovery data are to the reports from other laboratories because I don't know what criteria others have used; the large, easily recognizable megakaryocytes are generally all recovered by most isolation techniques and if only they are counted, very high recoveries would be reported. For example, I have not been able to find as high a recovery with Dr. Rabellino's modifications of these methods as he reports. The net recovery by our methods, which can surely be improved upon, has been 30–40% of the starting population in guinea pigs. By characterizing the isolated megakaryocytes carefully, we know just how representative they are of the marrow population, so that appropriate inferences can be properly drawn.

Evatt, Levine, and Williams, Editors
MEGAKARYOCYTE BIOLOGY AND PRECURSORS:
IN VITRO CLONING AND CELLULAR PROPERTIES

Analysis of Megakaryocyte DNA Content: Comparison of Feulgen Microdensitometry and Flow Cytometry

Paul A. Bunn, Jr., Richard F. Levine, Mark L. Schlam, and Karen C. Hazzard

NCI-VA Medical Oncology Branch, Division of Cancer Treatment, National Cancer Institute and Department of Medicine, Veterans Administration Medical Center, Washington, D. C.

Megakaryocytes are giant multinucleate cells from which platelets arise. Early microdensitometric studies quantifying deoxyribonucleic acid (DNA) content using the Feulgen technique confirmed that these cells were polyploid and contained 4, 8, 16, 32, and 64N amounts of DNA with 2N being the normal diploid amount of DNA (deLeval, 1964; Garcia, 1964; Odell et al., 1968; Paulus, 1968a; Weste and Penington, 1972). This suggested that megakaryocytes arise by a series of synchronous nuclear doublings without cytokinesis (endomitosis, endoreduplication). These studies demonstrated that, in unperturbed situations, 16N was the most frequent ploidy class and more than 95% of megakaryocytes were in the 8, 16, and 32N ploidy classes (deLeval, 1964; Garcia, 1964; Odell et al., 1976; Paulus, 1968a; Weste and Penington, 1972).

Regulation of platelet production is centered around changes in megakaryocyte number and ploidy. In response to situations of increased platelet demand (thrombocytopenia), increases in megakaryocyte number and ploidy precede a rise in platelet count by several days (Ebbe et al., 1968; Harker, 1968; Odell et al., 1965). Megakaryocyte size and the number of platelets produced per megakaryocyte are proportional to the megakaryocyte ploidy. On the other hand, situations with thrombocytosis and decrease in platelet demand are accompanied by a decrease in megakaryocyte number and ploidy (Odell et al., 1976; Penington et al., 1976).

Studies of megakaryocyte ploidy are few in part because of the infrequency of megakaryocytes in marrow suspensions (< 1% nucleated cells) and because of the tedious nature of the Feulgen microdensitometric procedure. Studies of the humoral regulation of megakaryopoiesis and hence thrombopoiesis have been hampered by these technical problems. Flow cytometry (FCM) provides rapid analysis of DNA content of large numbers of cells and can be used to measure megakaryocyte ploidy.

Recent studies have shown that megakaryocytes can be enriched from marrow suspensions through several physical separation techniques (Levine and Fedorko, 1976; Nachman et al., 1977; Nakeff and Maat, 1974; Rabellino et al., 1979), providing large numbers of megakaryocytes for analysis. The ability to grow mature megakaryocytes and megakaryocyte precursors *in vitro* has led to new avenues for studies on the regulation of megakaryopoiesis (Metcalf et al., 1975; Nakeff and Daniels-McQueen, 1976; Nachman et al., 1977; Williams and Jackson, 1978). Measurement of megakaryocyte ploidy will play an essential role in these studies. In this report, we compare and contrast flow cytometry and Feulgen microdensitometry for measurement of megakaryocyte ploidy.

Methods and Materials

Sample Collections

Bone marrow cell suspensions were obtained by scooping out the bones of the humeri and femora of two guinea pigs or one African green monkey. The marrow cells were placed into calcium and magnesium-free Hank's solution containing citrate, theophylline, and adenosine as previously described (Levine and Fedorko, 1976; Nachman et al., 1977). The gelatinous marrow was finely minced and disaggregated by pipetting and passage through a 100 μm diameter stainless steel sieve. When purification procedures were employed, marrow suspensions were subjected to a density gradient centrifugation followed by one or two velocity sedimentations as previously described (Levine and Fedorko, 1976; Nachman et al., 1977). At each step megakaryocytes and total cells in both the enriched fractions and discarded fractions were counted in hemocytometer chambers at 250 × and analyzed for DNA content by flow cytometry.

Flow Cytometry

The marrow cell suspensions were stained with 5 mg/100 ml propidium iodide (Sigma Chemical Co., St. Louis, Missouri) in 0.1% sodium citrate by the methods of Krishan (1975). In cell sorting experiments, the cells were fixed in 50% ethanol and stained with propidium iodide after RNase treatment according to the method of Crissman and Steinkamp (1973). The DNA content of up to 5×10^6 cells was measured in a Coulter TPS-1 cell sorter (Coulter Electronics, Hialeah, Florida). The electronics of the instruments were adjusted so that the diploid marrow cells were in channel 6 of 128 total channels allowing simultaneous analysis of the 2N to 32N ploidy classes. In some experiments 2N cells were placed in channel 3 so that 64N cells could also be analyzed. The relative fluorescence of the cells was displayed in a frequency distribution histogram. A computer-derived Gaussian fit program (M-Lab) was used to determine the proportion of cells in each ploidy class (Knott and Reece, 1972). In electronic sorting experiments, cells with 8N DNA content and cells with 16N or greater DNA content were sorted into separate plastic beakers. The cells were centrifuged onto glass slides and stained with Giemsa.

Feulgen Microdensitometry

Cell suspensions were centrifuged onto cover slips and air dried. The cells were subjected to acid hydrolysis with 5N hydrochloric acid for 40 min at room temperature and stained with a modified Feulgen reaction. Microdensitometric quantification of the Feulgen stain reaction by integrated optical density measurements over single nuclei was performed using an Artek 800 Image Analyzer (Artek Systems Corp., Farmingdale, New York) or with a Vickers M-85 Scanning Microdensitometer (Vickers instruments, Inc., Woburn, Massachusetts). For routine measurements, the slide was scanned for megakaryocyte nuclei which were identified by their size, shape, and density. For determination of the diploid 2N standard, 20–50 polymorphonuclear or band form neutrophils were measured in each sample.

In one experiment, an effort was made to obtain a more complete analysis of unpurified bone marrow suspensions. All cells which were not obvious polymorphonuclear or band form neutrophils or mature lymphocytes were measured.

Results and Discussion

Isolated Megakaryocytes

Megakaryocyte samples obtained after density and velocity sedimentations were split and analyzed by both flow cytometry and Feulgen microdensitometry. A representative histogram obtained by FCM is shown in Figure 1, and shows that the majority of cells nonmegakaryocytes in the 2N–4N ploidy classes, were found between channels 6–12. The 8N, 16N, and 32N ploidy classes are readily apparent at channels 24, 48, and 96, and contained 15%, 46%, and 39% of the megakaryocytes, respectively. This sample contained 66,435 cells of which 16,732 were in the 8N, 16N, and 32N ploidy classes. The DNA histogram obtained from the same sample analyzed by Feulgen microdensitometry is shown in Figure 2. The 8N, 16N, and 32N ploidy classes were easily recognizable and contained 13%, 64%, and 24% of the 200 megakaryocytes, respectively.

Direct comparison of flow cytometric and Feulgen microdensitometric analysis of 4 split specimens is shown in Table 1. The purity of the 4 specimens ranged from 28%–48% megakaryocytes and in each instance the results were extremely similar despite the fact that many more megakaryocytes were analyzed by flow cytometry. Analysis by flow cytometry was far more rapid; about 2 hr was required to measure the DNA content of 200 megakaryocytes by Feulgen microdensitometry. In contrast, the 16,000–64,000 megakaryocytes were analyzed by flow cytometry in less than 10 min.

Unseparated Bone Marrow

Isolated marrow suspensions were analyzed first because the enrichment procedures allowed more megakaryocytes to be measured in a shorter time. The results of analysis of split samples of unseparated marrow suspensions analyzed by FCM and by microdensitometry are shown in Table 2. The concentration of megakaryocytes

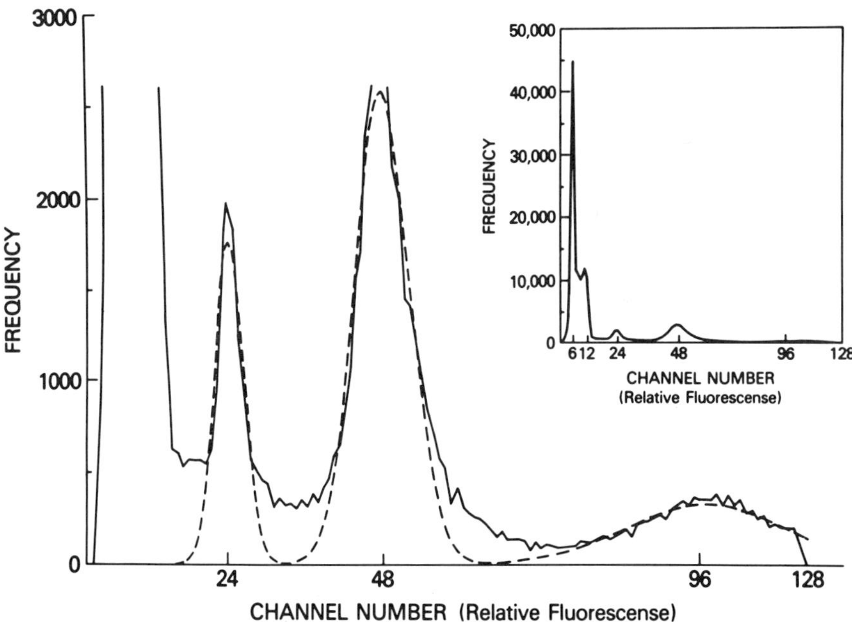

Fig. 1. DNA histogram of megakaryocyte ploidy in partially purified guinea pig bone marrow analyzed by flow cytometry. The insert shows the DNA distribution of 66,435 cells obtained after a density centrifugation and a velocity sedimentation. The largest peak at channel 6 is comprised of 2N diploid cells (G phase). The cells between channels 6 and 12 are cells undergoing DNA synthesis (S phase), while the cells at channel 12 have 4N DNA (G_2 + M cells). Three additional peaks centered at channels 24, 48, and 96 are apparent and are the 8N, 16N, and 32N megakaryocyte peaks. These peaks are shown more clearly by expanding the Y axis as shown in the body of the figure. The solid line represents the actual data, while the dotted line is the computer-derived Gaussian fit of this raw data. By Gaussian fit analysis, 15% of the megakaryocytes were 8N, 46% were 16N, and 39% were 32N.

in these samples ranged from 0.06%–0.26%. In contrast to the findings on purified suspensions, the two methods gave strikingly different distribution patterns. In each instance, flow cytometry found that 8N was the most frequent ploidy class; 16N and 32N contained a smaller percentage (8N > 16N > 32N). Microdensitometric results were quite dissimilar. 8N megakaryocytes were the least frequent ploidy class. In these experiments, about 10^6 marrow cells and 1,200–40,000 megakaryocytes were analyzed by flow cytometry in about 1 hr. Microdensitometry analyzed 80–131 megakaryocytes in about 3 hr. The differences in these results led us to examine whether the technique of scanning the unseparated marrow suspensions introduced a bias favoring the analysis of larger, more easily recognizable megakaryocytes.

In routine practice, one analyzes cells which are unquestionably megakaryocytes and since many microscopic fields had to be examined to analyze a single mega-

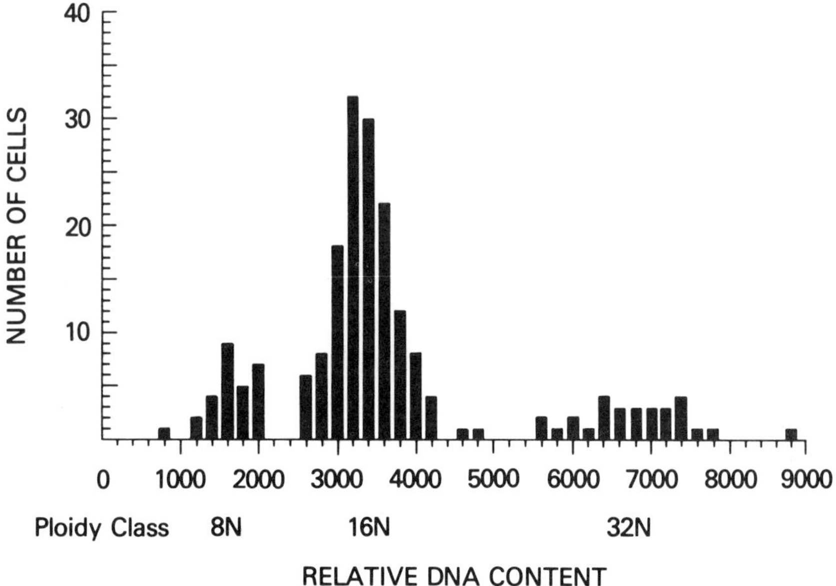

Fig. 2. DNA histogram of megakaryocyte ploidy in partially purified guinea pig bone marrow (same specimen as Fig. 1) analyzed by Feulgen microdensitometry (Artek 800 Image Analyzer). The DNA distribution of 200 megakaryocytes analyzed by scanning the slide for recognizable megakaryocytes is shown. The diploid standard polymorphonuclear neutrophils (not shown) had a relative DNA content of 400 arbitrary units. 8N megakaryocytes had a modal DNA content of 1600 arbitrary units, 16N megakaryocytes of 3200 arbitrary units, and 32N megakaryocytes of 6400 arbitrary units. By this technique 12% of the megakaryocytes were 8N, 64% were 16N, and 24% were 32N.

karyocyte, it was easy to overlook small, less mature megakaryocytes. Thus, to determine whether this selection approach omits analysis of smaller, less easily recognizable megakaryocytes, a more comprehensive sample of unseparated marrow cells was analyzed and the results are shown in Table 3. We measured the integrated optical density of 6000 cells which were not obvious neutrophils or mature lymphocytes. Megakaryocytes comprised 124/6000 or 0.21% of these cells and the distribution of these cells was similar to that found by FCM (8N > 16N > 32N). This type of analysis is impractical for routine purposes as 50 person-hours were required to analyze these 124 megakaryocytes.

To ensure that there was no bias favoring measurement of lower ploidy cells by FCM, we electronically sorted cells with 8N or greater DNA content and looked for evidence of cell clumping. Examples of sorted cells are shown in Figure 3. No clumps of nonmegakaryocytes were seen in 5 separate sorting experiments. Only a rare cell not morphologically a megakaryocyte was seen, attached to a morphologically identifiable megakaryocyte. As shown in the figure, higher ploidy megakaryocytes (16N–32N) were consistently larger than 8N megakaryocytes. Anal-

Table 1. A Comparison of the Ploidy of Isolated Megakaryocytes Analyzed by Flow Cytometry and by Feulgen Microdensitometry using a Vickers Microdensitometer and an Artek Image Analyzer.

	Flow cytometry			Vickers microdensitometry			Artek microdensitometry		
Sample	**8N**	**16N**	**32N**	**8N**	**16N**	**32N**	**8N**	**16N**	**32N**
Guinea Pig #1	15	46	39	13	57	29	10	58	32
#2	18	56	26	12	72	16	12	64	24
Monkey #1	9	42	49	3	53	44	4	51	45
#2	6	29	65	5	45	50	5	43	52
Sample sizes (megakaryocytes)	16,050-63,700			129-209			200-250		

Table 2. Comparison of the Ploidy Distribution of Megakaryocytes in Unseparated Marrow Cell Suspensions Analyzed by Flow Cytometry and Feulgen Microdensitometry Using a Scanning Approach.

		Vickers microdensitometry			Flow cytometry	
Sample	**8N**	**16N**	**32N**	**8N**	**16N**	**32N**
Guinea Pig #1	21	54	24	64	33	3
#2	21	57	22	68	26	6
Monkey #1	20	25	54	53	24	23
#2	21	34	46	50	29	20
Sample sizes	79-131 megakaryocytes			1,226-40,000 megakaryocytes		

ysis of the DNA histograms revealed clumping was not a significant problem. There were no discrete 6N, 10N, and 12N peaks which would be seen had clumping been present. In addition, clumping was not observed when the propidium iodide stained cells were examined by fluorescense microscopy.

By both methods, 8N megakaryocytes were more frequent in unseparated than in partially purified preparations. To examine these ploidy shifts in more detail, experiments were performed in which the ploidy distribution was analyzed both in the enriched and in the discarded fractions after each step in sequential purification procedure. The results are provided in Table 4. Overall, there was a 2 log enrichment of megakaryocytes from 0.53%–50%. During this enrichment the 8N fraction fell from 50%–21%, while the 32N fraction rose successively from 7%–21%. The 16N fraction increased slightly following the density centrifugation and then remained stable. Analysis of the discarded fraction shows that few 32N megakaryocytes were lost at any step accounting for the successive enrichment. The 8N megakaryocytes were the predominant cells lost during the density centrifugation accounting for the sharp decline in the 8N fraction at this step.

The megakaryocyte ploidy distributions were analyzed before and after partial purification in 2 monkeys, before purification in 8 guinea pigs and after partial purification in 12 guinea pigs and the results are provided in Table 5. In each instance, ploidy values were shifted toward higher values in monkeys compared to guinea pigs. In every instance, monkeys had more than twice as large a fraction of 32N megakaryocytes, as guinea pigs.

Analysis of megakaryocyte ploidy levels will play an important role in unraveling the mechanisms involved in megakaryopoiesis and thrombocytosis. Flow cytometry provides rapid objective measurement of large numbers of all marrow cells and is not biased by morphologic identification of cells. The validity of FCM analysis of megakaryocyte ploidy was established by direct comparison with Feulgen microdensitometry on split samples, and by morphologic examination of sorted populations. In this report, we have established that 8N is the predominant megakaryocyte

Table 3. Comparison of Ploidy Distribution of Megakaryocytes in Unseparated Guinea Pig Marrow Suspensions Using "Comprehensive Microdensitometry" and Flow Cytometry.

	Sample		Percentage			
	No. cells	No. megakaryocytes	8N	16N	32N	64N
Comprehensive microdensitometry	6,000	124	48	42	8	2
Flow cytometry	370,000	1,226	65	31	4	–

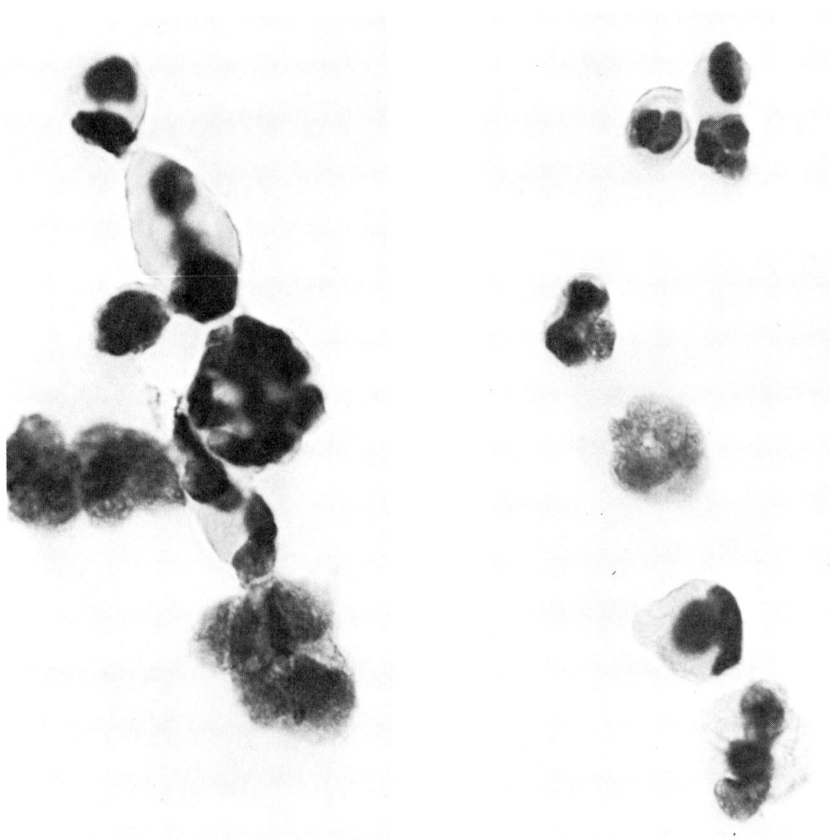

Fig. 3. Examples of megakaryocyte ploidy classes after electronic cell sorting (Giemsa × 1000). Figure 3a shows cells sorted on the basis of DNA content 8N or greater. Figure 3b shows cells with 8N DNA content. Figure 3c shows cells with 16N or greater DNA content. The sorted cells have the morphologic appearance of megakaryocytes except for a rare single myeloid cell which may have been splashed from the discard stream of lower ploidy cells. Note that the 16N and 32N megakaryocytes are larger than the 8N megakaryocytes.

ploidy class in guinea pig marrow, that density and velocity centrifugation procedures enrich megakaryocyte concentration by 2 logs but shift the ploidy distribution toward higher values, and that monkey marrows have a ploidy distribution shifted toward higher values than guinea pigs.

There are several advantages and disadvantages to FCM and Feulgen microdensitometry in analysis of megakaryocyte ploidy. FCM is rapid, measures all cells and, while not biased by morphologic criteria, has the capability to allow morphologic or biochemical evaluation of sorted cells. For example, we have confirmed that all cells with 8N or greater DNA content had the morphologic appearance of

Table 4. Shifts in Ploidy with Isolation of Megakaryocytes from Guinea Pig Marrow Cell Suspensions, Measured by Flow Microcytometry.

	Megakaryocyte in sample	Enriched percentage			Discarded percentage		
Purification step	%	8N	16N	32N	8N	16N	32N
Unseparated	0.5	50	43	7	58	37	5
After density centrifugation	6	31	55	14	46	50	4
After 1st velocity sedimentation	29	29	51	20	35	56	9
After 2nd velocity sedimentation	50	21	58	21	–	–	–

Table 5. Comparison of Megakaryocyte Ploidy Values in African Green Monkeys and in Guinea Pigs.

	Flow cytometry percentage			Feulgen microdensitometry percentage		
	8N	16N	32N	8N	16N	32N
Unseparated						
Monkey	52[a]	26	22	20[a]	30	50
Guinea pig	50[b]	40	10	21[a]	55	24
Partially purified						
Monkey	8[a]	35	57	4[a]	49	47
Guinea pig	21[c]	53	26	13	65	22

[a]Average of 2 experiments.
[b]Average of 8 experiments.
[c]Average of 12 experiments.

megakaryocytes. In addition, Nakeff et al (1979) have sorted the 2N and 4N populations of rodent marrow and stained them with acetylcholinesterase to determine the fraction of cells in the megakaryocytic series in these ploidy classes.

The staining procedure for FCM is short and simple. Large samples including those with a small fraction of megakaryocytes can be analyzed in relatively short periods of time (5–60 min). Since flow cytometry allows for rapid evaluation of megakaryocytes in guinea pigs and monkeys, the ploidy distributions in other animal species including man will be determined in future studies. It will be particularly important to measure the ploidy distribution in various disease states.

A major disadvantage of single parameter FCM studies is that 2N and 4N megakaryocytes and megakaryocyte precursors cannot be distinguished from other marrow cells. In this study, this was not a serious problem since 4N megakaryocytes never accounted for more than 1%–2% of the identifiable megakaryocyte population. However, the ability to analyze these early cells will be important in studies on factors regulating megakaryopoiesis. There are at least three solutions to this problem: (1) As mentioned previously, 2N and 4N cells can be sorted separately and analyzed by other biochemical (e.g., acetylcholinesterase), morphologic (e.g., electron microscopy) or fluorescent (e.g., anti-platelet antibody) techniques. (2) Megakaryocyte precursors can be cloned in soft agar (CFU-M) as demonstrated by several groups (Metcalf et al., 1975; Nakeff and Daniels-McQueen, 1976; Williams and Jackson, 1978). The ploidy distributions of individual cells in these colonies can then be analyzed serially and without various growth factors. We have shown elsewhere that FCM can readily analyze cells picked from soft agar colonies. (3) Flow cytometric instruments can simultaneously measure more than one parameter (Horan and Wheeler, 1977). Thus, cells can be stained with more than one fluorescent dye such as a fluorescent antibody as well as propidium iodide. In this way, the number of 2N and 4N cells binding specific fluorescent antibodies could be measured.

Feulgen microdensitometry is limited by the amount of time required for staining (up to 3 hr) and for analysis (1–3 hr for 100–200 megakaryocytes). Another serious drawback is the sampling bias introduced by the morphologic selection of cells, particularly when evaluating samples with a small percentage of megakaryocytes. This is especially important since the small, young megakaryocytes are the ones most likely to be overlooked. Recently developed laboratory techniques and improvements in image analyzers can avoid these limitations as well. Megakaryocyte ploidy can be analyzed in the small numbers of cells in colonies of CFU-M which contain only cells of the megakaryocyte series. The Artek 800 Image Analyzer can measure cell area, nuclear area and grain counts as well as integrated optical density. Cells stained with specific markers (e.g., fluorescent antibodies or biochemical cytoplasmic stains) can be analyzed for DNA content separately.

This study established that 8N is the major ploidy class in guinea pigs and that the 8N, 16N, and 32N ploidy classes have a normal or Gaussian distribution. The predominance of the 8N ploidy class is in contrast to several studies in the literature (deLeval, 1968; Garcia, 1964; Paulus, 1968a,b) but the Gaussian distribution is similar to the findings of Odell et al. (1970). Previous reports that 16N was the predominant ploidy assumed there was a faster turnover in 8N megakaryocytes compared to the 16N class (Odell et al., 1965, 1976). This study is the first time that megakaryocyte ploidy shifts due to purification procedures have been documented. Other experiments analyzing purified megakaryocyte populations must take these shifts into account when extrapolating to the *in vivo* population. This is especially important since certain megakaryocyte properties and functions might vary with ploidy level (Penington et al., 1976). The capacity for rapid, simultaneous measurements of megakaryocyte ploidy (and other parameters) should lead to routine studies of megakaryocyte patterns in human platelet disorders and in culture experiments examining growth factors and drugs which affect megakaryopoiesis.

Summary

Megakaryocyte ploidy levels are shifted by platelet demand but data on megakaryocyte ploidy levels are meager because of technical difficulties. We have compared semi-automated image analysis (SIA) and multiparameter flow cytometry (FCM), for their ability to analyze megakaryocyte DNA content and size. For SIA studies, samples were stained with Feulgen. Integrated nuclear density, cell volume, and nuclear volume were analyzed with an Artek Image Analyzer. For FCM studies, samples were stained directly with hypotonic propidium iodide (PI) or with isotonic PI after ethanol and RNase treatment. From 2×10^5 to 10^6 cells were analyzed for relative DNA content and relative cell size by light scatter in a Coulter TPS-1 cell sorter. Both techniques provided similar ploidy results on bone marrow samples enriched for megakaryocytes although FCM was more rapid, allowing analysis of more cells (16N > 32N > 8N). In unseparated marrow suspensions, FCM results found 8N > 16N > 32N while a traditional microdensitometric approach analyzing only easily recognizable megakaryocytes found 16N > 8N = 32N. However, where SIA analysis was performed on unselected marrow cells, many more smaller 8N megakaryocytes were found (8N > 16N > 32N) confirming the bias of the traditional sampling method. The FCM method was further validated by microscopic identification of megakaryocytes electronically sorted by ploidy level. The decrease in 8N megakaryocytes during enrichment was attributed to preferential sparing of 32N and 16N megakaryocytes during the enrichment procedures. Nuclear and cell volumes were proportional to ploidy levels although there was more overlap in volume than in ploidy. These techniques can be applied to the study of megakaryocyte ploidy in various disease states and to the study of megakaryocyte regulation by humoral factors.

References

Crissman, H. A., and J. A. Steinkamp. 1973. Rapid simultaneous measurement of DNA, protein, and cell volume in single cells from large mammalian cell populations. *J. Cell Biol.* 59:766–771.

deLeval, M. 1964. DNA levels in normal guinea pig megakaryocytes. *C. R. Soc. Biol.* 158:2198–2201.

deLeval, M. 1968. Quantitative cytochemical study of DNA in the course of megakaryocyte maturation. *Nouv. Rev. Fr. d'Hem.* 8:392–394.

Ebbe, S., F. Stohlman, Jr., J. Overcash, J. Donovan, and D. Howard. 1968. Megakaryocyte size in thrombocytopenic and normal rats. *Blood* 32:383–398.

Garcia, A. M. 1964. Fuelgen-DNA values in megakaryocytes. *J. Cell Biol.* 20:342–345.

Harker, L. A. 1968. Kinetics of thrombopoiesis. *J. Clin. Invest.* 47:458–465.

Horan, P. K. and L. L. Wheeless, Jr. 1977. Quantitative single cell analysis and sorting. *Science* 198:149–157.

Knott, G. D. and D. K. Reece. 1972. M-Lab: A civilized curve fitting system. In *Proceedings of Online '72 International Conference* (Vol. 1). Brunel Univ., England. pp. 497–526.

Krishan, A. 1975. Rapid flow cytofluorometric analysis of mammalian cell cycle by propidium iodide staining. *J. Cell Biol.* 66:188–193.

Levine, R. F. and M. E. Fedorko. 1976. Isolation of intact megakaryocytes from guinea pig femoral marrow. *J. Cell Biol.* 69:159–172.

Metcalf, D., H. R. McDonald, N. Odartchenko, and B. Sordat. 1975. Growth of mouse megakaryocyte colonies *in vitro*. *Proc. Nat. Acad. Sci. USA* 72:1744–1748.

Nachman, R., R. Levine, and E. A. Jaffe. 1977. Synthesis of Factor VIII antigen by cultured guniea pig megakaryocytes. *J. Clin. Invest.* 60:914–921.

Nakeff, A. and S. Daniels-McQueen. 1976. *In vitro* colony assay for a new class of megakaryocyte precursor: Colony-forming unit megakaryocyte (CFU-M). *Proc. Soc. Exp. Biol. Med.* 151:587–590.

Nakeff, A. and B. Maat. 1974. Separation of megakaryocytes from mouse bone marrow by velocity sedimentation. *Blood* 43:591-595.

Nakeff, A., F. Valeriote, J. W. Gray, and R. J. Grabske. 1979. Application of flow cytometry and cell sorting to megakaryocytopoiesis. *Blood* 53:732–745.

Odell, T. T., Jr., C. W. Jackson, and D. G. Gosslee. 1965. Maturation of rat megakaryocytes studied by microspectrophotometric measurement of DNA. *Proc. Soc. Exp. Biol. Med.* 119:1194–1199.

Odell, T. T., Jr., C. W. Jackson, and T. J. Friday. 1970. Megakaryocytopoiesis in rats with special reference to polyploidy. *Blood* 35:775–782.

Odell, T. T., Jr., J. R. Murphy, and C. W. Jackson. 1976. Stimulation of megakaryocytopoiesis by acute thrombocytopenia in rats. *Blood* 48:765–775.

Paulus, J. M. 1968a. Cytophotometric measurements of DNA in thrombopoietic megakaryocytes. *Exp. Cell Res.* 53:310–313.

Paulus, J. M. 1968b. Ultrastructural and microphotometric study of megakaryocyte maturation. *Nouv. Rev. Fr. d'Hem.* 8:394–397.

Penington, D. G. and T. E. Olsen. 1970. Megakaryocytes in states of altered platelet production: Cell number, size, and DNA content. *Br. J. Haematol.* 18:447–463.

Penington, D. G., K. Streatfield, and A. E. Roxburg. 1976. Megakaryocytes and the heterogeneity of circulating platelets. *Br. J. Hematol.* 34:639–653.

Rabellino, E. M., R. L. Nachman, N. Williams, R. J. Winchester and G. D. Ross. 1979. Human megakaryocytes. I. Characterization of the membrane and cytoplasmic components of isolated marrow megakaryocytes. *J. Exp. Med.* 149:1273–1287.

Weste, S. M. and D. G. Penington. 1972. Fluorometric measurement of deoxyribonucleic acid in bone marrow cells. The measurement of megakaryocytic deoxyribonucleic acid. *J. Histochem. Cytochem.* 20:627–633.

Williams, N. and H. Jackson. 1978. Regulation of the proliferation of murine megakaryocyte progenitor cells by cell cycle. *Blood* 52:163-170.

Williams, N., H. Jackson, A. P. C. Sheridan, M. J. Murphy, Jr., A. Elste, and M. A. S. Moore. 1978. Regulation of megakaryopoiesis in long-term murine bone marrow cultures. *Blood* 51:245–255.

Discussion

Hempling: What nozzle size are you using on your TPS-1?

Bunn: The TPS-1 has an orifice size of 70 μm. In our basic studies, we used hypotonic lysis, so we only ran the nucleus through. In the cell-sorting experiments, though, we used ethanol-fixed cells which, of course, shrinks the cells. However, with either technique, we have had no problems with the 70 μm orifice.

Hempling: When you use the whole marrow and then sort for 8N cells, what sort of contamination do you get with other cells?

Bunn: Essentially, none. We have seen a few nonmegakaryocytes. An occasional cell may splash from the central discard stream. Another problem is that the machine has to be calibrated exactly right, and it takes almost a day to get the machine ready to do a sorting experiment. In addition, if you get any kind of a plug in the orifice, it throws the sorting off and you have to start all over again. In my experience, sorting to enrich populations is not a good way to go, but it is satisfactory for getting small numbers of cells to examine.

Levine: I should like to mention that we saw a total of 5 granulocytic cells in all the experiments and that they were always in the 16N and 32N population, never in the 8N group.

Paulus: I would like to discuss the sampling problems that you mentioned and that are important in doing ploidy histograms. It is true that when you examine smears there are sampling problems. I think they come mostly from the process of smearing. A careful observer will not miss 8N megakaryocytes if he measures every doubtful cell. In your suspension procedure I think there is no doubt that some megakaryocytes are lost by mechanical disruption because of the procedure. You would expect that because it is a mechanical disruption it kills more of the larger megakaryocytes: in other words, more 32N. I think that there are sampling problems in both methods. Another way would be to make measurements on sections: Your first slide showed a ploidy histogram made on smears. I wouldn't say that there are absolutely no sampling problems on that, because 8N megakaryocytes would appear small on sections; however, there at least, you are sure you don't lose any megakaryocytes. So I think that the sampling problem must be addressed and there are sampling problems in absolutely every method.

Bunn: I agree, none of the methods is perfect. In each you have advantages and disadvantages. When we compared the Feulgen and the flow techniques in these experiments (they were on the same sample), the same sample was divided in half so that any bias that was introduced in obtaining the sample would be in both.

Levine: Additionally, when we did the unbiased look at every cell by Feulgen densitometry, we found 2 1/2 times as many 8N cells as we did looking at the same sample but only picking out obvious megakaryocytes, standard technique. I think that by the standard technique our result, averaging around 20% 8N, is at least as high as that reported in the literature. So I don't think we were remiss in finding them. There are a lot more unrecognizable or not easily recognizable small 8N cells that, with just a nuclear stain, one is very apt to pass over.

Bussel: Do you have any preliminary data on human marrow megakaryocytes?

Bunn: We have only run a couple of samples and the ploidy distribution seemed to be more similar to the guinea pig than to the monkey, but it is too preliminary to say anything about.

Mayer: In flow cytometry, how did you account for the problem of doublets or triplets in your cells suspension? What kind of fixation did you use, methanol or formaldehyde?

Bunn: We used methanol. We didn't have problems with doublets and triplets. First, when you use the purification procedures that Richard Levine has used, doublets and singles rarely occur. Second, with hypotonic lysis, cell stickiness is not a major problem. Third, we filter the sample through a 105 nylon mesh before we analyze the sample. We routinely look under a fluorescent microscope at what we are about to run through the photocytometer and essentially we never saw clumping. And, finally, if you look at the histogram, with clumping you would see 6N peaks, 10N peaks, 12N peaks, 14N peaks, and so on. So we don't think cell clumping was a problem.

Nakeff: The distrubution that you showed for the unseparated marrow showed a large number of cells prior to the 8N peak. What are those cells, if nonmegakaryocytic cells shouldn't have ploidy greater than 5.2N?

Bunn: In unseparated samples, the exact cut-off is a little more difficult to determine. You have to remember that the cells are less that 0.5% of the total marrow. From a Gaussian distribution, the tail of the 4N curve is going to have some cells overlapping into the megakaryocyte peaks. That is why using a Gaussian curve-fitting program to determine under each peak was a little better; but the cells you saw coming down were just the tail end of the 4N distribution. It is just that the 4N peak is so much larger than the 8N peak that its tail is going to overlap with some of the 8N cells.

Evatt, Levine, and Williams, Editors
MEGAKARYOCYTE BIOLOGY AND PRECURSORS:
IN VITRO CLONING AND CELLULAR PROPERTIES

Studies of Human Marrow Megakaryocytes

Enrique M. Rabellino, Leslie Goodwin, James B. Bussel, Richard B. Levene, and Ralph L. Nachman

Division of Hematology-Oncology, Cornell University Medical College, New York, New York

Megakaryocytes have been recognized as the progenitor cells of circulating platelets for many years. The processes by which megakaryocytes differentiate and produce platelets, however, have not been extensively investigated until recently (Ebbe, 1976; Ebbe and Phalen, 1979; Williams et al., 1978). Most of the information concerning the biology of megakaryocytes has been obtained from studies conducted on experimental animals and from analyses of megakaryocyte colonies grown *in vitro* (Levine and Fedorko, 1976; McDonald, 1978; Metcalf et al., 1975; Nakeff and Maat, 1974; Nakeff et al., 1975; Odell and Jackson, 1968; Williams et al., 1978). Studies of human megakaryocyte physiology, however, have been more limited because of the unavailability of methods to prepare pure populations of human megakaryocytes. Consequently, much of the knowledge of human megakaryocytes has been obtained indirectly from studies of patients with platelet-associated bleeding disorders (Queisser et al., 1971; Ridell and Branehog, 1976). Another primary source of information has been the cytochemical and cytogenetic analysis of marrow cells from patients with myeloproliferative disorders with associated megakaryocytic abnormalities (Breton-Gorius et al., 1978a,b; Efrati et al., 1979; Harker and Finch, 1969; Maldonado, 1975; Zucker-Franklin, 1975).

Pure preparations of human megakaryocytes are now available because of the recent development of techniques to isolate and clone marrow megakaryocytes (Rabellino et al., 1979; Vainchenker and Breton-Gorius, 1979). Using these pure preparations of cells, a number of platelet-associated proteins and differentiation markers have now been recognized (Rabellino et al., 1979).

This communication briefly reviews some of the recent advances in the isolation of human megakaryocytes and presents data on the characterization of various proteins associated with megakaryocytes. Preliminary data on the culture and protein synthetic capability of isolated marrow megakaryocytes as well as the use of cell antigens as markers for identification of human megakaryocyte progenitor cells are also reported.

Methods and Materials

Isolation of Human Bone Marrow Megakaryocytes

Human megakaryocytes were isolated by sequentially processing marrow cells through density gradient centrifugation and velocity sedimentation (Rabellino et al., 1979). The cell density profile of human megakaryocytes and other marrow cell types was first determined by density centrifugation in discontinous gradients of Percoll, a suspension of colloidal silica particles coated with polyvinyl pyrrolidone (Pharmacia Fine Chemicals, Piscataway, N.J.). Solutions were diluted in Ca^{++} Mg^{++}-free Hank's balanced salt solutions containing 0.0129 M sodium citrate, 10^{-3} M adenosine, 10^{-3} M theophylline, 25 mM Hepes (HBSS-CAT) and 50-80 μg/ml DNase at a final pH 7.0 and osmolarity of 290 mOsm. Marrow cells were obtained from ribs routinely removed from individuals undergoing thoracotomy. Rib fragments of 5 to 12 cm generally provided 5–13 × 10^8 nucleated marrow cells with very little blood cell contamination. Marrow cells were harvested within 10 min after removal of the ribs from the patients. These cells were obtained from ribs by injecting medium very gently into the bone cavity and rendered into single cell suspension by pipetting cells with a siliconized Pasteur pipette. Cells were then pelleted in 50 ml polypropylene tubes at 345 × g for 10 min at 20°C and resuspended in HBSS-CAT medium. Siliconized glassware and polypropylene plasticware were used throughout the procedure.

For the density centrifugation procedure, 7–8 × 10^7 cells were first resuspended in a 3 ml solution of Percoll and HBSS-CAT medium with a density of 1.050 g/cm^3 at 20°C and placed in 17 × 100 mm polypropylene tubes. Cells were then underlayered with 3 ml of a Percoll medium solution with a density of 1.085 g/cm^3 and subsequently overlayered with 3 ml of HBSS-CAT. After centrifuging gradients at 700 × g for 20 min at 20°C, cells were harvested in three fractions. Fraction 1 consisted of those cells from the upper medium layer and the interface (density < 1.050 g/cm^3). Fraction II was comprised of cells from the intermediate gradient layer (1.050 g/cm^3). Fraction III included cells from the lower interface and gradient layer as well as the pellet (>1.050 g/cm^3). All cell fractions were diluted with an equal volume of medium and centrifuged at 340 × g for 10 min at 20°C. Cells were counted and examined for various properties or further purified by velocity sedimentation.

For the velocity sedimentation, cells obtained from Fraction I of the density centrifugation gradient were resuspended in 1 ml of medium and layered onto a Percoll continuous gradient generated with two 6 ml solutions of Percoll in HBSS-CAT at density 1.010 and 1.020 g/cm^3, respectively. After sedimentation at 60 × g for 8 min at 20°C, megakaryocytes were collected from the fastest sedimenting fractions (FS) that consisted of the pellet and lowest 2 ml portions of the gradient.

The highest purity was obtained after processing this megakaryocyte enriched fraction through a second cycle of velocity sedimentation in a similar manner.

Mouse megakaryocytes were used occasionally and were obtained from three different sources. These included fresh bone marrow tissue, long-term bone marrow culture and colony-forming unit megakaryocytes (CFU-M) grown in agar (Williams et al., 1978, 1979). Marrow megakaryocytes were isolated from femurs of (DBA/2-C57BL/7) F_1 mice by the method described ahove. Cultured mouse megakaryocytes were generated by techniques described in more detail elsewhere (Williams et al., 1978). Platelet preparations were obtained from freshly drawn human and mouse blood collected in a 0.12 M sodium chloride solution containing 0.0129 M sodium citrate and 25 mM glucose at pH 6.8.

Culture of Isolated Human Megakaryocytes

Isolated marrow megakaryocytes were cultured in liquid medium using a microculture technique. For culturing megakaryocytes, marrow cells were separated by the technique previously described with the following modifications: (a) generation of a velocity sedimentation gradient with the culture medium instead of HBSS-CAT; (b) centrifugation of velocity sedimentation gradients at 60 $\times$ g for 6 min. After velocity sedimentation, megakaryocytes were collected from the FS fraction, counted and plated in siliconized, flat-bottom microtiter plates (Falcon #3040) which had been previously gas sterilized. Two hundred to 3000 megakaryocytes resuspended in 100 μl of alpha medium (Flow Laboratories) were plated in triplicate wells and incubated in a fully humidified incubator at 37°C, 5% CO_2. The alpha medium was supplemented with 10% FCS, 25 mM Hepes buffer and adjusted to pH 7.0 and 240 mOsm. FCS was adsorbed with aluminum hydroxide and subsequently heat inactivated. Multiple sets of cultures from each experiment were monitored for persistence and morphological appearance of megakaryocytes by examination under phase contrast microscopy at various periods of time. Occasionally, cultures were plated with other media such as RPMI-1640, Hank's balanced salt solution, as well as different concentrations of FCS.

^{3}H-leucine Incorporation

To investigate protein synthesis by megakaryocytes, cultures were pulse labeled with ^{3}H-leucine (1350 Ci/mmol, New England Nuclear). "Cold" leucine was removed from the culture system by using leucine-free Minimal Essential Medium (LF-MEM) and by dialyzing the FCS against LF-MEM (LF-FCS). For ^{3}H-leucine incorporation experiments, the velocity sedimentation gradient was harvested in three fractions: (a) the upper 7 ml portion of the gradient that contained less than 0.01% megakaryocytes and over 99.9% of the other nucleated marrow cell types (Nucleated cell-enriched fraction); (b) the middle 3 ml portion that contained 0.5%–5% megakaryocytes; and (c) the lowest 3 ml portion that contained 10%–50%megakaryocytes (Megakaryocyte-enriched or FS fraction). Culture mixing experiments were conducted by diluting cells from the Megakaryocyte-enriched fraction with cells obtained from the Nucleated-cell enriched fraction. Cultures were set at a constant volume of 100 μl of LF-MEM supplemented with 10% LF-FCS. After pulsing each culture with 2 μCi ^{3}H-leucine for 16 hr, cells were harvested with an automated cell harvester. After solubilizing filters in a NCS-Liquifluor-toluene cocktail, TCA precipitable ^{3}H-leucine was counted with a scintillation counter (Searle Mark III Scintillation Counter).

Immunofluorescence Assay

Monospecific antisera against human Factor VIII: AGN, fibrinogen, fibronectin, platelet factor 4, platelet myosin, platelet glycoproteins I, IIB and IIIa as well as human Ia and mouse Ia^k were prepared and tested for specificity as described previously (Kincade et al., 1978; Nachman et al., 1977; Moore et al., 1977; Rabellino et al., 1979; Winchester and Fu, 1977). For direct immunofluorescence, cells were incubated at 20°C for 25 min with either IgG $F(ab')_2$ fragments or with whole IgG preparations of the antibody treated with soluble *Staphylococcus aureus* Protein A (Pharmacia Fine Chemicals). These were conjugated to either fluorescein isothiocyanate (FITC) or tetramethylrhodamine isothiocyanate (TRITC) (Forsgren and Sjoquist, 1976; Rabellino et al., 1971, 1979; Winchester and Fu, 1976; Winchester et al., 1977). Occasionally, indirect immunofluorescence staining was performed and samples were incubated with the unlabeled antibody followed by the second antibody preparation conjugated to either FITC or TRITC. For fluorescence staining of membrane-restricted components, viable cells were resuspended directly with the antisera, whereas for staining of cytoplasmic elements, cells were incubated with the antisera after being smeared and fixed with acetone or methanol for 15 min.

Complement Receptor Assay

The two types of complement (C) receptors, complement receptor type one, the immune adherence (C4b-C3b) receptor (CR_1) and complement receptor type two, the C3d-receptor (CR_2), were detected by rosette formation with sheep erythrocytes-rabbit IgM antibody complement complexes (EAC) prepared and carried out as described (Rabellino et al., 1978a,b; Ross and Polley, 1976).

Detection of Fc IgG Receptor (FcR)

FcR cell membrane receptors for the FC portion of IgG molecules were detected either by rosette formation with sheep erythrocytes sensitized with rabbit IgG antibody (EA_{IgG}) or by fluorescence assay with soluble immune complexes conjugated to FITC. The rosette and fluorescence assays were performed as previously described (Abbas and Unanue, 1975; Halberg et al., 1973).

Results and Discussion

Isolation of Human Bone Marrow Megakaryocytes

Human marrow megakaryocytes were isolated with high purity and yield by processing marrow cells sequentially through density centrifugation and velocity sedimentation. Initial studies on density separation of marrow cells with Percoll gradients revealed that maximal separation of megakaryocytes from other nucleated marrow cell types was found within the low density fractions. Table 1 shows results of a typical experiment in which the cell density distribution of bone marrow cells was analyzed in Percoll medium of various densities. Over 90% of all morphologically recognizable megakaryocytes and only 10% of all other marrow cell types were less dense than 1.050 g/cm³ Percoll. Based on these distribution analyses, a simplified procedure was devised to isolate human megakaryocytes using Percoll as the separating medium. The procedure was established as described in Materials and Methods and allowed the isolation of megakaryocytes with high recovery. Table 2 shows results of a typical experiment. After Percoll gradient centrifugation, Fraction I, containing cells with a density lower than 1.050 g/cm³, had 1.77% megakaryocytes representing a 17.7-fold enrichment over unseparated marrow cells. The megakaryocytes recovered in this fraction accounted for over 80% of the starting sample. Fractions II and III contained only a minor proportion of the megakaryocytes, 0.05% and < 0.01%, respectively.

Megakaryocytes were further purified by processing density gradient derived Fraction I cells by velocity sedimentation. Optimal separation was obtained by centrifuging cells in 1.010–1.020 g/cm³ Percoll at 60 × g for 8 min at 20°C. The vast majority of megakaryocytes were harvested from the fastest sedimenting fraction (FS), together with only 1%–2% of other marrow cells (Table 2). After two sedimentation cycles, purity was increased to 681-fold over the unseparated sample and megakaryocytes represented 63% of the total cells.

Culture of Isolated Human Marrow Megakaryocytes: ³H-Leucine Incorporation

Originally, studies were oriented towards establishing favorable conditions to sustain isolated megakaryocytes in culture. Several types of media supplemented with different concentrations of FCS were tested and thus far the optimal medium was

Table 1. Density Distribution of Human Marrow Cell as determined by Discontinuous Gradient Centrifugation in Percoll.

	Relative number of cells[a]	
Medium density (g/cm³)	**Megakaryocytes** %	**Other cell types[b]** %
1.022	60	2
1.044	30	7
1.050	9	5
1.062	1	8
1.085	< 0.001	22
>1.085	< 0.001	56

[a]Percent of megakaryocytes and other marrow cell types from each density fraction as determined by both number and differential analysis.

[b]Other cell types included: blastoid cells, erythroid precursors, and small mononuclear cells.

Table 2. Purification of Human Marrow Megakaryocytes by Density Centrifugation and Velocity Sedimentation in Percoll Gradients.

		Megakaryocytes			
		Yield			
Cell suspension	**Medium density**	**Number × 10^3**	**Percent[a]**	**Purity**	**Enrichment[b]**
Unseparated marrow cells		285[c]		0.10	
After density centrifugation					
Fraction I	<1.050	239	84	1.77	17.7
Fraction II	≅1.050	20	7	0.05	
Fraction III	>1.050			<0.01	
After velocity sedimentation					
1 cycle: FS		208	73	13.6	136.0
2 cycle: FS		180	63	68.1	681.0

[a]Percent recovery from unseparated marrow cells.
[b]Fold enrichment over unseparated marrow cells.
[c]Data from one typical experiment.

found to be alpha medium with 10% FCS at pH 7.0, 290 mOsm. In addition, conditions for velocity sedimentation were modified in order to minimize cell handling and washing which in turn improved cell yield. Cultured cells were monitored by examination under phase contrast microscopy. Morphologically recognizable megakaryocytes were counted and their cell integrity was assessed by the continuity and the refractive appearance of the cell membrane as well as the distribution of cytoplasmic granules (Paulus and Mel, 1967). Persistence of megakaryocytes in culture was at least 70% of the original number plated without apparent loss of cell integrity for as long as 16 days. It is noteworthy that isolated megakaryocytes were stable in 0.2% agar cultures during 2 weeks incubation, indicating that the conditions established to manipulate these cells were adequate to preserve viable megakaryocytes for extended periods of time. Attempts to evaluate cell viability by the dye exclusion test and supravital staining methods proved to be futile since cell staining was not associated with cell integrity. For instance, using methylene blue and Janus green–neutral red, all megakaryocytes as well as megakaryocyte fragments were stained. Trypan blue uptake by megakaryocytes was not discriminative since isolated intact megakaryocytes but not disrupted megakaryocytes were stained. Moreover, colony megakaryocytes grown in agar were deeply stained *in situ* when trypan blue was added directly to the plate. In examining the cultures, it was observed that the vast majority of megakaryocytes plated on polystyrene or glass surfaces underwent rapid adherence, spreading and degranulation. This phenomenon was minimized in our culture system by siliconizing the polystyrene culture plates.

To investigate the integrity of the megakaryocyte protein biosynthetic apparatus, marrow cell preparations containing various proportions of megakaryocytes were studied for their ability to incorporate ^{3}H-leucine into TCA precipitable protein using the microculture technique described above. Table 3 summarizes a typical ^{3}H-leucine incorporation experiment in which megakaryocyte enriched marrow cells obtained from the fastest velocity sedimentation fraction (FS) were diluted with megakaryocyte depleted marrow cells harvested from the slowest sedimenting fraction (SS). Change in the number of nucleated cells per culture from 84,000 to 1,500 was paralleled by a decrease of only 50% in the total counts per min. Since the number of megakaryocytes per culture increased as the number of other nucleated cells decreased, it seemed probable that some ^{3}H-leucine may have been incorporated by megakaryocytes into protein. Furthermore, a positive linear correlation was found between the estimated ^{3}H-leucine incorporated by megakaryocytes and the proportional and absolute number of megakaryocytes in each culture (Table 3, Fig. 1). Additional calculations on the estimated ^{3}H-leucine incorporated by megakaryocytes as compared with the estimated ^{3}H-leucine incorporated by other marrow nucleated cells indicated that megakaryocytes incorporated at least one hundred times more ^{3}H-leucine/cell than the other marrow cell types as measured by cpm/cell (Fig. 2). Over six experiments, the estimated ^{3}H-leucine incorporation per megakaryocyte was 24.82 cpm/cell (range: 8–122), whereas the estimated ^{3}H-

Table 3. ^{3}H-Leucine Incorporation by Isolated Marrow Megakaryocytes in Short-Term Liquid Cultures.

Number of cells per culture		Incorporated ^{3}H-leucine per culture	
Me Megakaryocytes	Other nucleated cells	Total cpm	Cpm-megakaryocytes[a]
49	84,000	4535	0[b]
92	75,000	4305	330
113	65,000	3800	335
145	56,000	3556	588
178	47,000	3202	711
211	38,000	3217	1201
243	29,000	2918	1381
275	20,000	3077	2017
309	10,500	2674	2118
343	1,500	2200	2115

[a]Cpm-Mk = An estimation of ^{3}H-Leucine incorporated by megakaryocytes. Cpm-Mk/culture = total cpm/culture – cpm-N/culture. Cpm-N = estimation of ^{3}H-Leucine incorporated by other types of nucleated cells. Cpm-N = (Number of nucleated cells/culture) × (average cpm-N.

[b]$$\text{Average cpm-N} = \frac{\text{total cpm (>99\% nucleated cells)}}{\text{number of nucleated cells}} = \frac{4{,}535}{84{,}000}.$$

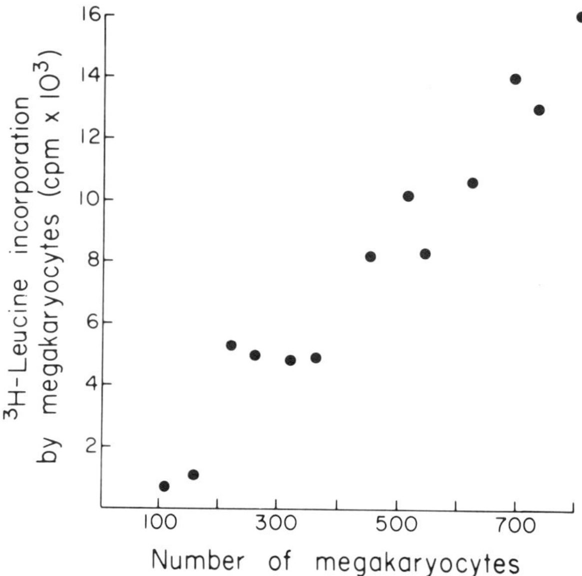

Fig. 1. Relationship between the estimated ^{3}H-leucine incorporated by megakaryocytes and the number of megakaryocytes in each culture. Data from a typical experiment.

leucine incorporation within other marrow cell types was only 0.034 cpm/cell (range: 0.014–0.053).

Identification of Platelet Proteins in Isolated Marrow Megakaryocytes

In other experiments, isolated human megakaryocytes and cultured mouse megakaryocytes were tested for the presence of various membrane and cytoplasmic components by either fluorescence or rosette techniques. Fibrinogen, Factor VIII:AGN, platelet myosin, platelet glycoprotein I, IIb and IIIa, fibronectin as well as platelet factor 4 were detected by immunofluorescence on the membrane and in the cytoplasm of about 80–90% of the marrow megakarocytes (Table 4).

Identification of Mononuclear Marrow Cells Bearing Factor VIII: AGN and Platelet Factor 4

Since the vast majority of morphologically recognizable marrow megakaryocytes contained a variety of the platelet associated proteins, it was of interest to investigate whether other marrow cells, not easily recognizable as megakaryocytes, bore some of these platelet markers. These studies were undertaken in an attempt to search for marrow cells that may represent putative megakaryocyte precursors. Analysis of cells by immunofluorescence from unseparated marrow samples as well as density gradient fractions revealed that there were a minor proportion of small mononuclear

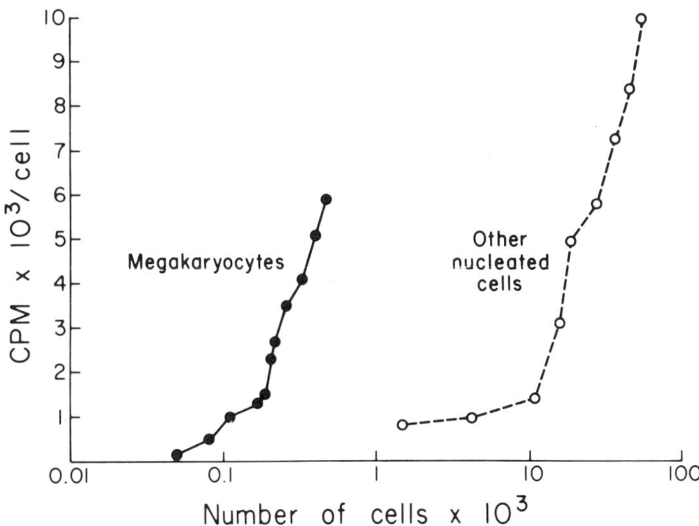

Fig. 2. Relationship between the estimated ^{3}H-leucine incorporated per megakaryocyte and the estimated ^{3}H-leucine incorporated per nucleated cell in each culture. Data from a typical experiment.

marrow cells expressing some of the platelet proteins. Factor VIII:AGN and platelet factor 4 bearing mononuclear marrow cells represented 0.005%–0.01% of the unseparated nucleated marrow cells and between 1%–5% of the cells harvested from density centrifugation of Fraction I. Specificity studies carried out by preincubating monospecific antisera with highly purified antigen disclosed that these mononuclear cells were indeed specifically stained with the two antisera used. Stained mononuclear cells had a lymphoid-like appearance and had diameters ranging between 8–15 μm as compared with that of red cells and lymphocytes present in the same preparation. Stained cells contained a centrally located round nucleus surrounded with smooth cytoplasmic mass containing very few granules. Also, a small proportion of small and medium sized cells from density Fractions I and III was found to stain non-specifically with all antisera used including antisera to non-related antigens such as BSA, KLH and ovalbumin. These cells were easily distinguishable from the specifically stained cells because of their intense granular appearance under phase microscopy. Moreover, preincubation of these cells with any source of protein preparation substantially decreased the intensity of the fluorescent staining.

Expression of Hematopoietic Differentiation Markers in Megakaryocytes and Platelets

Various cell markers known to be expressed at different stages of differentiation in the myeloid, monocytic and erythroid series were studied in both human and

Table 4. Identification of Platelet Associated Proteins in Human Megakaryocytes by Immunofluorescence.

Proteins	Megakaryocytes	
	Membrane	Cytoplasm
Fibrinogen	+[a]	+
Factor VIII:AGN	+	+
α_2-macroglobulin	–	–
Platelet-myosin	+	+
Platelet glycoprotein I	+	+
Platelet glycoprotein IIb	+	+
Platelet glycoprotein IIIa	+	+
Fibronectin	+	+
Platelet factor 4	+	+

[a]Positive (+) indicates that over 80% of morphologically recognizable megakaryocytes were stained.

mouse megakaryocytes and platelets with the expectation that they may be useful in further defining the processes of megakaryocytopoiesis. Fc Receptors (FcR) for IgG, two types of complement receptors (CR_1 and CR_2), and Ia antigen were detected by immunofluorescence and/or rosette assay techniques (Table 5). FcR were found on over 90% of isolated human megakaryocytes with either fluoresceinated antigen-antibody complexes or IgG antibody-sensitized erythrocytes. Ia antigens were demonstrated on only small proportions of isolated human megakaryocytes and, significantly, were not present on platelets. By contrast, FcR and Ia antigens were not detected on freshly isolated or cultured mouse megakaryocytes. Human megakaryocytes were devoid of complement receptors, whereas mouse megakaryocytes expressed CR_1 but not CR_2. Studies of platelets and their homologous megakaryocytes revealed a parallel expression of FcR and complement receptors.

This communication has reviewed some of our current studies on human megakaryocytes. With the development of a method to isolate human marrow megakaryocytes, studies were undertaken to investigate a variety of their properties. These included the identification of various platelet associated proteins and several differentiation markers as well as the development of culture systems for isolated megakaryocytes. Purified preparations of megakaryocytes were obtained by a slightly modified version of the density centrifugation and velocity sedimentation techniques previously described (Rabellino et al., 1979). The modification consisted of the replacement of the upper density gradient layer of Percoll (1.020 g/cm^3) by medium. Additionally, Percoll, rather than Ficoll, was used to generate the velocity sedimentation gradient. These two modifications greatly facilitated the separation procedure since Percoll has a very low viscosity and can be readily washed away

Table 5. Differentiation Markers in Human and Mouse Megakaryocytes and Platelets.

	Percent of cells bearing markers[a]			
	Human		Mouse	
Marker	Megakaryocytes	Platelets	Megakaryocytes	Platelets
FcR[b]	92	97	<0.4	0
CR_1[c]	<0.1	0	10-22	1-3
CR_2[d]	<0.1	0	<0.4	0
Ia antigen	8-24	0	<0.4	0

[a]Percentage represents mean value of at least six experiments.

[b]FcR for IgG molecules.

[c]CR_1 was assayed with EAC complexes containing human or mouse C3 in its C3b form.

[d]CR_2 was assayed with EAC complexes containing human or mouse C3 in its C3d form.

from cells. The uniform use of a single type of gradient solution throughout the entire purification procedure was believed to minimize the detrimental effect that varying gradient solutions might have on the cell membrane structure.

Human megakaryocytes behaved uniquely in Percoll density gradients since all megakaryocytes were much less dense in Percoll solutions as compared to the buoyant density distribution of megakaryocytes in other gradient media such as BSA and Ficoll (Rabellino et al., 1979). The special behavior of megakaryocytes in Percoll was not attributed exclusively to the density effect since the density distribution of other types of marrow cells in Percoll was similar to those found in BSA and Ficoll. These particular properties of marrow megakaryocytes in Percoll were essential for establishing the separation procedure since it provided maximal separation of megakaryocytes from other marrow cell types. Moreover, this may be explained by a combination of the megakaryocytes morphological features in conjunction with some properties of the Percoll solutions. Megakaryocytes, for instance, have a demarcation membrane system that is continuous with the outer membrane of the cell. With such a system, it might be anticipated that these cells could incorporate variable amounts of the uppermost density medium and thus become less sensitive to centrifugation forces. Percoll, on the other hand, will not penetrate the cells thereby conferring upon this system the advantages that only the salt solution from the separating medium could be taken up and subsequently influence the cell density.

Persistence of megakaryocytes in culture was monitored and evaluated by phase contrast microscopy following the criteria established by Paulus and Mel (1967). As previously indicated, two principal conventional methods for ascertaining nucleated cell viability, exclusion of dyes and supravital staining, were found to be unreliable for megakaryocytes. It was determined that an accurate viability profile of other nucleated marrow cells could be assured by observation under phase contrast microscopy following an incubation period of these cells with a 0.5% solution of trypan blue. This same criterion, however, was not valid for megakaryocytes. Frequently, colony megakaryocytes whose viability had been established by their proliferation in agar culture were found to be deeply stained with the trypan blue solution, whereas grossly disrupted megakaryocytes and cell fragments remained unstained. Similarly, inconsistent staining patterns were observed with the two supravital stains used. Disrupted megakaryocytes frequently were observed to be deeply stained with methylene blue and Janus green–neutral red. However, freshly isolated and colony megakaryocytes remained unstained. The mechanisms underlaying the behavior of megakaryocytes in the two staining systems still remain unclear. It is suspected that this erratic behavior may be attributed to a combination of properties uniquely expressed by the megakaryocyte membrane. These include the demarcation membrane system and a high, non-specific adsorbent capacity of the membrane associated with a dynamic transmembrane incorporation capability.

Isolated megakaryocytes were found to undergo rapid adherence, spreading and degranulation when incubated on polysytrene or glass surfaces. This phenomenon was minimized by siliconization of surfaces with water-soluble silicon.

Isolated megakaryocytes were cultured in a liquid medium system for up to 16 days. Cultured megakaryocytes were able to synthesize proteins as assessed by the ^{3}H-leucine incorporation studies. Newly synthesized protein was readily detected during the first 24 hr of culture. The estimated ^{3}H-leucine incorporated by megakaryocytes in each culture was found to be positively correlated with the number of megakaryocytes plated. Furthermore, the ^{3}H-leucine incorporation rate per cell was at least 100 times higher for megakaryocytes than other types of marrow cells.

Human megakaryocytes were studied for various platelet proteins. Using the immunofluorescence technique, the vast majority of morphologically identifiable megakaryocytes were found to express membrane and cytoplasmic fibrinogen, Factor VIII:AGN, platelet factor 4, platelet myosin, fibronectin as well as platelet glycoproteins I, IIb and IIIa. This indicates that these platelet-related molecules are intrinsically associated with megakaryocytes and that their appearance may occur during the early stages of megakaryocyte development. From early experiments, it was learned that human megakaryocytes expressed membrane receptors for the Fc portion of IgG (Rabellino et al., 1979). Consequently, for all future studies of megakaryocyte membrane antigens only antibody preparations free of reactive Fc regions were used. For this purpose, antibody preparations were treated with either pepsin or soluble *Staphylococcus aureus* Protein A, in order to generate $F(ab')_2$ fragments or to inactivate the Fc portion of the IgG molecules, respectively. These treatments were effective in preventing non-specific binding of antibodies via Fc receptors (Forsgren and Sjoquist, 1966).

More recently, Factor VIII:AGN and platelet factor 4 have been detected in the cytoplasm of a number of small mononuclear marrow cells. These cells represented between 0.005%–0.01% of the entire population of nucleated marrow cells as detected by immunofluorescence. Factor VIII:AGN bearing mononuclear cells have low cell density and can be enriched up to 1%–5% by Percoll density centrifugation. These studies have clearly demonstrated that there is a proportion of small mononuclear marrow cells that stained specifically for Factor VIII:AGN and platelet factor 4. This raises the intriguing possibility that this population of mononuclear cells may represent immature or precursor megakaryocytes. This population may well represent analogous cells to the mononuclear cells that contain acetylcholinesterase activity that have been described in the rat and mouse bone marrow as early megakaryocyte (Jackson, 1973; Long and Williams, 1980). Further studies are required to define the true nature of these mononuclear marrow cells and to establish whether these cells do indeed represent early human megakaryocytes.

Because platelets contain a growth promoting factor (PDGF) for fibroblasts, megakaryocytes were studied for their ability to induce proliferation of cultured fibroblasts (Antoniades et al., 1979; Ross et al., 1979). Cell homogenates obtained from purified preparations of megakaryocytes were tested for their capacity to stimulate cell proliferation. Cell proliferation was assessed by measuring the ^{3}H-TdR uptake of cultured mouse 3T3 cells and human bone marrow fibroblasts (Castro-Malaspina et al., 1980). Growth promoting activity was detected in extracts obtained from highly enriched cell preparations of megakaryocytes. These studies are discussed in more detail in Chapter 31 of this book.

Parallel studies of megakaryocytes and platelets from human and mouse sources revealed that there were differences in the expression of FcR and two types of C receptors. A proportion of human megakaryocytes, but not homologous platelets, bore the Ia antigen, indicating that the expression of this marker may be restricted to a subclass of megakaryocytes. This latter type of cell may represent a more primitive megakaryocyte class similar to those early cells expressing Ia antigen in the granulocytic and erythroid series (Rabellino et al., 1978b; Winchester et al., 1978). Alternatively, this megakaryocyte class may reflect the development of separate megakaryocyte lineages, with subsequent restricted platelet functions (Corash et al., 1977; Penington et al., 1976). Heterogeneity in the expression of membrane-associated differentiation markers has also been observed with mouse megakaryocytes. Our studies demonstrated that about a third of mouse megakaryocytes expressed CR_1 along with a small proportion of blood platelets. This confirms our prediction that this marker may be expressed at a late stage in development (Rabellino et al., 1979). It is anticipated that analysis of the acquisition and loss of membrane and cytoplasmic antigens on highly purified megakaryocyte populations will provide new insights into the mechanisms of platelet production and the events of platelet release. Furthermore, the availability of highly purified populations of megakaryocytes maintained in culture should help define the microenvironmental conditions that influence the processes of thrombopoiesis.

Summary

Human marrow megakaryocytes were isolated by processing bone marrow cells sequentially through Percoll gradients, firstly by density centrifugation and subsequently by velocity sedimentation. Analysis of isolated cells for various platelet-associated components by immunofluorescence demonstrated fibrinogen, platelet glycoprotein Ib and IIIa, Factor VIII:AGN, platelet myosin, fibronectin and platelet factor 4 in over 85% of marrow megakaryocytes. Factor VIII:AGN and platelet factor 4 antigens were also found in a population of small mononuclear cells that represents only 0.005%–01% of the entire marrow nucleated cells. Factor VIII:AGN and PF4 bearing mononuclear cells have low density and they can be enriched up to 1%–2% by centrifugation in BSA or Percoll density gradients. Parallel studies on human and mouse megakaryocytes and platelets for IgG receptor (FcR), complement receptor type one (CR_1) (C3b receptor), complement receptor type two (CR_2) (C3d receptor), and Ia antigen by fluorescence and (or) rosette formation methods were performed. FcR were present on most human megakaryocytes and platelets. The Ia antigen was detected on a proportion (10%–15%) of human megakaryocytes but it was undetectable on human platelets. CR_1 was found on 20%–40% of mouse megakaryocytes and also on a proportion of mouse platelets. These differentiation markers may be used in monitoring megakaryocyte maturation.

ACKNOWLEDGMENTS

E.M. Rabellino is a recipient of a National Institute of Health Research Career Development Award Ca 00518/01. This work was supported by grants from the National Institute of Health (HL-18828), The New York Community Trust, and The Irma T. Hirschl Charitable Trust.

References

Abbas, A. K. and E. R. Unanue. 1975. Interrelationships of surface immunoglobulin and Fc receptors on mouse B lymphocytes. *J. Immunol.* 115:1665–1671.

Antoniades, H. N., C. D. Scher, and C. D. Stiles. 1979. Purification of human platelet-derived growth factor. *Proc. Nat. Acad. Sci. USA* 76:1809–1813.

Breton-Gorius, J., F. Reyes, J. P. Vernant, M. Tulliez, and B. Dreyfus. 1978a. The blast crisis of chronic granulocytic leukemia: Megakaryoblastic nature of cells as revealed by the presence of platelet peroxidase, a cytochemical ultrastructural system. *Br. J. Haematol.* 39:295–303.

Breton-Gorius, J., F. Reyes, G. Duhamel, A. Najman, and N. C. Gorin. 1978b. Megakaryoblastic acute leukemia: identification by the ultrastructural demonstration of platelet peroxidase. *Blood* 51:45–59.

Castro-Malaspina, H. E. M. Rabellino, A. Yen, R. B. Levene, R. L. Nachman, and M. A. S. Moore. 1980. Megakaryocyte-derived gfrowth factor. Studies on 3T3 cells and bone marrow fibroblasts. *Megakaryocytes In Vitro.* pp. 234-250.

Corash, L., H. Tan, and H. R. Gralnick. 1977. Heterogeneity of human blood-platelet subpopulations. I. Relationship between buoyant density, cell volume and ultrastructure. *Blood* 49:71–87.

Ebbe, S. 1976. Biology of megakaryocytes. *Prog. Hemostasis Thromb.* 3:211–229.

Ebbe, S. and E. Phalen. 1979. Does autoregulation of megakaryocytopoiesis occur? *Blood Cells* 5:123–138.

Ebbe, S., E. Phalen, P. D'Amore, and D. Howard. 1978. Megakaryocyte responses to thrombocytopenia and thrombocytosis in $S1/S1^d$ mice. *Exp. Hematol.* 6:201–212.

Efrati, P., E. Nir, A. Yaari, A. Berrebi, H. Kaplan, and A. Dvilanski. 1979. Myeloproliferative disorders terminating in acute micromegakaryoblastic leukemia. *Br. J. Haematol.* 43:79–86.

Forsgren, A. and J. Sjoquist. 1966. Protein A from *S. aureus*. I. Pseudo immune reaction with human gamma-globulin. *J. Immunol.* 97:822–827.

Halberg, T., B. W. Gurner, and R. R. A. Coombs. 1973. Opsonic adherence of sensitized ox red cells to human lymphocytes as measured by rosette formation. *Int. Arch. Allergy Appl. Immunol.* 44:500–513.

Harker, L. A. and C. A. Finch. 1969. Thrombokinetics in man. *J. Clin. Invest.* 48:963-974.

Jackson, C. W. 1973. Cholinesterase as a possible marker of early cells of the megakaryocyte series. *Blood* 42:413-421.

Kincade, P. W., C. Y. Page, R. M. E. Parkhaus, and G. Lee. 1978. Characterization of murine colony B cells. I. Distribution, resistance to anti-immunoglobulin antibodies and expression of Ia antigens. *J. Immunol.* 120:1289–1296.

Levine, R. F. and M. E. Fedorko. 1976. Isolation of intact megakaryocytes from guinea pig femoral marrow. *J. Cell Biol.* 69:159–172.

Long, M.W. and N. Williams. 1980. Relationship of small acetylcholinesterase positive cells to megakaryocytes and clonable megakaryocytic progenitor cells. In *Megakaryocytes In Vitro*. Eds.: R. Levine, N. Williams and B. Evatt. pp. 59–75.

McDonald, T. P. 1978. A comparison of platelet production in mice made thrombocytopenic by hypoxia and by platelet specific antisera. *Br. J. Haematol.* 40:299–309.

Maldonado, J. E. 1975. The ultrastructure of the platelets in refractory anemia (preleukemia) and myelomonocytic leukemia. *Ser. Haematol.* 8:1–125.

Metcalf, D., H. R. MacDonald, N. Ordartchenko, and B. Sordat. 1975. Growth of mouse megakaryocyte colonies *in vitro*. *Proc. Natl. Acad. Sci. USA* 72:1744–1748.

Moore, A., E. A. Jaffe, C. G. Becker, and R. L. Nachman. 1977. Myosin in cultured human endothelial cells. *Br. J. Haematol.* 35:71–79.

Nachman, R. L., R. Levine, and E. A. Jaffe. 1977. Synthesis of factor VIII antigen by cultured guinea pig megakaryocytes. *J. Clin. Invest.* 60:914–921.

Nakeff, A. and B. Maat. 1974. Separation of megakaryocytes from mouse bone marrow by velocity sedimentation. *Blood* 43:591–595.

Nakeff, A., K. A. Dicke, and M. J. Van Noord. 1975. Megakaryocytes in agar culture of mouse bone marrow. *Ser. Haematol.* 8:1–21.

Odell, T. T. and C. W. Jackson. 1968. Polyploidy and maturation of rat megakaryocytes. *Blood* 32:102–110.

Paulus, J. M. and H. C. Mel. 1967. Viability studies on megakaryocytes in mechanically and enzymatically suspended rat bone marrow. *Exp. Cell Res.* 48:27–38.

Penington, D. G., K. Streatfield, and A. E. Roxburgh. 1976. Megakaryocytes and the heterogeneity of circulating platelets. *Br. J. Haematol.* 34:639–653.

Queisser, V., W. Queisser, and B. Spiertz. 1971. Polyploidization of megakaryocytes in normal humans and in patients with idiopathic thrombocytopenia and with pernicious anaemia. *Br. J. Haematol.* 20:489–501.

Rabellino, E. M., R. L. Nachman, N. Williams, R. J. Winchester, and G. D. Ross. 1979. Human megakaryocytes. I. Characterization of the membrane and cytoplasmic components of isolated marrow megakaryocytes. *J. Exp. Med.* 149:1273–1287.

Rabellino, E. M., S. Colon, H. M. Grey, and E. R. Unanue. 1971. Immunoglobulins on the surface of lymphocytes. I. Distribution and quantitation. *J. Exp. Med.* 133:156–167.

Rabellino, E. M., G. D. Ross, and M. J. Polley. 1978a. Membrane receptors of mouse leukocytes. I. Two types of complement receptors for different regions of C3. *J. Immunol.* 120:879–885.

Rabellino, E. M., G. D. Ross, H. T. K. Trang, N. Williams, and D. Metcalf. 1978b. Membrane receptors of mouse leukocytes. II. Sequential expression of membrane receptors and phagocytic capacity during leukocyte differentiation. *J. Exp. Med.* 147:434–445.

Ridell, B. and I. Branehog. 1976. The ultrastructure of the megakaryocytes in idiopathic thrombocytopenic purpura (ITP) in relations to thrombokinetics. *Pathol. Eur.* 11:179–187.

Ross, G. D. and M. J. Polley. 1976. Assay for the two different types of lymphocyte complement receptors. *Scand. J. Immunol.* 5 (Suppl. 5):99–115.

Ross, R., A. Vogel, P. Davies, E. Raines, R. Kariya, M. J. Rivest, C. Gustafson, and J. Glomset. 1979. The platelet-derived growth factor and plasma control cell proliferation. In *Hormones and Cell Culture*. Sato, G. H. and Ross, R., eds. Cold Spring Harbor Conferences on Cell Proliferation (Vol. 6). pp. 3–16.

Vainchenker, W., J. Guichard, and J. Breton-Gorius. 1979. Growth of human megakaryocyte colonies in culture from fetal, neonatal, and adult peripheral blood cells: Ultrastructural analysis. *Blood Cells* 5:25–42.

Williams, N., T. P. McDonald, and E. M. Rabellino. 1979. Maturation and regulation of megakaryocytopoiesis. *Blood Cells* 5:43–55.

Williams, N., H. Jackson, A. P. C. Sheridan, M. J. Murphy, A. Elste, and M. A. S. Moore. 1978. Regulation of megakaryocytopoiesis in long-term murine bone marrow cultures. *Blood* 51:245–255.

Winchester, R. T. and S. M. Fu. 1976. Lymphocyte surface membrane immunoglobulin. *Scand. J. Immunol.* 5 (Suppl. 5):77–82.

Winchester, R. J., G. D. Ross, D. I. Jarowsky, C. Y. Wang, J. Halper, and H. Broxmeyer. 1977. Expression of Ia-like antigen molecules on human granulocytes during early phases of differentiation. *Proc. Nat. Acad. Sci. USA* 74:4012–4016.

Winchester, R. J., P. A. Meyers, H. E. Broxmeyer, C. Y. Wang, M. A. S. Moore, and H. G. Kunkel. 1978. Inhibition of human erythopoietic colony formation in culture by treatment with Ia antisera. *J. Exp. Med.* 148:613–618.

Zucker-Franklin, D. 1975. Ultrastructural studies of hematopoietic elements in relation to the myelofibrosis-osteosclerosis syndrome, megakaryocytes and platelets (MMM or MOS). *Adv. Biosciences.* 16:127–140.

Discussion

Levine: Have you quantified any of these substances that you have been finding?

Rabellino: We are in the process of doing that.

Levine: How small are your small cells? Have you measured them?

Rabellino: I have measured them in relation to other cells in the same preparation. I would guess that they are between 7 to 11 μm in diameter.

Levine: People who don't work with isolated megakaryocytes should know that it is a little bit difficult to get consistent counts. I found it necessary to do 8 hemocytometer counts to get consistent data so that the results from opposite femurs were exactly the same. How can you accurately count 100 or 800 megakaryocytes in a sample? What was the size of your aliquot and how did you do it?

Rabellino: When we culture cells, we count the cell suspension first, then again in the microtiter plates *in situ* with an inverted microscope.

Dombrose: Were your studies with the Factor VII antigen done with Fab fragments?

Rabellino: Yes, we used $F(ab')_2$ antibody and on occasion we used the whole IgG antibody, coupled with Protein A to block the Fc portion.

Dombrose: I think your studies on the platelet factor 4 are very interesting. If you grow megakaryocytes in serum, you can anticipate very high levels of platelet factor 4, so that your study with marrow cells being washed right out is, I think, valid, but it could be a problem to those who are trying to grow megakaryocytes in any kind of media that has serum.

Levine: I should like to point out that in these kinds of isolation procedures there may be a dozen washes from the time the cells are removed from the marrow until they are plated. Kathy Kellar and I culture these cells overnight with albumin instead of serum, thereby obviating that problem.

Leven: You mentioned that 80% of the cells you looked at showed these different factors by immunofluorescence. Is this true of the myosin as well; that you only saw 80% of all megakaryocytes fluorescing?

Rabellino: Yes.

Leven: Can you tell me what kind of fixation you used?

Rabellino: We used acetone or methanol.

Leven: It is very surprising that you would see any cells that don't have myosin. In my studies I have never seen any cells that didn't. Also, we have found that certain siliconizing agents cause megakaryocytes to become very adherent, especially a compound called Prosil 28 (PCR Research Chemicals, Inc.).

Rabellino: We were using the Clay-Adams product, Siliclad, and later switched to Prosil 28.

Levine: Jim George has found that the best substance to coat surfaces to prevent platelet sticking was polyethylene glycol.

Mayer: In your immunofluorescence investigations, you showed that 85% of megakaryocytes showed platelet factors on the surface or in the cytoplasm. Amongst these cells were immature cells. What is the nature of the remaining 15% of megakaryocytes?

Rabellino: I think that the great majority of the cells do bear those markers, the balance being cells that, for instance, are not well stained because of artifact or are dead megakaryocytes that do not stain well. I don't know if the nonstaining megakaryocytes represent any kind of subset; it is not possible to speculate on that possibility at this time.

Hoffmann: Do other nucleated bone marrow cells have fibronectin on their cell surface?

Rabellino: Yes. Monocytes and the precursor cells have it, as it appears to actively participate in phagocytosis. The finding of fibronectin on the surface appears to be common for different cell types.

Hoffmann: Is fibronectin on the surface of megakaryocytes an absorption phenomenon or a production phenomenon?

Rabellino: I do not have that information.

Levine: That is very important because fibronectin may have to do with platelets sticking to collagen.

Evatt: Your Percoll separation showed a very nice megakaryocyte purification. Have you done ploidy analyses to determine if cells of lower ploidy, as described by Paul Bunn, were obtained in that fraction?

Rabellino: No, we have not done that.

Breton-Gorius: Have you observed any difference in the intensity of fluorescence in the staining of different glycoproteins between young and fully mature megakaryocytes?

Rabellino: That study has not been done yet.

Hempling: Are all your media Ca^{++}-,Mg^{++}-free?

Rabellino: It is the same medium as described by Richard Levine for guinea pig megakaryocytes. The osmolarity was reduced to 290 mOsm, the same as that of human plasma. That seemed to influence the yield of human marrow cells. DNAse is added in the early separation steps.

Evatt, Levine, and Williams, Editors
MEGAKARYOCYTE BIOLOGY AND PRECURSORS:
IN VITRO CLONING AND CELLULAR PROPERTIES

Megakaryocyte Lipids

Barbara P. Schick and Paul K. Schick

Department of Physiology and Biochemistry, Medical College of Pennsylvania, Philadelphia, Pennsylvania and the Specialized Center for Thrombosis Research, Temple University Health Sciences Center, Philadelphia, Pennsylvania

We have undertaken studies to compare lipid composition and lipid synthesis in guinea pig megakaryocytes and platelets in order to define the role of the megakaryocyte in determining the lipid composition of the platelet. The megakaryocyte may be largely responsible for the regulation of the molecular content and organization of the platelet. For example, evidence has been presented that guinea pig megakaryoctyes synthesize Factor VIII antigen (Nachman et al., 1977) and actin (Nachman et al., 1978) and contain (and presumably synthesize) platelet growth factor (Chernoff et al., 1980), and rat megakaryoctyes can produce prostaglandins (Demers et al., 1980). It is reasonable to expect also that the essential characteristics of the platelet membrane, and, therefore, the composition and orientation of the membrane lipids, are determined at the level of the megakaryocyte.

The lipid composition of platelets is well established (Marcus et al., 1969; Cohen and Derksen, 1969; Nordy and Lund, 1968), and there is considerable evidence that lipids are organized in a specific manner in the platelet membrane (Chap et al., 1977; Otnaess and Holm, 1976; Schick, 1978; Schick et al., 1976; Perret et al., 1979). While it is clear that platelets have the ability to synthesize phospholipids from a variety of precursors (Cohen et al., 1970; Deykin and Desser, 1968; Hennes and Awai, 1965; Lewis and Majerus, 1969; Majerus et al., 1969; Okuma et al., 1971; Spector et al., 1970), we do not think that the pathways which have been demonstrated are sufficient to explain or determine the overall phospholipid structure of the platelet. Furthermore, platelets have little or no ability to synthesize cholesterol (Deykin and Desser, 1968; Okuma et al., 1971; Derksen and Cohen, 1973; Derksen et al., 1976). Therefore, the overall regulation of platelet lipid composition is not understood.

Much work has been done to characterize platelet lipid metabolism. It is our intention in this paper to review the literature with regard to the role of phospholipids and cholesterol in platelet aggregation, the evidence for specific organization of lipid in platelet, and the *in vitro* capacity of platelets for lipid synthesis as a background and rationale for studying megakaryoctye lipids. We will discuss briefly our current work concerning lipid compositions and lipid biosynthesis in guinea pig megakaryoctyes and platelets.

The reader is referred to more extensive reviews of platelet lipid metabolism for additional information (Schick, 1979; Marcus, 1978; Zwaal, 1978).

The Role of Lipids in Platelet Function

Platelet lipids have been implicated in a number of ways in platelet function. Platelet factor 3 activity has long been thought to be related to platelet phospholipids, although this relationship is no longer clear (Broekman et al., 1976) and may involve instead the interactions of a complex lipid-protein surface (Marcus, 1978). Phosphatidylcholine (PC) may be involved in platelet activity, since hydrolysis of platelet PC by phospolipase C from *Clostridium perfringens* results in a release reaction similar to that produced by thrombin (Schick and Yu, 1974) and stimulates platelet aggregation (Chap and Douste-Blazy, 1974). A small pool of phosphatidylethanolamine (PE) also may be physiologically important, since treatment of platelets with thrombin permits labeling of a small amount of PE with the chemical reagent trinitrobenzenesulfonate (TNBS) that is not available to this reagent in control platelets (Schick et al., 1976). The source of this newly-exposed PE is not known, but may represent fusion of granule membranes with the plasma membrane with exposure of granule membrane on the platelet surface, transfer of PE from the inner to the outer surface of the plasma membrane, or unmasking of PE already present on the surface by means of hydrolysis of glycoproteins. Perhaps the most intriguing functional role for phospholipids which is currently being investigated is their ability to provide arachidonic acid for thromboxane and prostaglandin synthesis (Bills et al., 1976, 1977; Derksen and Cohen, 1975; Jessie and Cohen, 1976; Rittenhouse-Simmons et al., 1976, 1977; Rittenhouse Simmons, 1979; Bell and Majerus, 1980; Lapetina and Cuatrecasas, 1979).

Platelet cholesterol content and localization may be an important determinant of platelet reactivity. Platelets made hypercholesterolemic by incubation with cholesterol-rich liposomes become hyperreactive to ADP (Shattil et al., 1975) and produce greater amount of thromboxane B_2 than do control platelets upon thrombin stimulation (Stuart et al., 1980). Platelets from patients with Type IIa hyperlipidemia contain excess cholesterol in their plasma membranes (Shattil et al., 1975) and are, like the platelets made hypercholesterolemic *in vitro*, hyperreactive to ADP (Carvalho et al., 1974; Shattil et al., 1977). This platelet hyperreactivity may contribute to the severe atherosclerosis found in these patients. It has been postulated that cholesterol enrichment of the platelet may exert an effect on the adenylate cyclase system (Sinha et al., 1977). Excess cholesterol decreases the fluidity of platelet membranes (Shattil and Cooper, 1976) and may thereby profoundly affect the ability of the platelet to function. In this regard, a recent study in which platelets were monitored with a fluorescent probe demonstrated that platelet membranes become more rigid during thrombin-induced aggregation (Nathan et al., 1980). This increased rigidity was thought to be either a primary or secondary effect of the platelet lipids. Cholesterol enrichment of the platelet may, therefore, mimic the rigidity induced by thrombin. Conversely, depletion of a small amount of platelet cholesterol reduces platelet reactivity to ADP and epinephrine (Shattil et al., 1975) and decreases platelet microviscosity (Shattil and Cooper, 1976). Thus a precise regulation of the amount and localization of cholesterol in the platelet membrane may be highly significant for platelet function, but the mechanism of this regulation is unknown.

Organization of Lipids in Platelet Membranes

Evidence has been presented that phospholipids are arranged in a highly specific fashion in the platelet plasma membrane (Chap et al., 1977; Otnaess and Holm, 1976; Schick, 1978; Schick et al., 1976; Perret et al., 1979). These studies, which have probed the platelet surface with phospholipases and the non-penetrating chemical probe trinitrobenzenesulfonic acid, suggest that the phospholipids of the outer surface of the membrane are primarily PC and sphingomyelin (SM), with considerably lesser amounts of PE and phosphatidylinositol (PI). One study suggests that much PE is on the surface (Otnaess and Holm, 1976), although most of the experimental evidence suggests that PE is located primarily in the interior of the platelet. The plasmalogen species of PE is distributed in the same manner as total PE (Rittenhouse-Simmons et al., 1976, 1977). Phosphatidylserine (PS), which has remained largely undetected by the external labeling reagents, is thought to be localized in the inner portion of the plasma membrane, although a recent study has detected 9% of platelet PS on the surface (Perret et al., 1979). There is little data on the distribution of fatty acids between the inner and outer surfaces of the platelet. Arachidonic acid, however, may be distributed asymmetrically, with only 10% of the platelet arachidonic acid on the exterior surface of the plasma membrane (Perret et al., 1979).

Cholesterol is found in both the plasma membrane and granule membranes in approximately the same molar ratio to phospholipids in both (Marcus et al., 1969). The cholesterol content of platelets from Type IIa hyperlipidemic patients has been found to be elevated, but only in the plasma membrane (Shattil et al., 1977). Incubation of platelets with cholesterol-rich liposomes, on the other hand, results in increased content of cholesterol in the granules as well as the plasma membrane, although the increase in cholesterol in the granule membranes was only one-third that of the plasma membrane (Shattil et al., 1975). The ability of the platelet to obtain cholesterol from plasma lipoproteins is not clear, since normal platelets incubated in Type IIa hyperlipidemic plasma did not increase their cholesterol content (Shattil et al., 1977). Thus, the physiologic mechanism for incorporation of cholesterol into platelets is not understood.

Lipid Synthesis in Platelets—De Novo Lipid Synthesis and Utilization of Exogenous Fatty Acids

Platelets have been shown to be capable of lipid synthesis from glycerol (Lewis and Majerus, 1969), acetate (Deykin and Desser, 1968; Hennes and Awai, 1965; Okuma et al., 1971), and phosphate (Cohen et al., 1971; Lloyd et al., 1974; Lloyd and Mustard, 1974; Leung et al., 1974). In addition, the presence of the two enzymes critical for *de novo* fatty acid synthesis, acetyl coenzyme A carboxylase and fatty acid synthetase, has been demonstrated (Majerus et al., 1969). The acetate incorporation studies have demonstrated that platelets have the ability to chain elongate preformed fatty acids, but it is not known whether platelets have fatty acid desaturases. When human and rat platelets were radiolabeled with acetate, most of the lipid-incorporated radioactivity was associated with PC, ceramide, and unesterified fatty acids (Deykin and Desser, 1968; Okuma et al., 1971). The radioactivity was not distributed among the lipids in proportion to the amounts of each lipid in the cell; indeed, ceramide and free fatty acids comprise a very small proportion of platelet lipids. The uneven distribution of radioactivity among the phospholipids might be anticipated, since most of the radioactivity from acetate is incorporated into the fatty acid moieties of the phospholipids rather than the glycerol backbone (Deykin and Desser, 1968). The phospholipid radioactivity thus would reflect only the fatty acids synthesized or elongated within the platelet, and phospholipids which contain predominantly fatty acids obtained from extracellular sources would contain little radioactivity. However, when platelets were incubated with ^{14}C glycerol (Lewis and Majerus, 1969; Cohen et al., 1970), PC was again the most heavily radiolabeled phospholipid, followed by PI, which is only about 4% of platelet phospholipids, and phosphatidic acid (PA), which is present only in trace amounts. Exogenous $^{32}PO_4$ is incorporated into PA and PI in short incubations (Cohen et al., 1971; Lloyd et al., 1974) and is incorporated into other phospholipids only after many hours of incubation (Leung et al., 1974). Furthermore, the five major platelet fatty acids (palmitate, stearate, oleate, linoleate, and arachidonate), when supplied to the platelets *in vitro* as either complexes with albumin or as lipid dispersions in buffer, are incorporated predominantly into PC and PI (Deykin and Desser, 1968; Spector et al., 1970; Cohen et al., 1970; Bills et al., 1976). The degree of incorporation of exogenous fatty acid into platelet lipids appears to depend upon the fatty acid/albumin ratio in the medium (Spector et al., 1970). The capacity of the platelet for esterification appears to be limited, since elevation of the free fatty acid/albumin ratio resulted in accumulation of free fatty acid in the platelet (Spector et al., 1970). Similarly, in a different approach to fatty acid incorporation into platelet phospholipids, Bewreziat et al. (1978) studied exchange of ^{14}C arachidonate and ^{14}C linoleate from synthetically prepared $2-{}^{14}C$ acyl PC complexed to plasma lipoproteins. The ^{14}C linoleate exchanged almost exclusively into platelet PC, with a few percent into PI, and about half of the ^{14}C arachidonate which exchanged into the platelet was found in PC and about one-fifth into PI. These investigators estimated that the exchangeable pool represented only 2–3% of the total platelet PC, and was exchangeable in 30 minutes. About 14–16% of the arachidonyl-PC and 20–35% of the linoleyl-PC were implicated in the exchange. The studies on platelet phospholipid metabolism, on the whole, suggest a high rate of metabolism of PC and PI in platelets. The physiologic significance of these pathways is not established, and the pools involved may be small.

It is notable that PE and PS are labeled to only a small degree by the agents used thus far to investigate platelet lipid synthesis. Of particular interest is the inability of exogenous arachidonic acid to label PE, since that phospholipid is extremely rich in arachidonic acid (Marcus et al., 1969). The general trend of the synthetic pathways which have been demonstrated in platelets, suggests to us that the fatty acid composition of PE and PS is not determined by the platelet itself because the potential precursors are incorporated to a very small degree into these phospholipids.

Human (Deykin and Desser, 1968; Derksen and Cohen, 1973), monkey (Derksen et al., 1976), and rat (Okuma et al., 1971) platelets have little or no ability to synthesize cholesterol. Thus, platelets are not able to establish their cholesterol content by endogenous synthesis.

Megakaryocyte Lipids

The experiments described above concerning lipid synthesis in platelets have revealed that PC and PI are metabolically very active, but in comparison, PE, PS, and SM exhibit relatively minimal involvement in the platelet synthetic pathways. Furthermore, platelets have almost no capacity to synthesize sterols. The apparent inability of the platelet to regulate its overall lipid composition has led us to postulate that the megakaryoctye may be the most important determinant of the ultimate composition and organization of the platelet lipids.

Our experiments have compared the composition and lipid synthetic capabilities of guinea pig megakaryocytes and platelets. We are indebted to Drs. R. F. Levine and M. E. Fedorko (1976) for development of a technique for isolation of megakaryocytes from guinea pigs which has made this work possible.

We anticipated that a comparison of the lipid content of megakaryocytes and platelets would shed light on the possibility that the megakaryocytes might determine the lipid composition of the platelets. As we have reported (Schick and Schick, 1979), the phospholipid distribution in the two cells is similar, as is the distribution of the major phospholipid fatty acids. Furthermore, the fatty acid composition of the individual phospholipids of both cells is remarkably similar. We predict on the basis of the similarities which we have observed that the megakaryoctye may predetermine the general pattern of platelet phospholipid fatty acids. The details of this work will be presented elsewhere when complete.

We have compared the ability of the guinea pig megakaryocytes and platelets to synthesize cholesterol, phospholipids, and ceramide from acetate (Schick and Schick, 1979). Both cell types were incubated with $U - {}^{14}C$ acetate for 1.5 hours, the cells were washed, the lipids were extracted, and the lipid classes were analyzed in several thin layer chromatographic systems to determine incorporation of radioactivity. The megakaryocytes committed about half of their lipid synthesis from acetate to cholesterol synthesis, whereas the guinea pig platelet, like platelets of other species (Deykin and Desser, 1968; Okuma et al., 1971) demonstrated negligible incorporation of acetate into sterols. Megakaryocytes were also capable of phospholipid and ceramide synthesis from acetate, as were the guinea pig platelets. Cholesterol synthesis in the megakaryoctye appears to be a significant factor in control of cholesterol content of the platelet. We cannot calculate from our data precisely how much cholesterol synthesis occurs, but we can compare cholesterol synthesis in the megakaryoctye to cholesterol synthesis from acetate in other guinea pig tissues (Swann et al., 1975) on the basis of nmol acetate incorporated per mg protein. The megakaryoctye incorporates acetate into cholesterol at a rate equal to acetate incorporation into sterols in guinea pig ileum, the most active sterol-synthesizing tissue in the guinea pig, and at a rate 5–20 times greater than that reported for the other guinea pig tissues studied. We predict, on the basis of our experiments, that the megakaryoctye may establish a great deal of platelet cholesterol by endogenous cholesterol synthesis. The role of plasma lipoproteins as a source of megakaryoctye cholesterol has not yet been established. Incorporation of acetate into phospholipids was more rapid per mg cell protein in megakaryocytes than in platelets (Schick and Schick, 1979), and the pattern of labeling suggested that the megakaryoctye would need to rely to some degree on plasma fatty acid for phospholipid synthesis. We predict that utilization of plasma fatty acid by the megakaryoctye rather than by the platelet is the more important determinant of platelet phospholipid fatty acid composition.

We expect that further studies of megakaryoctye lipid metabolism will yield exciting new information on the regulation of platelet lipid composition and the importance of the metabolic pathways available to the platelet.

Summary

Platelet lipids appear to play a central role in platelet aggregation, and, therefore, in blood coagulation. In order to understand the regulation of platelet lipid composition, we have compared the lipid composition and lipid biosynthetic capabilities of guinea pig megakaryocytes and platelets. Megakaryocytes were isolated from guinea pig marrow to 85% purity by the method of Levine and Fedorko (1976). The overall phospholipid and fatty acid composition and the fatty acid composition of the individual phospholipids of both cells were similar, except that megakaryocytes had a somewhat higher percentage of phosphatidylinositol and oleic acid and lower percentage of arachidonic acid than platelets. The cholesterol/phospholipid molar ratio was lower in megakaryocytes than in platelets. The differences in lipid composition between the two cells may reflect loss of the nucleus and endoplasmic reticulum upon platelet shedding. Lipid biosynthesis was determined by incubating megakaryocytes or platelets with ($U-{}^{14}C$) acetate. Megakaryocytes synthesized cholesterol as the major product of lipid synthesis from acetate, but platelets did not synthesize cholesterol. The pattern of acetate incorporation into phospholipids and ceramide was similar in both cells, but, per approximate equivalent cell volume, the megarkaryocytes were several times more active than platelets. We hypothesize, on the basis of the similarity of lipid composition of megakaryocytes and platelets, the ability of megakaryoctes but not platelets to synthesize cholesterol, and the several-fold greater capability of megakaryocytes to synthesize phospholipids and ceramide, that the megakaryocyte regulates the lipid composition of platelets and determines platelet lipid synthetic pathways.

ACKNOWLEDGMENTS

Supported in part by USPHS Grant HL 22633 from the National Institutes of Health, Veterans Admin-itration Merit Review Grant, and a grant from the Pennsylvania Heart Association. B. Schick is supported by a National Research Service Award from the National Heart, Lung and Blood Institute.

References

Bell, R. L. and P. W. Majerus. 1980. Thrombin-induced hydrolysis of phosphatidylinositol in human platelets. *J. Biol. Chem.* 255:1790–1792.

Bereziat, G., J. Chambaz, G. Trugnan, D. Pepin, and J. Polonovski. 1978. Turnover of phospholipid linoleic and arachidonic acid in human platelets from plasma lecithins. *J. Lipid Res.*, 19:495–500.

Bills, T. K., J. B. Smith, and M. J. Silver. 1976. Metabolism of ^{14}C arachidonic acid by human platelets. *Biochim. Biophys. Acta* 424:303–313.

Bills, T. K., J. B. Smith, and M. J. Silver. 1977. Selective release of arachidonic acid from the phospholipids of human platelets in response to thrombin. *J. Clin. Invest.* 60:1–6.

Broekman, M. J., R. I. Handin, A. Derksen, and P. Cohen. 1976. Distribution of phospholipids, fatty acids, and platelet factor 3 among subcellular fractions of human platelets. *Blood* 47:963–971.

Carvalho, A. C. A., Colman, R. W., and R. S. Lees. 1974. Platelet function in hyperlipoproteinemia. *N. Eng. J. Med.* 290:434–438.

Chap, H. and L. Douste-Blazy. 1974. Reaction de liberation plaquettaire induite par la phospholipase C. *Eur. J. Biochem.* 48:351–355.

Chap, H. J., R. F. A. Zwaal, and L. L. M. Van Deenen. 1977. Action of highly purified phospholipases on blood platelets. Evidence for an asymmetrical distribution of phospholipids in the surface membrane. *Biochim. Biophys. Acta* 467:146–164.

Chernoff, A., R. F. Levine, and D. W. Goodman. 1980. Origin of platelet-derived growth factor in megakaryocytes in guinea pigs. *J. Clin. Invest.* 65:926–930.

Cohen, P. and A. Derksen. 1969. Comparison of phospholipid and fatty acid composition of human erythrocytes and platelets. *Br. J. Haematol.* 17:359–371.

Cohen, P., A. Derksen, and H. van den Bosch. 1970. Pathways of fatty acid metabolism in human platelets. *J. Clin. Invest.* 49:128–139.

Cohen, P., M. J. Broekman, A. Verkley, J. W. W. Lisman, and A. Derksen. 1971. Quantification of human platelet phosphoinositides and the influence of ionic environment on their incorporation of orthophosphate ^{32}P. *J. Clin. Invest.* 50:762–770.

Demers, L. M., R. E. Budin, and B. S. Shaikh. 1980. The effects of aspirin on megakaryocyte prostaglandin production. *Proc. Soc. Exp. Biol. Med.* 163:24–29.

Derksen, A. and P. Cohen. 1973. Extensive incorporation of $2-{}^{14}$C mevalonic acid into cholesterol precursors by human platelets *in vitro*. *J. Biol. Chem.* 248:7396–7403.

Derksen, A. and P. Cohen. 1975. Patterns of fatty acid release from endogenous substrates by human platelet homogenates and membranes. *J. Biol. Chem.* 250:9342–9347.

Derksen, A., M. M. Meguid, and P. Cohen. 1976. Non-human primates and arterial tissue cannot convert preformed ^{14}C lanosterol into ^{14}C cholesterol *in vivo*. *Biochem. J.* 158:157–159.

Deykin, D. and R. K. Desser. 1968. The incorporation of acetate and palmitate into lipids by human platelets. *J. Clin. Invest.* 47:1590–1602.

Hennes, A. R. and K. Awai. 1965. Studies on incorporation of radioactivity into lipids by human blood. IV. Abnormal incorporation of acetate-1-C-14 by whole blood and platelets from insulin-dependent diabetics. *Diabetes* 14:709–715.

Jesse, R. L. and P. Cohen. 1976. Arachidonic acid release from diacyl phosphatidylethanolamine by human platelet membranes. *Biochem. J.* 158:283–287.

Lapetina, E. G. and P. Cuatrecasas. 1979. Stimulation of phosphatidic acid production in platelets precedes the formation of arachidonic acid and parallels the release of serotonin. *Biochim. Biophys. Acta* 573:394–402.

Leung, N. L., R. L. Kinlough-Rathbone, and J. F. Mustard. 1974. Incorporation of $^{32}PO_4$ into phospholipids of blood platelets. *Br. J. Haematol.* 36:417–425.

Levine, R. F. and M. E. Fedorko. 1976. Isolation of intact megakaryocytes from guinea pig femoral marrow. Successful harvest made possible with inhibitors of platelet aggregation; enrichment achieved with a two-step separation technique. *J. Cell Biol.* 69:159–172.

Lewis, N. and P. W. Majerus. 1969. Lipid metabolism in human platelets. II. *De novo* phospholipid synthesis and the effect of thrombin on the pattern of synthesis. *J. Clin. Invest.* 48:2114–2123.

Lloyd, J. V. and J. F. Mustard. 1974. Changes in ^{32}P content of phosphatidic acid and the phosphoinositides of rabbit platelets during aggregation induced by collagen or thrombin. *Br. J. Haematol.* 26:243–253.

Lloyd, J. V., E. E. Nishizawa, J. Halder, and J. F. Mustard. 1972. Changes in ^{32}P labeling of platelet phospholipids in response to ADP. *Br. J. Haematol.* 23:571–585.

Majerus, P. W., M. B. Smith, and G. H. Clamon. 1969. Lipid metabolism in human platelets. I. Evidence for a complete fatty acid synthesizing system. *J. Clin. Invest.* 48:156–164.

Marcus, A. J. 1978. The role of lipids in platelet function: With particular reference to the arachidonic acid pathway. *J. Lipid Res.*, 19:793–826.

Marcus, A. J., H. L. Ullman, and L. B. Safier. 1969. Lipid composition of subcellular particles of human blood platelets. *J. Lipid Res.* 10:108–114.

Nachman, R., R. Levine, and E. A. Jaffe. 1978. Synthesis of actin by guinea pig megakaryocytes. Complex formation with fibrin. *Biochim. Biophys. Acta* 543:91–105.

Nachman, R., R. Levine, and E. A. Jaffe. 1977. Synthesis of Factor VIII antigen by cultured guinea pig megakaryocytes. *J. Clin. Invest.* 60:914–921.

Nathan, I., G. Fleischer, A. Livine, A. Dvilansky, and A. H. Parola. 1980. Membrane microenvironmental changes during activation of human blood platelets by thrombin. *J. Biol. Chem.* 254:9822–9828.

Nordoy, A. and S. Lund. 1968. Platelet factor 3 activity, platelet phospholipids, and their fatty acid and aldehyde pattern in normal male subjects. *Scand. J. Clin. Lab. Invest.* 22:328–338.

Okuma, M., M. Steiner, and M. G. Baldini. 1971. Incorporation of acetate and fatty acids into lipids of rat platelets. *Proc. Soc. Exp. Biol. Med.* 136:842–847.

Otnaess, H. B. and T. Holm. 1976. The effect of phospholipase C on human blood platelets. *J. Clin. Invest.* 57:1419–1425.

Perret, B., H. J. Chap, and L. Douste-Blazy. 1979. Asymmetric distribution of arachidonic acid in the plasma membrane of human platelets. A determination using purified phospholipases and a rapid method for membrane isolation. *Biochim. Biophys. Acta* 556:434–446.

Rittenhouse-Simmons, S. 1979. Production of diglyceride from phosphatidylinositol in activated human platelets. *J. Clin. Invest.* 63:580–587.

Rittenhouse-Simmons, S., F. A. Russell, and D. Deykin. 1977. Mobilization of arachidonic acid in human platelets. Kinetics and Ca^{2+} dependency. *Biochim. Biophys. Acta* 488:370–380.

Rittenhouse-Simmons, S., F. A. Russell, and D. Deykin. 1976. Transfer of arachidonic acid to human platelet plasmalogen in response to thrombin. *Biochem. Biophys. Res. Comm.* 70:295–301.

Schick, B. P. and P. K. Schick. 1979. Lipid composition and lipid synthesis in guinea pig megakaryocytes and platelets. *Thrombos. Haemostas.* 42:283.

Schick, P. K. 1978. The organization of aminophospholipids in human platelets. Selective changes induced by thrombin. *J. Lab. Clin. Med.* 91:802–810.

Schick, P. K. 1979. The role of platelet membrane lipids in platelet hemostatic activities. *Sem. Hematol.* 16:221–233.

Schick, P. K. and B. P. Yu. 1974. The role of platelet membrane phospholipids in the platelet release reaction. *J. Clin. Invest.* 54:1032–1039.

Schick, P. K., K. B. Kurica, and G. K. Chacko. 1976. Location of phosphatidylethanolamine and phosphatidylserine in the human platelet plasma membrane. *J. Clin. Invest.* 57:1221–1226.

Shattil, S. J. and R. A. Cooper. 1976. Membrane microviscosity and human platelet function. *Biochem.* 15:4832–4837.

Shattil, S. J., J. S. Bennett, R. W. Colman, and R. A. Cooper. 1977. Abnormalities of cholesterol and phospholipid composition in platelets and low-density lipoproteins of human hyperbetalipoproteinemia. *J. Lab. Clin. Med.* 89:341–353.

Shattil, S. J., R. Anaya-Galindo, J. Bennett, R. W. Colman, and R. A. Cooper. 1975. Platelet hypersensitivity induced by cholesterol incorporation. *J. Clin. Invest.* 55:636–643.

Sinha, A. K., S. J. Shattil, and R. W. Colman. 1977. Cyclic AMP metabolism in cholesterol-rich platelets. *J. Biol. Chem.* 252:3310–3314.

Spector, A. A., J. C. Hoak, E. D. Warner, and G. L. Fry. 1970. Utilization of long-chain free fatty acids by human platelets. *J. Clin. Invest.* 49:1489–1496.

Stuart, M. J., J. M. Gerrard, and J. A. White. 1980. Effect of cholesterol on production of thromboxane B_2 by platelets *in itro*. *N. Eng. J. Med.* 302:6–10.

Swann, A., M. H. Wiley, and M. D. Siperstein. 1975. Tissue distribution of cholesterol feedback control in the guinea pig. *J. Lipid Res.* 16:360–366.

Zwaal, R. F. 1978. Membrane lipid involvement in blood coagulation. *Biochim. Biophys. Acta* 515:163–205.

Discussion

Ebbe: Could you restate how you decided that each megakaryocyte makes 1000 platelets?

B. Schick: This is an average over the entire population. We had calculated the amount of protein per platelet and the amount of protein per megakaryocyte. This would be a low estimate, however, since the population includes many immature megakaryocytes.

Ebbe: Several years ago, Cronkite made an estimate of 4000 platelets/megakaryocyte.

B. Schick: The estimate is 2000–4000. In a recent paper (*J. Clin. Invest.* 65:926, 1980), Levine analyzed one very large megakaryocyte and reported 2500 clearly demarcated platelets.

Dombrose: Because of the ubiquity of phospholipids on cell membranes, do you think the minor lipids, such as the sulfatides, may be more important?

B. Schick: I don't know that they have been detected in platelets.

Dombrose: That is what I meant. Do you think they might be present and play a role as receptors for certain classes of plasma proteins that are throught to arise as a result of platelet activation? Perhaps certain glycolipids come to the surface upon platelet activation.

B. Schick: I think that if we could isolate very large masses of megakaryocytes that study might be possible. But you may be right. The minor lipids may play a very important role and the major lipids may have a more structural role. Large quantities of lipids would have to be analyzed to detect the minor ones.

Dombrose: Did you say that there are no sulfalipids and sulfatides in platelets?

B. Schick: I haven't seen any reported.

Dombrose: Maybe no one has looked for them.

B. Schick: It may be.

Hempling: It is reasonable to hypothesize that the reason for the large turnover of cholesterol and cell mass is in synthesis of membrane?

B. Schick: Yes.

Evatt, Levine, and Williams, Editors
MEGAKARYOCYTE BIOLOGY AND PRECURSORS:
IN VITRO CLONING AND CELLULAR PROPERTIES

The Response of Cultured Megakaryocytes to Adenosine Diphosphate

R. M. Leven and V. T. Nachmias

Department of Anatomy, School of Medicine/G3, University of Pennsylvania, Philadelphia, Pennsylvania

It is 70 years since it was realized that megakaryocytes give rise to platelets by fragmentation of their cytoplasm (Wright, 1910). Yet only recently have techniques been developed to separate megakaryocytes from bone marrow suspensions so that relatively pure cultures can be obtained. This makes it possible to study the properties of this remarkable cell type under controlled conditions. Since the end point of megakaryocyte maturation is the circulating platelet, one set of developmental problems centers around questions of when receptors are formed and whether megakaryocytes can respond to stimuli which activate platelets. If the stimulus-response mechanism develops early, megakaryocytes are more favorable for study than platelets because of their large size and ease of manipulation. Studies of megakaryocytes may in turn lead to better understanding of how platelets are activated.

With this approach in mind, we tested the response of cultured guinea pig megakaryocytes to adenosine diphosphate (ADP), which causes shape change and aggregation of platelets (Born, 1970). We examined cells by phase microscopy for morphological changes and by indirect immunofluorescence for changes in tubulin distribution. Once the response was observed, a set of criteria was made to estimate the magnitude of spreading. It was then possible to study the ionic dependence of ADP stimulation.

Methods and Material

Megakaryocytes were prepared by modifications of the methods of Levine and Fedorko (1976) and Rabellino et al. (1979). The yield from two guinea pigs was 150,000–350,000 megakaryocytes of 50%–90% purity. The cells were cultured in McCoy's 5A medium supplemented with 10% fetal calf serum, 2 mM glutamine and 50 U/ml penicillin-streptomycin in 5% CO_2 – 95% 0_2 at 37°C. The cells were cultured for 18–24 hr prior to an experiment. Antitubulin antibody has been previously characterized (Tucker et al., 1977). The cells were treated, fixed, and stained by the method of Osborn and Weber (1977). Photographs were taken with an Olympus Vanox microscope with epifluorescent illumination and Zeiss Plan-apochromat objectives. Quantitation of spreading was made by counting at least 100 cells per dish and assigning each to one of three categories. Stage O – cell spherical and unspread; Stage 1 – central, raised perinuclear region is greater than half the total cell diameter; Stage 2 – central raised perinuclear region is half the total cell diameter; Stage 3 – central raised perinuclear region is less than half the total cell diameter.

Results and Discussion

Response to ADP

When megakaryocytes are maintained in liquid culture, they are spherical and adhere only weakly to the culture dish (Fig. 1). When ADP (1–100 μM final concentration) is added, the cells begin to form an actively ruffling membrane, adhere to the substrate, and gradually spread out to five times their original size. The process takes about 30 min; at the end of that time the whole peripheral cytoplasm appears very thin and the ruffling has subsided (Fig. 2).

Cells were cultured on a variety of substrates: glass, plastic, rat-tail collagen, poly-1-lysine and pure or crude fibronectin; none altered the spreading response.

Fig. 1. Untreated cultured megakaryocytes. Bar equals 50 μm.

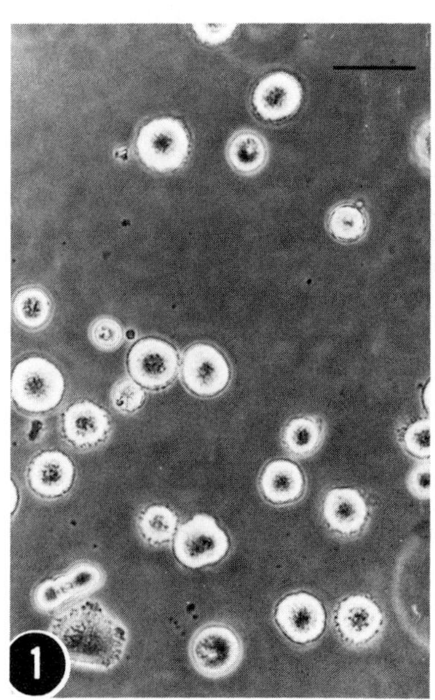

Spreading did not require serum, but proceeded as well in buffered salt solutions. Spreading appeared to be an energy requiring process as it was blocked by 1 mM sodium azide and 1 mM sodium fluoride in glucose free medium.

Several other platelet agonists were tested. Thrombin (0.3 U/ml), epinephrine (10 μM), A23187 (2 μM) and collagen surfaces were all ineffective in altering cell shape. Arachidonic acid (0.5 mM) caused spreading but to a lesser extent than with ADP.

Since ADP induced platelet shape change is prevented or reversed by increased levels of cyclic AMP (Haslam et al., 1978) we tested the effects of 1.0 mM dibutyrl cyclic AMP with 1.0 mM isobutyl methylxanthine (a potent phosphodiesterase inhibitor). The combination would be expected to cause increased levels of cyclic AMP in the cells. This combination prevented spreading if added before ADP, or reversed it if added after cells were fully spread. Remarkably, the cells rounded up completely in about 15 to 45 min. If such reversed cells were put into fresh medium they could be respread with ADP. Cytochalasin D was tested because it inhibits actin polymerization *in vitro* (Flanagan and Lin, 1980). It blocked spreading completely at 2 μg/ml, but did not reverse cells previously spread.

Localization of Tubulin

Since platelets contain an unusual coiled microtubule (White, 1976; Nachmias et al., 1977), it was of interest to examine the distribution of microtubules in megakaryocyte cytoplasm and to see whether any change occurred during spreading.

Fig. 2. Cultured megakaryocytes exposed to 10 μm ADP for 30 min. Arrow marks active ruffled border. Bar equals 50 μm.

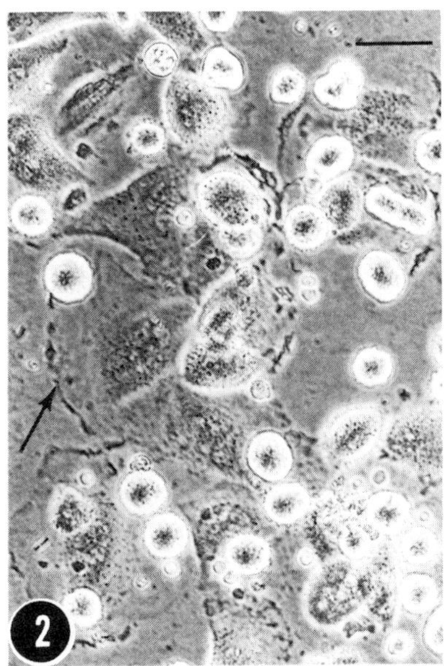

In untreated, rounded megakaryocytes, microtubules were observed by indirect immunofluorescence as a tightly packed array (Fig. 3). When cells were examined at the onset of spreading, there appeared to be a rim in which the microtubules were absent. After ADP induced spreading, the distribution is essentially the same as in controls, although the array is more loosely arranged (Fig. 4). In some of the spread cells, distinctive ring shaped structures can be seen (Fig. 4 and inset). These rings measure 1.5–2.5 μm in diameter. Megakaryocyte microtubules were depolymerized by either colchicine (1 mg/ml) or incubation at 4°C. Colchicine, however, did not appear to affect spreading.

Ionic Dependence of ADP-Induced Spreading

Much work has suggested, though not rigorously proved, that increased levels of free calcium are present in activated platelets (White et al., 1974; LeBreton et al., 1976; Kaser-Glazmann et al., 1977). On the other hand, work with other cell types has shown that the formation of microfilaments from amorphous precursor material can be induced by increases in the cellular pH (Tilney et al., 1978; Begg and Rebhun, 1979). Platelet activation results in microfilament bundles in the filopodia (Nachmias et al., 1977; Zucker-Franklin et al., 1967). Actin ruffling and spreading by megakaryocytes probably involves the formation or rearrangement of microfilaments and not microtubules since it is blocked by cytochalasin D, which is known to prevent actin polymerization (White, 1976), but not affected by colchicine. Therefore we tested agents which should increase megakaryocyte free Ca^{++} or pH

Fig. 3. Untreated cultured megakaryocyte fixed and stained with antitubulin antibody. Bar equals 10 μm.

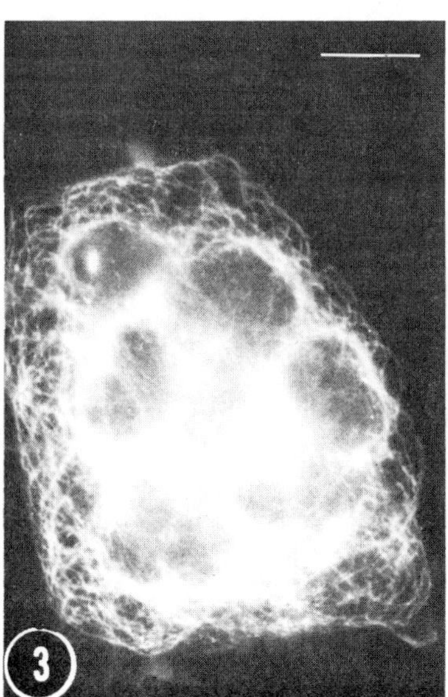

for their ability to induce spreading. As previously mentioned, the ionophore A23187 which is relatively specific for calcium transport and causes platelet activation, did not cause any spreading at 2 μM. If left in contact with the cells for 1–2 hr, cell lysis was observed. We then incubated cells in 5 mM methylamine, an agent which can cross cell membranes and increase cytoplasmic pH (Gillies and Deanes, 1979). Occasionally some spreading occurrred, but the response was inconsistent. When, however, both agents were added together to the cultures, virtually every megakaryocyte responded with a spreading reaction. The spreading of individual cells was not as extensive as with ADP, but the membrane ruffling appeared very active.

Since this experiment suggests that a rise in both cellular pH and free Ca^{++} are necessary for spreading, it was of interest to ask how ADP might cause such changes. One possible link is through proton or calcium exchange with sodium across the cell membrane. Such exchanges have been found in other cell types (Blaustein, 1974). To test this hypothesis, cells were incubated in calcium- and magnesium-free Hank's salt solution (CMFH) in which all sodium was replaced with either potassium or choline. The results of this experiment are diagrammed in Table 1. The abscissa points, 0–3, represent the different stages of spreading (see Methods). The ordinate represents the difference between the percent of the total population in each spreading stage before and after exposure to ADP. In this experiment, cells were exposed to 10 μM ADP for 30 min. The top two panels

Fig. 4. Cultured megakaryocyte exposed to 10 μm ADP for 30 min, fixed and stained with antibulin antibody. Bar equals 10 μm. Inset is 1.5 magnification of same cell showing microtubule rings.

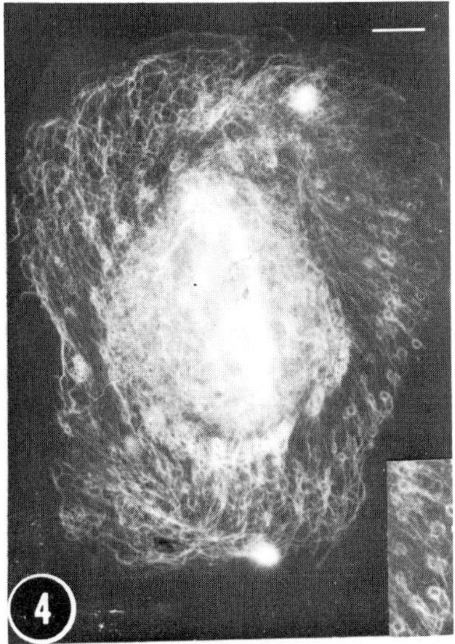

show the response to ADP in normal or CMFH media. The center two panels illustrate identical cultures incubated in CMFH with Na^+ replaced by K^+ or choline. The megakaryocytes do not respond under either of these conditions. When the cultures from the center two panels were returned to fresh medium and exposed again to 10 μM ADP, they responded again, as shown in the bottom two panels. Not shown here is the finding that Li^+ was not able to replace Na^+ in ADP-induced spreading.

Table 1.

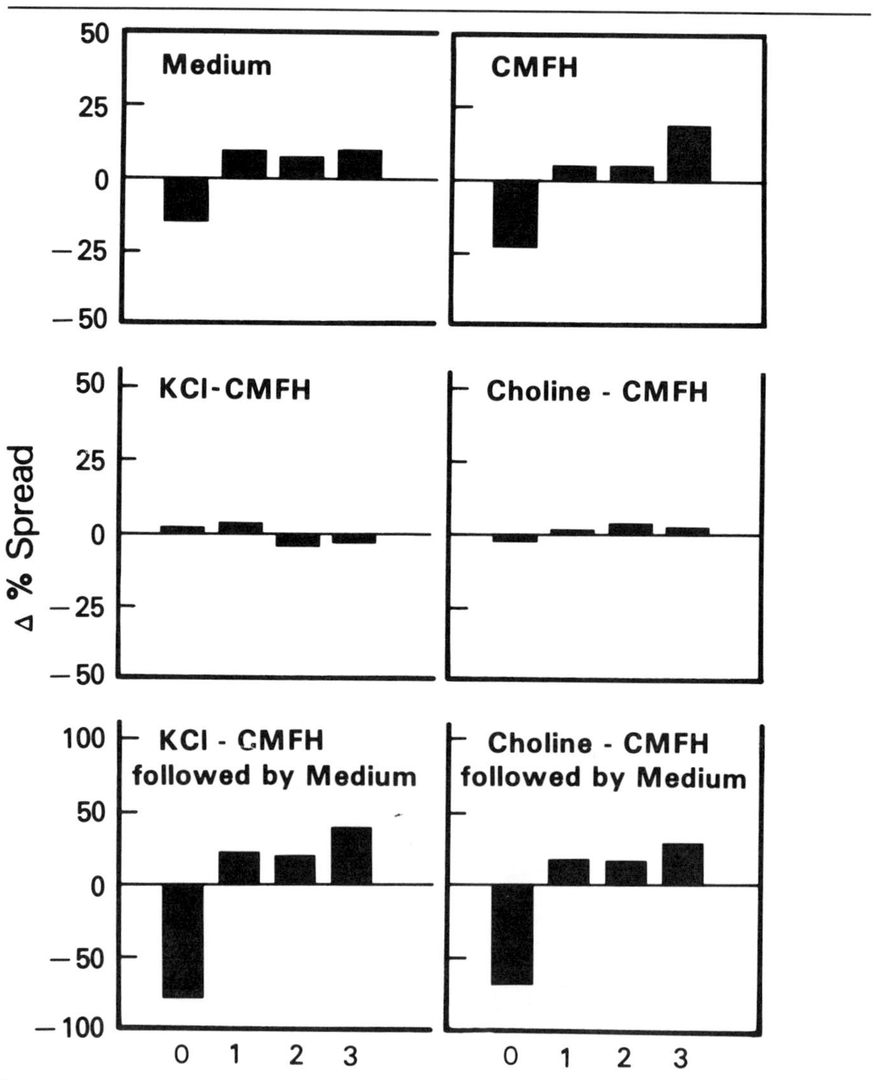

The spreading response of megakaryocytes described here is similar to platelet shape change in several respects. It is inhibited by cytochalasin D and reversed by agents which cause an increase in cytoplasmic cyclic AMP. It is induced by ADP (but not by other diphosphonucleotides) and by arachidonic acid. On the other hand, it differs from the platelet response in several other respects. It is slow (minutes rather than seconds), occurs in buffered salt solutions, and involves, as a rule, the formation of ruffles rather than filopodia. Also, megakaryocytes in this system are not responsive to several platelet activating agents. It may be that cofactors are required for these agents which are not present in the cultures; it may be that the changes in platelets induced by different agents are not the same, and that only ADP and arachidonic acid cause changes that can be expressed as spreading; or finally, it may be that receptors for thrombin, epinephrine or collagen are not expressed on the cultured cells.

Our data show that there is no overall breakdown of microtubules during the response to ADP. It is possible that ADP causes a transient breakdown in the region of the ruffled membrane, where microtubules are absent, and that subsequent reassembly occurs, or it may be that there is a lag period before the central microtubules move or grow out into the spreading cytoplasm. The rings, seen with antitubulin antibody in some spread cells, are identical in size to the microtubule coil present in platelets. There are three possible explanations for the appearance of the rings. It may be that they are obscured in the nonspread cells. It may be that the physical effect of spreading itself causes ring formation. Or, it may be that ADP in addition to inducing spreading also has an effect on the arrangement of microtubules. Though very speculative, it is an intriguing idea that the formation of platelets could be stimulated by the products released when platelets are activated.

In respect to the ionic mechanism of ADP stimulation, the data are still too scanty to present a definite model. It is clear that in the relative absence of Ca^{++} and Mg^{++}, sodium is required for the response. This finding, together with the dramatic effect of methylamine plus A23187 in induction of ruffling and spreading, suggests that there is an exchange of sodium in ADP induced spreading. This could be an exchange for H^+. An internal rise in pH with or without increased sodium levels could trigger Ca^{++} release from internal stores. This has been reported for cardiac sarcoplasmic reticulum (Dunnet and Nayler, 1979). One or both of these changes may affect microfilament polymerization (which we postulate to be an early step in spreading) and possibly, further interactions of the microfilaments with myosin. Experiments to test these proposals are underway.

Summary

Guinea pig bone marrow megakaryocytes have been isolated by modifications of published procedures and maintained in liquid culture. The cells are normally spherical. It was discovered that ADP (10 μM) causes the spherical cells to spread and flatten to several times their normal size in the presence or absence of serum. Spreading is identical when cells are cultured on glass, plastic, poly-1-lysine or fibronectin. By time lapse cinematography, treated cells have actively ruffling membranes and some filopodia as they spread. Using antitubulin antibody, the distribution of microtubules in spread and unspread cells was compared by indirect immunofluorescence. Both have a tightly packed, highly convoluted array of microtubules throughout their cytoplasm. Colchicine breaks down all visible microtubules but does not inhibit spreading even at 1 mg/ml. Incubation at 4°C breaks down microtubules which reform after reincubation at 37°C. Spreading requires metabolic energy. Cytochalasin D (0.5 μg/ml) prevents but does not reverse spreading. Spreading can, however, be reversed by dibutyryl cAMP (1 mM) and isobutylmethyl xanthine (1 mM) together. A23187 (2 μM) does not cause spreading. Replacement of extracellular NA^+ by choline, potassium or lithium completely prevents ADP induced spreading. Thus, spreading shares some characteristics of platelet activation and relaxation and probably requires sodium influx for ADP stimulation.

ACKNOWLEDGMENTS

We thank Dr. Fred Frankel for the generous gift of antitubulin antibody and Dr. Joseph Sanger for the use of his fluorescence microscope. Supported by Grants HL 15835, AM 17492, and T32-HD 07067.

References

Begg, D. A. and L. I. Rebhun. 1979. PH regulates the polymerization of actin in the sea urchin egg cortex. *J. Cell Biol.* 83:241–248.

Blaustein, M. P. 1974. The interrelationship between sodium and calcium fluxes across cell membranes. *Rev. Physiol. Biochem. Pharmacol.* 70:33–82.

Born, G. V. R. 1970. Observations on the change in shape of blood platelets brought about by adenosine diphosphate. *J. Physiol.* 209:487–511.

Dunnet, J. and W. Nayler. 1979. Effect of pH on calcium accumulation and release by isolated fragments of cardiac and skeletal muscle sarcoplasmic reticulum. *Arch. Biochem. Biophys.*, 198:434–438.

Flanagan, M. D. and S. Lin. 1980. Cytochalasins block actin filament elongation by binding to high affinity sites associated with F-actin. *J. Biol. Chem.* 255:835–838.

Gillies, R. J. and D. W. Deamer. 1979. Intracellular pH: Methods and applications. *Curr. Top. in Bioenergetics* 9:63–87.

Haslam, R. J., M. M. Davidson, T. Davies, J. A. Lynham, and M. D. McClenaghan. 1978. Regulation of blood platelet function by cyclic nucleotides. *Adv. Cyclic Nucleotide Res.* 9:533–552.

Käser-Glanzmann, R., M. Jakabova, J. N. George and E. F. Lüscher. 1977. Stimulation of calcium uptake in platelet membrane vesicles by adenosine 3′, 5′ – cyclic monophosphate and protein kinase. *Biochim. Biophys. Acta* 466:429–440.

LeBreton, G. C., R. J. Dinerstein, L. J. Roth, and H. Feinberg. 1976. Direct evidence for intracellular divalent cation redistribution associated with platelet shape change. *Biochem. Biophys. Res. Commun.* 71:362–370.

Levine, R. F. and M. E. Fedorko. 1976. Isolation of intact megakaryocytes from guinea pig femoral marrow. Successful harvest made possible with inhibitors of platelet aggregation; enrichment achieved with a two-step separation technique. *J. Cell Biol.* 69:159–172.

Nachmias, V., J. Sullender, and A. Asch. 1977. Shape and cytoplasmic filaments in control and lidocaine-treated human platelets. *Blood* 50:39–53.

Osborn, M. and K. Weber. 1977. The display of microtubules in transformed cells. *Cell* 12:561–571.

Rabellino, E. M., R. L. Nachman, N. Williams, R. J. Winchester, and G. D. Ross. 1979. Human megakaryocytes. I. Characterization of the membrane and cytoplasmic components of isolated marrow megakaryocytes. *J. Exp. Med.* 149:1273–1287.

Tilney, L. G., D. P. Kiehart, C. Sardet, and M. Tilney. 1978. Polymerization of actin. IV. Role of Ca^{++} and H^{+} in the assembly of actin and in membrane fusion in the acrosomal reaction of echinoderm sperm. *J. Cell Biol.* 77:536–550.

Tucker, R. W., K. K. Sanford and R. Frankel. 1978. Tubulin and actin in paired nonneoplastic and spontaneously transformed neoplastic cell lines *in vitro*: Fluorescent antibody studies. *Cell* 13:629–642.

White, J. G. and J. M. Gerrard. 1976. Ultrastructural features of abnormal blood platelets. *Am. J. Pathol.* 83:589–632.

White, J. G., G. H. Rao, and J. M. Gerrard. 1974. Effects of the ionophore A23187 on blood platelets. I. Influence on aggregation and secretion. *Am. J. Pathol.* 77:135–149.

Wright, J. H. 1910. The histogenesis of the blood platelets. *J. Morphol.* 21:263-278.

Zucker-Franklin, D., R. L. Nachman, and A. J. Marcus. 1967. Ultrastructure of thrombosthenin, the contractile protein of human blood platelets. *Science* 157:945–946.

Discussion

Miller: You had spreading in Ca^{++}-, MG^{++}-free hanks, the same as in control media. What effect does the presence of EDTA or EGTA have?

Leven: If Ca^{++} and Mg^{++} are vigorously excluded, there is no adherence, and this is a prerequisite for spreading in Ca^{++}-, Mg^{++}-free experiments. I am sure there are still micromolar amounts there.

Miller: Could you tell us about the time course of the cAMP experiments? Do the cells revert to normal ADP sensitivity later on?

Leven: The time course of the reversal is variable. Some cells round up in about 15 min, the time required for spreading, other cells take up to 40–45 min. If you then give the cells ADP, they will apread again.

Levine: How long does the recovery from the spreading take?

Leven: Between 15–45 min.

Walz: You get no spreading with thrombin?

Leven: That is correct.

Walz: Do you have any evidence that thrombin will bind to megakaryocytes?

Leven: I have not looked at binding.

Levine: Thrombin will cause dramatic degranulation of megakaryocytes (*Lab. Invest.* 36:321, 1977). In fact, there are several ways in which platelet behavior has been demonstrated in megakaryocytes. I have seen secretion of alpha granules and formation of actin filaments after thrombin treatment.

Evatt, Levine, and Williams, Editors
MEGAKARYOCYTE BIOLOGY AND PRECURSORS:
IN VITRO CLONING AND CELLULAR PROPERTIES

Megakaryocyte Maturation Indicated by Methanol Inhibition of an Acid Phosphatase Shared by Megakaryocytes and Platelets

Olivera Saljinska Markovic and N. Raphael Shulman

National Institutes of Arthritis, Metabolism and Digestive Diseases, Bethesda, Maryland

Morphologic evaluation of megakaryocyte maturity based on conventional bone marrow staining is not precise. Histochemical techniques particularly for glycogen or DNA have been applied in attempts to distinguish human megakaryocyte stages (Forteza and Baguena, 1960; Wintrobe et al., 1974), but all stages contain PAS-positive material and may have the same degree of polyploidy. Acetylcholinesterase found in rat megakaryocyte precursors is not present in human megakaryocyte cell lines. Acid phosphatase (AP) has been found in the earliest recognizable rat megakaryocyte by electron microscopy (Bentfeld and Bainton, 1975), and a tartrate-inhibitable as well as non-inhibitable AP has been found in human blood platelets (Kaulen and Gross, 1971); but systematic studies of changes in AP with megakaryocyte maturation in human beings have not been done and soluble cytoplasmic megakaryocyte or platelet AP isoenzymes have not been noted.

During application of cytochemical techniques for AP we found that different stages of megakaryocytes, identified by conventional morphologic criteria, show differences in their AP activity and that peripheral blood platelets have AP characteristics similar to the cytoplasm of mature megakaryocytes. This study is concerned with the basis for observed differences in megakaryocyte AP activity and for the correlation between megakaryocyte AP activity and platelet formation.

Identification of megakaryocyte stages by morphologic criteria was facilitated by studying sequential bone marrow samples from a patient with a rare form of cyclic thrombocytopenia associated with megakaryocytopoiesis (Engström et al., 1966; Wilkinson and Firkin, 1966). Figure 1 shows the times at which bone marrows were obtained during two cycles. The earliest recognizable megakaryocytes (Stage I) appeared *de novo* in the marrow on day 35; approximately 10% promegakaryocytes (Stage II) and predominantly early mature megakaryocytes (Stage IIa) were

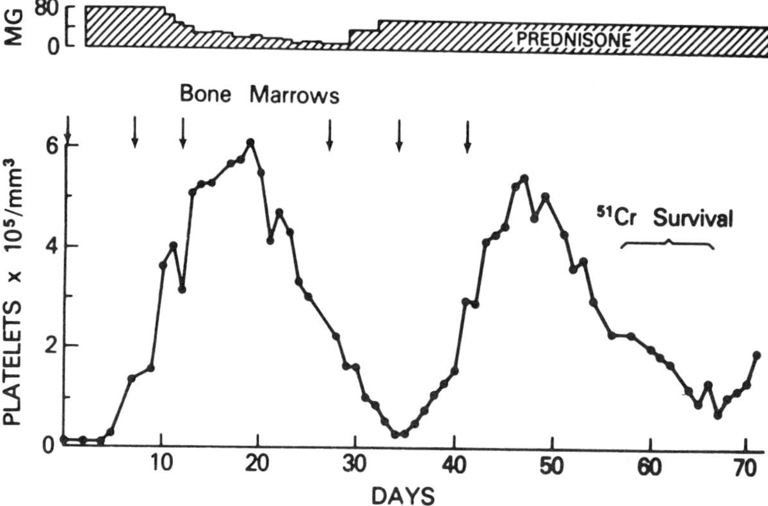

Fig. 1. Observation of a patient with idiopathic cyclic thrombocytopenia. Day 0 was the day of admission to the Clinical Center at NIH.

present on days 7 and 41 when platelets were rising; and when platelets reached normal levels (day 12) megakaryocytes were mainly mature (Stage III). Results on this patient's megakaryocytes were verified by studies of the megakaryocytes of 80 cases of acute leukemia at different stages of remission following chemotherapy and of the megakaryocytes of two hematologically normal individuals from whom marrow was obtained for other purposes. In normal marrow, there were 4%–10% Stage I megakaryocytes, 10%–25% Stage II, 50%–70% Stages IIa and III, and approximately 10% naked nuclei.

In the standard technique, fixation of cells on glass slides was performed in a mixture of 10% formalin in absolute methanol for 20 sec at 0°–4°C. Other conditions of fixation tried were as follows: 60% citrate-buffered acetone, pH 2.0, for 20 sec at room temperature; 10% formalin in 0.1 M acetate buffer, pH 5.0 for 20 sec at 0°–4°C; 3% glutaraldehyde in 0.1 M phosphate buffer, pH 7.0 for 20 sec at 0°–4°C. The standard method for cytochemical investigation of megakaryocyte AP or platelet AP (PAP) was a simultaneous azo-dye coupling technique with the α-naphthol AS-MX phosphate fast blue RR system, according to Burstone (1958, 1961) and Markovic (1976). After the AP reaction, smears were counterstained with a 1% solution of methylene blue. Counterstaining was performed over the pH range 6.0–7.5 in 0.1 M phosphate buffer for 5–15-min periods at room temperature. Best results were obtained at pH 7.0 for 10 min, which was used routinely.

The intensity of AP was estimated by the size, density, distribution, and color of granules representing the final reaction product to the AP reaction. Degrees of activity could be estimated from 0–4 + . Lowest activity, 1 + , appeared as an island of up to 10 small discrete, fine light blue granules localized in one part of the cytoplasm. Strong activity, 3+, consisted of many coarse single granules, larger than those of 2+ activity, uniformly distributed through the cytoplasm. The strongest activity, 4+, appeared as coarse, dense, intensely purple granulation with a tendency for granules to clump together and obscure the nucleus. For semiquantitative scoring the sum of the estimated enzyme activity per 100 cells was used.

Platelets were prepared from blood anticoagulated with acidified acid-citrate-dextrose and centrifuged at 1100 × g for 3 min. Smears of platelet-rich plasma, prepared in the same manner as peripheral blood smears, were used either unfixed or after fixation in acetone, formalin, methanol, or formalin-methanol with the same cytochemical procedures applied to megakaryocytes.

The same standard cytochemical methods were also applied on electophoretically separated isoenzymes of platelet acid phosphatase (PAP). To obtain soluble PAP, a button of platelets was washed to remove plasma proteins, suspended in 0.15 M NaC1 at a concentration of 10^6/ml and treated with 5% Triton X-100 in water for 30 min at 4°C. The disrupted cell suspension was then centrifuged at 100,000 × g to obtain the supernatant soluble protein fraction. To separate PAP isoenzymes, the protein fraction was electrophoresed in 7.5% polyacrylamide gel containing 0.5% Triton X-100 at pH 4.5 using 6–8 mA per tube for 45 min at 4°C (Ornstein and Davis, 1962; Markovic, 1974). Gels were rinsed in 0.1 M acetate buffer, pH 5.0, and reacted with the same AP substrate described above for megakaryocytes either directly or after fixation in formalin-methanol, formalin, methanol, or acetone at the same concentration used for fixing the cell preparations. Densitometric recordings of the AP reaction in the gels were made with a Gilford densitometer.

Results and Discussion

Megakaryocyte Acid Phosphatase

In unfixed smears all stages of megakaryocytes showed AP activity (Fig. 2), the density of the reaction increasing parallel with the degree of maturation, as shown by examples of individual cells in Figure 2 and on the basis of AP score in Figure 3 (0). When smears were fixed in formalin-methanol (Fig. 4B, D), Stage III cells showed markedly decreased AP activity (Fig. 4D) compared to the unfixed smears, in which their activity was the highest (Fig. 4C).

Figure 5 shows that megakaryocytes in smears stained unfixed (0) or fixed in glutaraldehyde (GA/B), or formalin (F/B) had essentially the same AP activity, which roughly paralleled the degree of maturity. The activity of cells in smears fixed in acetone was somewhat decreased, but the degree of decrease was similar for all stages of maturity. All megakaryocytes in smears fixed in formalin-methanol (F/M) or buffered methanol (M/B) showed decreased AP activity, the decrease being most marked in Stage III cells, in which activity was essentially absent. Thus methanol appeared to have a marked inhibitory effect specifically on AP of Stage III megakaryocytes. The inhibitory effect of methanol was time and concentration dependent. The optimal differential effect between Stage III and other megakary-

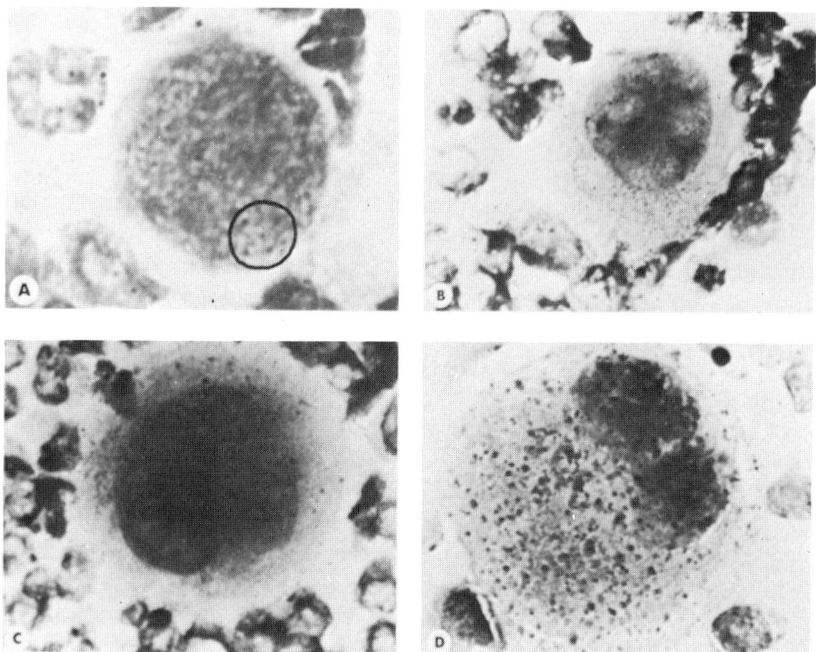

Fig. 2. Acid phosphatase activity of megakaryocytes from the patient whose clinical course is shown in Fig. 1. Stage I, II, IIa, and III megakaryocytes as defined by the morphologic criteria described in the text are shown. The dark granules of acid phosphatase reaction are light blue to dark purple. Degree of activity evaluated as described in text. (A) Stage 1 shows 1+ activity (an island of granules is encircled). (B) Stage II shows 2+ activity. (C) Stage IIa shows 3+ activity. (D) Stage III shows 4+ activity.

ocytes was obtained by a 20–30-sec exposure to 90% or absolute methanol. Formalin was included in the standard method to improve fixation.

Platelet Acid Phosphatase

In smears of PRP that were used either unfixed or fixed in glutaraldehyde, formalin, or acetone, PAP activity appeared as dense, purple granules in almost all cells. PAP had the same characteristics as AP in Stage III megakaryocytes. When formalin-methanol or methanol was used to fix PRP smears, PAP was demonstrable in only about 5% of platelets. Figure 4 shows the difference between unfixed (Fig. 4E) and formalin-methanol fixed (Fig. 4F) PRP.

Soluble platelet proteins electrophoresed in polyacrylamide gels and reacted with AP substrate showed two clearly separated bands of activity (Fig. 5). The fast moving fraction (Fraction 1; Fig. 5A-a) was 10.5–13.5 mm from the origin and represented approximately 2% of the total activity. The slow moving fraction, 2.5–7.0 mm from the origin, represented 98% of the total activity (Fig. 5C-a). The

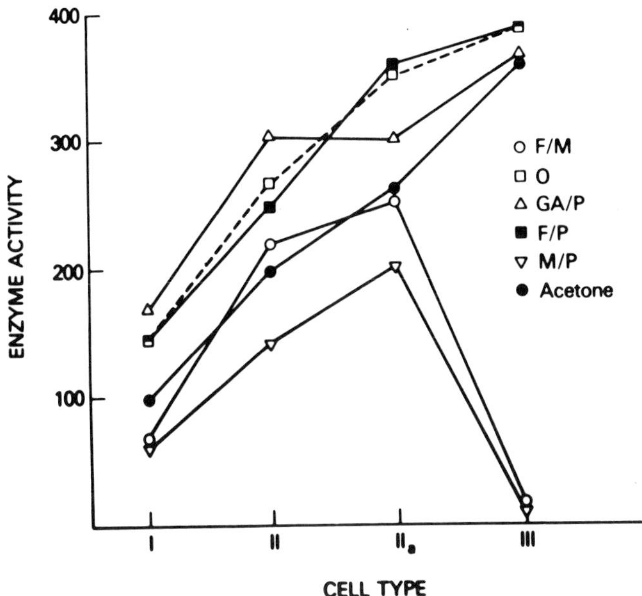

Fig. 3. Degree of activity scored 0–4+ as described in the text. Megakaryocyte stages shown on abscissa. Symbols shown on the graph represent results with formalin-methanol (F/M), unfixed (0), with buffered glutaraldehyde (GA/B), with buffered formalin (F/B), with buffered methanol (M/B), and with acetone. Values are based on scoring a total of 900 Stage III cells, 350 Stage II and IIa cells, and 60 Stage I cells. Content of fixatives described in the text.

slow moving fraction was not homogeneous, for a third fraction close to the origin became visible with prolonged electrophoresis (3 hr) which represented about 10% of the activity (Fig. 5B, D). Since the fast moving fraction became less distinct with prolonged electrophoresis and because fractions 2 and 3 had the same response to inhibitors, a 45-min electrophoresis was optimal for distinguishing methanol sensitivity of PAP.

When the gels were treated with 10% formalin-methanol before the AP reaction, the density of the slow moving fractions decreased by 50%–60% in 30 min, 80%–90% in 45 min (Fig. 5A-b, C-b), and 99% in 75 min (Fig. 5A-c, C-c); after further exposure it could not be detected. The density of the fast moving fraction decreased by only about 10%–20% in 75 min.

Buffered formalin or buffered acetone produced approximately 15% inhibition of all AP fractions in 45 min, and 0.04 M $CuSO_4$ completely inhibited all AP activity in the gels in 45 min.

We have found that methanol, the usual fixative applied in phosphatase cytochemistry, markedly inhibits an AP component, most likely an isoenzyme, that is present in the cytoplasm of mature megakaryocytes and in platelets, but does not

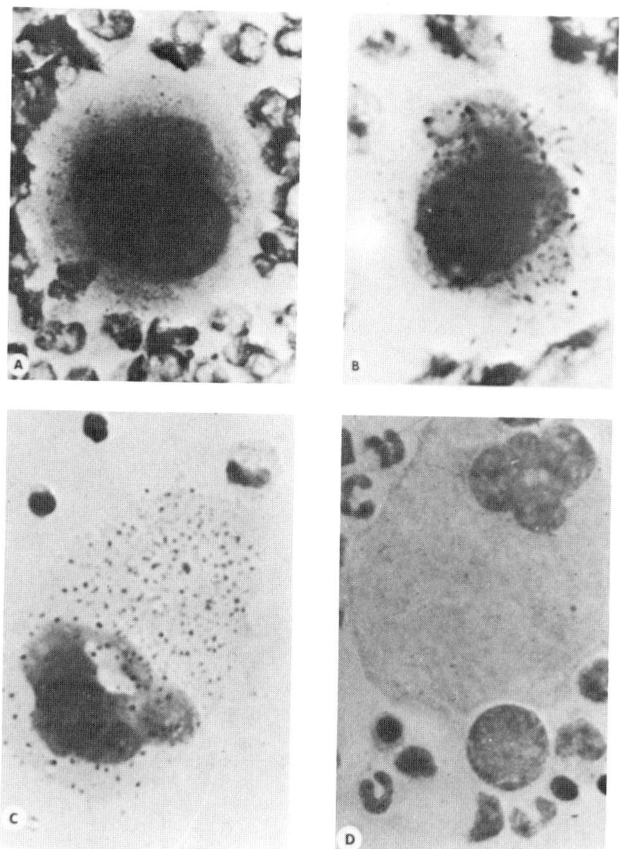

significantly affect another AP component, or isoenzyme, present in the cytoplasm of immature megakaryocytes. Since mature megakaryocytes and platelets contain a relatively high concentration of methanol-sensitive AP, the presence of this isoenzyme in megakaryocytes may be an indication of platelet formation. Comparison of megakaryocyte AP reactions of smears fixed by methanol and formalin-methanol with those of unfixed smears, or smears fixed by acetone, glutaraldehyde, or formalin, should provide, along with standard morphology based on panoptic staining, an indication of the number of mature megakaryocytes present. This information may prove to be a useful index of platelet production.

It is conceivable that in certain clinical conditions, for example, those involving rapid platelet turnover, there may be discrepancies between morphologic and biochemical criteria of megakaryocyte maturity which would be helpful diagnostically. Comparison of megakaryocyte methanol-sensitive and -resistant AP in disorders such as idiopathic thrombocytopenic purpura and other thrombocytopenic syndromes may provide clues to differentiating between defects involving platelet

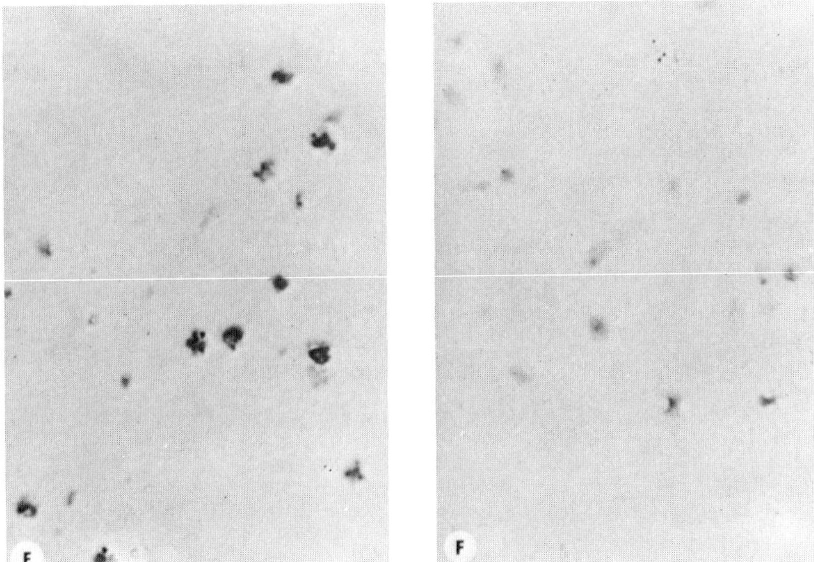

Fig. 4. Acid phosphatase activity in megakaryocytes and platelets on unfixed and formalin-methanol fixed smears. (A) Megakaryocyte Stage IIa, unfixed, activity 3+. (B) Megakaryocyte Stage IIa, fixed by formalin-methanol, activity 3+. (C) Megakaryocyte Stage III, unfixed, activity 4+. (D) Megakaryocyte Stage III, fixed by formalin-methanol, activity 0. (E) Platelet-rich plasma smear, unfixed, activity present. (F) Platelet-rich plasma smear, fixed by formalin-methanol, activity absent.

destruction or production. It is also conceivable that, under pathologic conditions, platelets may be formed by less mature megakaryocytes, which may be reflected in the relative content of methanol-sensitive and -resistant PAP measurable quantitatively by electrophoresis.

Three bands of PAP were seen on electrophoresis. The two slow moving bands, accounting for 98% of total activity, could be inactivated by methanol, while the fast moving band, accounting for approximately 2% of the total, was resistant to methanol. Since AP that predominates in younger megakaryocytes was not sensitive to methanol, methanol-resistant PAP may reflect the amount of that isoenzyme left in megakaryocytes when platelets form. The two slow moving PAP bands had the same sensitivity to methanol; it is possible that the slowest moving band, 3, is simply a dimer of band 2.

Summary

Optimal conditions necessary for the cytochemical demonstration of megakaryocyte and platelet acid phosphatase (AP) were determined. Methanol, a constituent of fixatives commonly used in AP cytochemistry, was found to inhibit megakaryocyte

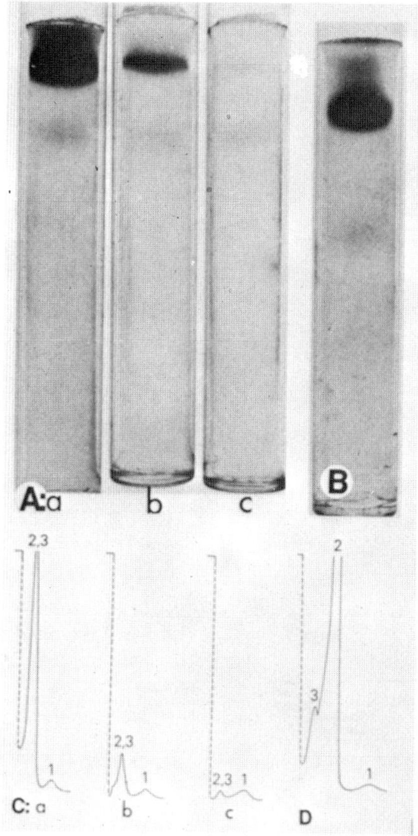

Fig. 5. Soluble platelet protein prepared, electrophoresed, and reacted with acid phosphatase substrate as described in text. (A) Electrophoretic conditions optimized to increase prominence of band 1 with bands 2 and 3 not separated distinctly. Inhibition of acid phosphatase by formalin-methanol after exposure for 45 min (b) and 75 min (c) compared to untreated gel (a). (B) Electrophoretic time increased to 3 hr for clearer separation of band 3. (C) Densitometric scan of the gels of (A). (D) Scan of the gel of (B).

AP, and the degree of inhibition varied with megakaryocyte maturity. Immature megakaryocytes contained predominantly methanol-resistant AP, and mature megakaryocytes, predominantly methanol-sensitive AP. Platelets contained methanol-sensitive AP similar to mature megakaryocytes, suggesting that this enzyme provides an index of platelet formation by megakaryocytes. Soluble platelet AP showed three bands on polyacrylamide gel electrophoresis, visualized by the same reactions applied cytochemically. Two bands, accounting for 98% of the platelet AP activity, were slow moving and methanol-sensitive; and one fast moving band accounting for 2% of activity was methanol-resistant. Measurement of megakaryocyte and platelet AP isoenzymes may prove to have applications in evaluating megakaryocyte function.

References

Axline, G. 1968. Isoenzymes of acid phosphatase in normal and Calmette-Guerin bacillus induced rabbit alveolar macrophages. *J. Exp. Med.* 128:1031–1048.

Bentfield, M. and D. Bainton. 1975. Cytochemical localization of lysosomal enzymes in rat megakaryocytes and platelets. *J. Clin. Invest.* 56:1635–1649.

Burstone, M. 1958. Histochemical demonstration of acid phosphatase with naphthol AS phosphate. *J. Nat. Cancer Inst.* 21:523–540.

Burstone, M. S. 1961. Histochemical demonstration of phosphatases in frozen sections with naphthol AS-phosphates. *J. Histochem. Cytochem.* 9:146–153.

Daniell, H. W. 1959. Studies of megakaryocyte glycogen. I. A semiquantitative method of measurement: Effect of phlebotomies in young adults. *Blood* 14:61–73.

Engström, K., A. Lindquist, and N. Söderström. 1966. Periodic thrombocytopenia or tidal platelet dysgenesis in a man. *Scand. J. Haematol.* 3:290–294.

Forteza, B. G. and C. R. Baguena. 1960. Valor del contenido en glucogeno megakariocitico en las megakariocitos de la medula osea. *Rev. Clin. Esp.* 78:17–24.

Kaulen, H. D. and R. Gross. 1971. The differentiation of acid phosphatases of human blood platelets. *Thromb. Diath. Haemorrh.* 26:353–361.

Kolarz, G., H. Ottmeier, F. Singer, and H. Pietschmann. 1975. Cytochemical studies in megakaryocytes in hematologic diseases. *Folia Haematol.* (Leipz) 102:507–514.

Markovic, O. and N. R. Shulman. 1977. Megakaryocyte maturation indicated by methanol inhibition of an acid phosphatase shared by megakaryocytes and platelets. *Blood* 50:905–914.

Markovic, O. and N. Markovic, 1975. Cytochemistry of megakaryocytes, II, Glycogen, *IV Kongres medicinskih biochemicara Jugoslavije.* Sarajevo, Jugoslavije. pp. 98-103.

Markovic, O. 1976. Cytochemistry of megakaryocytes. II. Acid phosphatase. 1. Identification of the enzyme. *God. Zb. Med. Fak. Skopje* 22:91–97.

Markovic, O. 1976. Cytochemistry of megakaryocytes. II. Acid phosphatase. 2. Selection of the basic method for cytochemical detection. *God. Zb. Med. Fak. Skopje* 22:99–105.

Markovic, O. 1974. Platelet acid phosphatase isoenzymes. *Zbornik II Kongresa hematologa Jugoslavije* 1:801–812.

Ornstein, L. and B. Davis. 1962. In *Disc Electrophoresis*. Rochester, NY: Distillery Products Industries.

Reisfeld, A., G. Levis, and E. Williams. 1962. Disc electrophoresis of basic proteins and peptides on polyacrylamide gels. *Nature*, 195:281–283.

Wilkinson, T. and B. Firkin. 1966. Idiopathic cyclical acute thrombocytopenic purpura. *Med. J. Aust.* 1:217–219.

Wintrobe, M. M., G. R. Lee, D. R. Boggs, T. C. Bithell, J. W. Adams, and J. Foerster. 1974. In *Clinical Hematology*. Philadelphia: Lea and Febiger. p. 383.

Discussion

Dombrose: I just want to make a comment on acid phosphatase in relation to the age of the megakaryocytes. Dr. B. Schick gave a nice presentation on the lipid profiles. It's been discussed in the literature that as cells age they begin to accumulate more sphingomyelin. I noticed that on some of the distributions of the sphingomyelin that they were changing between megakaryocytes and platelets. I wonder if you think that shift might be an indicator for megakaryocyte age.

Levine: It would require a better fractionation for young megakaryocytes to answer that. It should be remembered that the megakaryocytes are young platelets; they are probably all young and not aged cells in the term of the life span of that cytoplasm.

Mayer: Did you look at the ploidy stages of your youngest megakaryocytes in that which the acid phosphatase appears?

Shulman: No, this was essentially a clinical study with bone marrow smears. Does anyone know whether the acid phosphatase could be considered to be in the alpha granule?

Bentfeld-Barker: Dorothy Bainton and I did a cytochemical/electron microscopy study, indicating that the acid phosphatase was not in the alpha granules, but in separate vesicles. Unfortunately, we couldn't demonstrate acid phosphatase in the mature cells by electron microscopy, like you have in marrow smears. There are other kinds of evidence for that. There is a storage pool disease in which some of the patients have normal levels of acid phosphatase and other lysozomal enzymes and are lacking alpha granules. Some of the cell fractionation studies also indicate that the peaks for lysozomal enzyme activities are not in the alpha granule fraction (*J. Clin. Invest.* 56:1635, 1975).

Evatt, Levine, and Williams, Editors
MEGAKARYOCYTE BIOLOGY AND PRECURSORS:
IN VITRO CLONING AND CELLULAR PROPERTIES

Identification of Two Classes of Human Megakaryocyte Progenitor Cells

Eric M. Mazur, Ronald Hoffman, Edward Bruno, Sally Marchesi, and Joel Chasis

Hematology Section, Department of Internal Medicine, Yale University School of Medicine, New Haven, Connecticut

The study of human megakaryopoiesis has suffered from the lack of specific histochemical or immunochemical markers for cells of megakaryocytic lineage. Increasing attention has recently been focused upon platelet glycoproteins (PGP) as platelet specific membrance components. Using a direct immunofluorescent technique, Rabellino et al. (1979) have shown that 97–98% of human megakaryocytes also contain PGP while other bone marrow hematopoietic elements do not. Thus, PGP may be used as a specific immunochemical marker for cells of the megakaryocyte series. We have applied this information in studying megakaryocytic precursor cells. A highly reactive antiserum was raised in rabbits immunized with a purified PGP preparation. This antiserum, in conjunction with indirect immunofluorescent staining, specifically and unequivocally labeled human bone marrow megakaryocytes, a population of small bone marrow mononuclear cells, and megakaryocyte colonies cloned *in vitro*.

Methods and Materials

Bone Marrow Smears

Bone marrow aspirations were collected in EDTA and smeared directly onto glass slides. They were fixed with acetone:methanol (9:1) for 20 min, washed, air dried, and stored frozen at −20°C.

Cell Cultures

Peripheral blood mononuclear cells from five normal volunteers were obtained by Ficoll-Hypaque density centrifugation. Cells were cultured in 1 ml volumes in 35 mm Petri dishes using a modified plasma clot technique as described by McLeod et al. (1976) in which heat inactivated human AB serum replaced fetal calf serum and alpha media (GIBCO, Grand Island, New York) replaced NCTC 109 and supplemented HMEM. Culture additives consisted of erythropoietin (Step III, Connaught Laboratories, Willowdale, Ontario), human embryonic kidney cell conditioned medium (donated by Dr. T. P. McDonald) (McDonald, et al., 1975) and a T-lymphoblast cell line conditioned medium (donated by Dr. David Golde) (Golde et al., 1980). After incubation at 37°C in a humidified atmosphere of 5% CO_2 in air, plasma clots were fixed *in situ* with methanol:acetone (1:3) for 20 min, washed, air dried and stored frozen at −20°C.

Antiserum Preparation

Purified human PGP was prepared by lithium diiodosalicylate-phenol extraction of pooled platelet concentrates as described by Marchesi and Chasis (1979). New Zealand white rabbits were immunized by subcutaneous or intravenous injections of 1 mg of glycoprotein emulsified in Freund's adjuvant at 0, 2, and 4 weeks. Antiserum was harvested at six weeks, and stored at −80°C.

Immunofluorescent Staining

Bone marrow or peripheral blood smears and culture dishes were incubated for 30–60 min at 37°C in 5% CO_2 humidified air with rabbit anti-PGP antiserum diluted in phosphate buffered saline (1/200). Specimens were re-incubated with fluorescein conjugated goat anti-rabbit IgG (Meloy, Springfield, Virginia), counterstained with 0.125% Evan's Blue, and coverslips mounted with isotonic barbital buffer, pH 8.6, in glycerol (1:3).

Bone marrow megakaryocytes were identified by their large size and intense fluorescence. Small fluorescein positive PGP-bearing cells were defined as intensely fluorescent cells with diameters between those of red cells and neutrophils with a nucleus visible by either fluorescent or phase microscopy. Between 150 and 250 fluorescein positive cells were counted on each bone marrow smear except where noted. Megakaryocytes were estimated as normal, increased, or decreased in numbers using specimens stained for conventional light microscopy. *In vitro* plasma clot cultures were scored *in situ*. The 35 mm Petri dish was inverted and completely scanned under epifluorescent illumination. Clusters of three or more strongly fluorescein labeled cells were defined as a megakaryocyte colony.

Results and Discussion

Antibody Specificity

When normal bone marrow and peripheral blood smears were examined, only platelets and megakaryocytes fluoresced intensely. Red cells, granulocytes, monocytes, and lymphocytes fluoresced little or not at all. An infrequent population of bone marrow cells the size of lymphocytes also showed significant fluorescence. Blast cells obtained from patients with several different lymphoid and myeloid leukemias as well as cells derived from the human leukemic blast cell lines HL-60 and K562 demonstrated no fluorescence. HL-60 is a promyelocytic leukemia cell line (Collins et al., 1978) while the K562 cell line has recently been shown to possess a number of erythroid phenotypic markers (Benz et al., 1980).

In Vivo Small Platelet Glycoprotein-Bearing Cells

Small nucleated bone marrow cells, approximately the size of lymphocytes, demonstrated intense fluorescence when labeled with anti-PGP antibody (Fig. 1). These cells were most commonly located in or near the marrow stromal particles, areas also demonstrating the highest density of large megakaryocytes.

A direct correlation existed between the estimated megakaryocyte number and the percentage of small PGP-bearing cells expressed as the fraction of total nucleated fluorescein positive cells. When normal numbers of megakaryocytes were present, 22.9 ± 2.0 %($\times$ $\pm$ SEM) of the total number of fluorescein positive cells were small cells. With increased numbers of megakaryocytes, these small PGP-bearing cells comprised $42.1 \pm 3.8\%$ of the total (Fig. 2). Bone marrow specimens with decreased numbers of megakaryocytes were hard to quantitate since the total numbers of PGP-bearing cells were so small as to make the numerator unreliable.

In Vitro Megakaryocyte Cultures

Intensely fluorescent PGP-bearing cellular colonies were successfully cultured from the peripheral blood of five normal volunteers (Fig. 3). 13 ± 6 megakaryocyte colonies were cloned per 10^6 mononuclear cells cultured. The number of mega-

Fig. 1. Anti-platelet glycoprotein labeled megakaryocyte and a small cell the size of a lymphocyte present on a human bone marrow smear (fluorescent illumination, 400×).

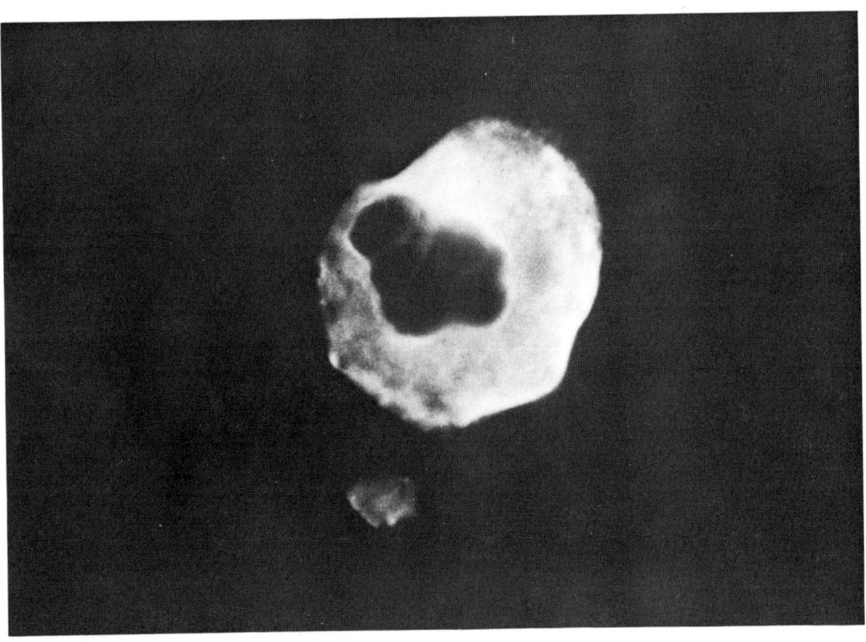

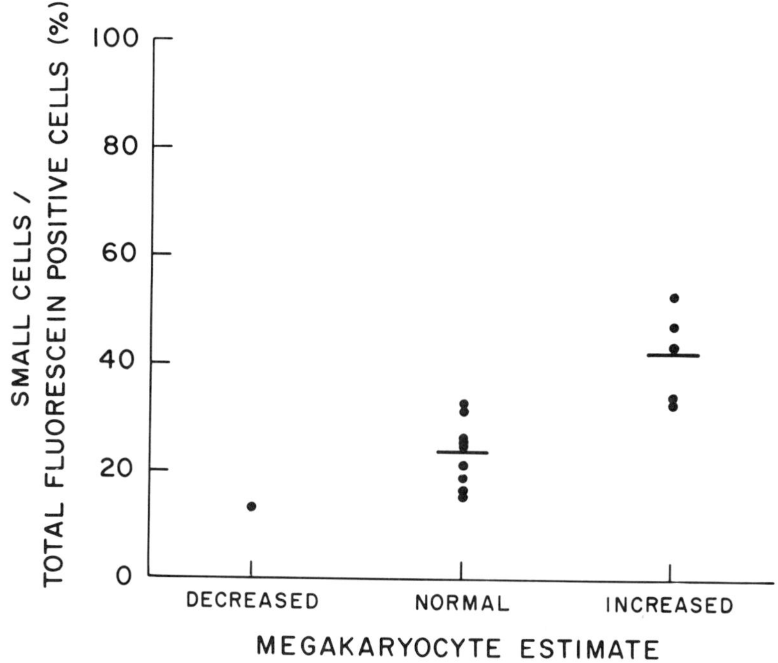

Fig. 2. Small platelet glycoprotein-bearing cells as the percentage of total fluorescein positive cells on human bone marrow smears. Each point represents a single patient. The single point for decreased megakaryocyte estimate is based upon a count of 8 cells.

karyocyte colonies demonstrated marked individual variation from donor to donor. Erythropoietin (Epo) (0.5–4 IU/ml) and human embryonic kidney cell medium (HEKM), a source of thrombopoietic activity (75–600 μg protein/ ml) (McDonald et al., 1975), did not further augment megakaryocyte colony formation (Fig. 4). Suppression of colony formation was observed at the higher concentrations of both substances.

The addition of medium conditioned by a T-lymphoblast cell line (TC-CM, final concentration 10%), a known source of burst promoting activity and colony stimulating activity (Golde et al., 1980), also failed to augment megakaryocyte colony formation (Fig. 4).

The number of cells present in each megakaryocyte colony was usually between 3 and 6. However, colonies consisting of up to 20 cells were also observed. Colonies were comprised of both large and small fluorescein positive cells, both in combination and alone. They were generally not mixed with other unlabeled cells (granulocytes, normoblasts, macrophages), although occasional mixed colonies of both fluorescein labeled and unlabeled cells were present.

Light microscopy of Giemsa-stained fluorescent colonies revealed both large cells with multilobulated nuclei which resembled megakaryocytes and small cells resembling mature lymphocytes. Pure colonies of both types of cells were seen.

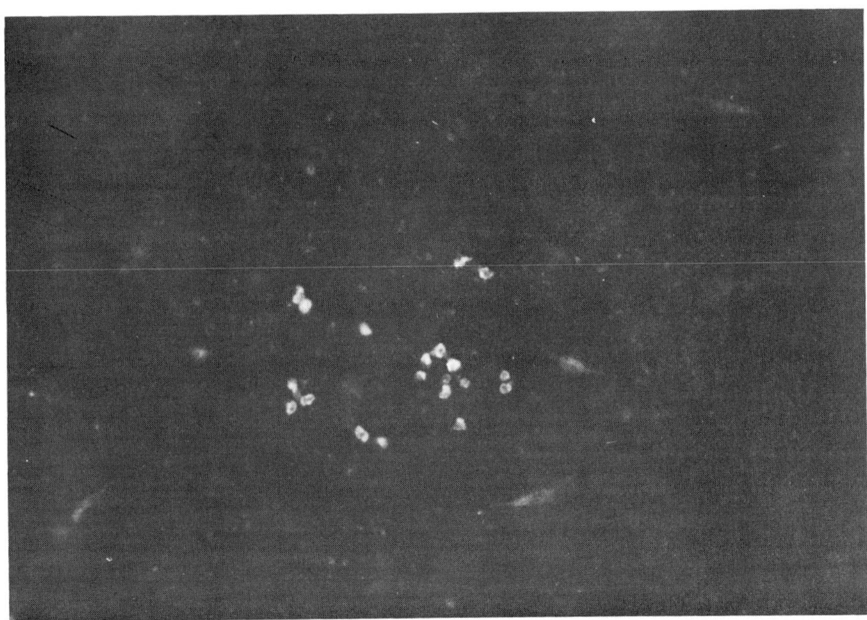

Fig. 3. A large, 20-cell fluorescent colony labeled with anti-platelet glycoprotein antibody grown in plasma clot culture (fluorescent illumination, 100×).

Antibody prepared to purified PGP provided a convenient and specific marker for cells of megakaryocytic lineage in man. When used in conjunction with indirect immunofluorescent staining, anti-PGP antibody identified a population of small nucleated cells not otherwise classifiable as megakaryocytes. Increased numbers of these small PGP-bearing cells were present when bone marrow smears manifesting megakaryocytic hyperplasia are examined. We believe these cells are analogous to the small acetylcholinesterase-positive cells previously described in rats (Jackson, 1973,1974). In rats, these cells are felt to represent late precursors of morphologic megakaryocytes whose numbers respond to perturbations in the peripheral platelet count (Jackson, 1973, Long and Henry, 1979). In man, whether these cells are true megakaryocyte precursors or merely represent 2N megakaryocytes is yet to be determined. However, since megakaryocyte ploidy tends to increase in the face of thrombopoietic stimulation (Odell et al., 1976), the increased numbers of small PGP-bearing cells found in conjunction with megakaryocyte hyperplasia suggests their identity is that of a megakaryocyte precursor.

PGP has also been useful as a probe for identifying *in vitro* megakaryocyte colonies grown in plasma clot cultures. Similar colonies grown from peripheral blood mononuclear cells have been previously described by Vainchenker and Breton-Gorius (1979). However, in that study, colony identification was based entirely on megakaryocyte morphology. The present study is free of the ambiguities inherent in such a morphological definition of megakaryocyte colonies. Sources of eryth-

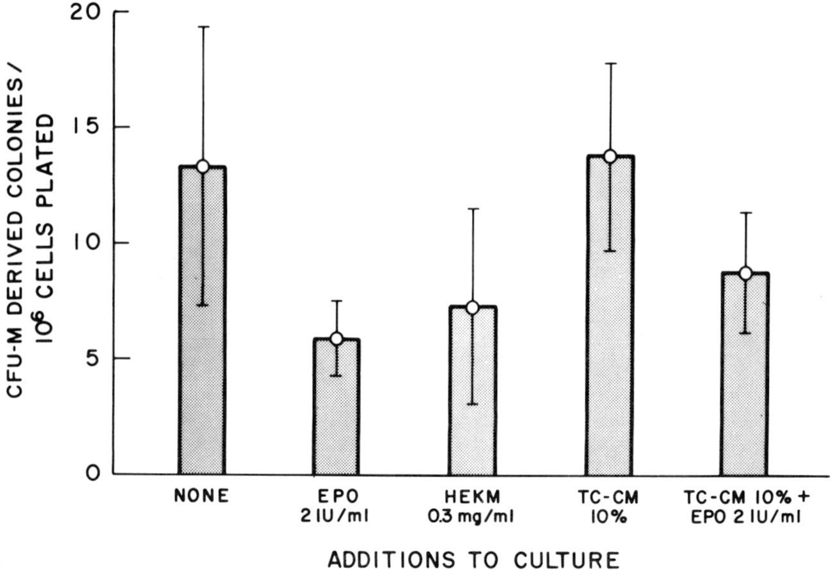

Fig. 4. CFU-M derived colony numbers cloned with and without culture additions as noted. Values represent the mean ± SEM.

ropoietin, thrombopoietin, burst promoting activity, and colony stimulating activity failed to significantly augment megakaryocyte colony formation. This suggests that these substances are not important physiologic regulators at the level of the CFU-M.

Indirect immunofluorescent staining of PGP-bearing cells in human bone marrow and *in vitro* plasma clot cultures has successfully demonstrated two types of megakaryocyte progenitor cells. The first is the human analogue of the small acetylcholinesterase-positive cells in rats. These small nucleated PGP-bearing cells identified in normal marrow are probably the derivatives of the CFU-M that gradually differentiate into megakaryocytes. The second is the human CFU-M whose progeny can be unequivocally identified and quantitated in culture. The ability to detect and quantitate these two megakaryocyte progenitor cells will be helpful in defining regulatory factors at different levels of megakaryocyte differentiation in man.

Summary

Identification of human megakaryocyte progenitor cells has been delayed by the lack of a specific histochemical marker. For this purpose we have raised an antibody in rabbits directed against a purified human platelet glycoprotein (PGP) preparation. Using indirect immunofluorescent staining of human bone marrow (BM) and peripheral blood (PB), megakaryocytes fluoresced intensely and uniquely except for an infrequent population of small mononuclear cells limited to the BM. The frequency of these small cells varied directly with the number of megakaryocytes present. We believe they are the human counterpart of similar cells described in rats which represent late precursors of morphologically identifiable megakaryoblasts. A second megakaryocyte progenitor cell PC (CFU-M) was identified by culturing PB mononuclear cells from normal volunteers in plasma clot. Plates were harvested after 7–14 days and doubly labeled with rabbit anti-PGP antiserum and fluorescein tagged goat anti-rabbit IgG. CFU-M derived colonies containing small and large fluorescein positive cells were present at day 7 peaking between days 10 and 12. Positive colonies were defined as those fluorescing with moderate to bright intensity. Two types of positive colonies were observed: pure colonies of fluorescein labeled cells, and mixed colonies containing both labeled and unlabeled cells. Many negative colonies were present in all plates. The ability to identify these two megakaryocyte progenitor cells will be helpful in defining regulatory factors at different levels of megakaryocyte differentiation in man.

ACKNOWLEDGMENTS

Supported in part by USPHS Grants CA22697 and HL07262 from the National Institutes of Health. Dr. Hoffman is a recipient of a Research Career Development Award.

References

Benz, E. J., Jr., M. J. Murnane, B. L. Tonkonow, B. W. Berman, E. M. Mazur, C. Cavallesco, T. Jenko, E. L. Snyder, B. G. Forget, and R. Hoffman. 1980. Embryonic-fetal erythroid characteristics of a human leukemia cell line. *Proc. Nat. Acad. Sci. USA* 77:3509–3513.

Collins, S. J., F. W. Ruscetti, R. E. Gallagher, and R. C. Gallo. 1978. Terminal differentiation of human promyelocytic leukemia cells induced by dimethyl sulfoxide and other polar compounds. *Proc. Nat. Acad. Sci. USA* 75:2458–2462.

Golde, D. W., N. Bersch, S. G. Quan, and A. J. Lusis. 1980. Production of erythroid-potentiating activity by a human T-lymphoblast cell line. *Proc. Nat. Acad. Sci. USA* 77:593–596.

Jackson, C. W. 1973. Cholinesterase as a possible marker for early cells of the megakaryocytic series. *Blood* 42:413–421.

Jackson, C. W. 1974. Some characteristics of rat megakaryocyte precursors identified using cholinesterase as a marker. In *Platelets: Production, Function, Transfusion, and Storage*. Baldini, M. G., and Ebbe, S., eds. New York: Grune and Stratton. pp. 33–40.

Long, M. W. and R. L. Henry. 1979. Thrombocytosis-induced suppression of small acetylcholinesterase-positive cells in bone marrow of rats. *Blood* 54:1338–1346.

McDonald, T. P., R. Clift, R. D. Lange, C. Nolan, I. I. E. Tribby, and G. H. Barlow. 1975. Thrombopoietin production by human kidney cells in culture. *J. Lab. Clin. Med.* 85:59–66.

McLeod, D. L., M. M. Shreeve, and A. A. Axelrad. 1976. Induction of megakaryocyte colonies with platelet formation *in vitro*. *Nature* 261:492–494.

Marchesi, S. L. and J. A. Chasis. 1979. Isolation of human platelet glycoproteins. *Biochim. Biophys. Acta*. 555:442–459.

Odell, T. T., J. R. Murphy, and C. W. Jackson. 1976. Stimulation of megakaryocytopoiesis by acute thrombocytopenia in rats. *Blood* 48:765–775.

Rabellino, E. M., R. L. Nachman, N. Williams, R. J. Wincester, and G. D. Ross. 1979. Human megakaryocytes. I. Characterization of the membrane and cytoplasmic components of isolated marrow megakaryocytes. *J. Exp. Med.* 149:1273–1287.

Vainchenker, W. and J. Breton-Gorius. 1979. Differentiation and maturation *in vitro* of human megakaryocytes from blood and bone marrow precursors. In *Cell Lineage, Stem Cells, and Cell Determination*. INSERM Symposium No. 10. Le Douarin, N., ed. New York: Elsevier North-Holland. pp. 215–226.

Vainchenker, W., J. Guichard, and J. Breton-Gorius. 1979. Growth of human megakaryocyte colonies in culture from fetal, neonatal, and adult peripheral blood cells: Ultrastructural analysis. *Blood Cells* 5:25–42.

Discussion

Breton-Gorius: Have you looked with your antibody staining at cases of blast crisis of chronic myeloid leukemia, to determine if such small cells are present, as we have shown with platelet peroxidase staining?

Mazur: We have looked to date at only one patient with a transformed myeloproliferative disorder which was not chronic myelogenous leukemia. The preliminary data is that it does appear that there are small circulating cells which contain platelet glycoprotein. However, we are not sufficiently confident of those results yet to make a definite statement.

Breton-Gorius: It is very strange that you have not obtained increased numbers of megakaryocyte colonies after addition of erythropoietin. In our experience, we have tested different sources of erythropoietin, and each time we have found a dose response to different amounts of erythropoietin with the number of megakaryocyte colonies.

Mazur: We were well aware of your results on the dose responsiveness to erythropoietin and I think one possible explanation may be the stage of megakaryocyte differentiation at which we are able to identify the megakaryocyte colonies. We only require the acquisition of platelet glycoprotein in the component cells; therefore, we are able to detect colonies which are otherwise not morphologically identifiable as megakaryocytes, so perhaps we are detecting colony cells which are then influenced in late differentiation by erythropoietin. That would explain your results, as well as ours.

Zuckerman: I wonder if there may be differences in the serum lots between yours and Dr. Breton-Gorius'. Another possibility is that she cultures nonadherent cells. One could postulate that maybe you are looking at a maximally stimulated state and that with other cells there that are able to make stimulators, no further increase can be detected.

Mazur: In the absence of erythropoietin, we see absolutely no erythroid colony formation, so that we certainly do not have from the plasma sufficient erythropoietin to cause that kind of differentiation. One of the differences in our assay system is that our culture media was enriched with nutrients, including additional amounts of L glutamine, minimal essential amino acids; we only saw megakaryocyte colony formation when there was exuberant growth, spontaneous growth, of other colony types, especially granulocyte-macrophage.

Mayer: Did you use platelets of one donor?

Mazur: They were actually blood bank-derived pooled platelets, which were then put through this complicated extraction procedure.

Mayer: Did you use further purification or absorption steps for nonspecific antibodies?

Mazur: We did at times and found it made no difference.

Weil: How did you control for nonspecific binding through the Fc receptor of both your first and second antibodies?

Mazur: We control by looking at a nonimmune rabbit serum on human bone marrow and found that there was no staining of megakaryocytes in the system we used. Interestingly, when we examined patients who had antiplatelet antibodies, we did find significant nonspecific binding with nonimmune human serum, but apparantly the species difference allowed us to circumvent that problem.

Long: Did the megakaryocytic progenitor cell that you isolated in the peripheral blood stain with the immune serum?

Mazur: No, it did not. We looked at several cytospins of the mononuclear cells which were subsequently cultured and did not see these cells. However, their frequency is so low that I am not sure that they weren't there. We would like to examine that possibility with a cell sorter.

Long: Are there lots of serums that do not show spontaneous colonies? Is this a phenomenon of all serum batches?

Mazur: We have tried several different human AB serums. We had at least 6 different donors and used a single donor for a batch of experiments and we see spontaneous colony formation in all of them. What seems to be the common denominator is that once we enriched the media we began to see this spontaneous colony formation.

Long: Have you done anything about trying to isolate and stimulate the small cells that you detect in the bone marrow to see if you can get them to either proliferate or mature into megakaryoctyes?

Mazur: Not yet.

Rabellino: How do you establish the specificity of your antiserum in terms of the different types of platelet glycoproteins?

Mazur: This is done by Dr. Chasis, who had been interested in this antiserum for other reasons. The immunogen forms three major glycoprotein bands on acrylamide gel. What she has done is to incubate the antiserum with acrylamide gels of this glycoprotein preparation and then add on top of that, radioiodinated Staph Protein A, and then (she) performed autoradiographs. Three platelet glycoprotein bands were labeled, using this antiserum.

Rabellino: Have you estimated the incidence of your small mononuclear cells bearing glycoproteins?

Mazur: Well, only by adding some of your results and ours. We found that in the normal human bone marrow — and these are very crude estimates, admittedly — about 20% of all fluorescent positive cells or platelet glycoproteins bearing cells were small cells. However, if the incidence is 0.5% of all bone marrow cells is megakaryocytes, we come up with an incidence of about 0.1%. This is consistent with the work of others.

Levin: What do you mean by two classes of human megakaryocyte colonies?

Mazur: Pure and mixed colony types. Mixed colonies were rare, but megakaryocytes were found with granulocytes and macrophages. I think that we should emphasize that, subsequent to submitting our abstract, the vast majority of the colonies that we saw were pure and we specifically did look for the megakaryocyte-erythroid mixed colonies described by others and we haven't been able to find them.

Levin: Could the mixed colony just be the superimposition of a megakaryocyte and granulocyte colony?

Mazur: Yes.

Levin: What did you mean by the two classes of human megakaryocytes?

Mazur: Well, we are talking about the first one being the small cell platelet glycoprotein-bearing cell that we identified in bone marrow, and I know others have done a great deal of work with it in the rat and I think that this is the human counterpart. The second one is assayed as a colony of identifiable cells.

Levine: Do you think that either of those are young megakaryocytes or something that is before the differentiation step?

Mazur: You mean the small platelet glycoprotein-bearing cells? We can only hypothesize. I think that the fact that these cells are increased in states of megakaryocyte hypoplasia would suggest that they are a precursor cell rather than, say, a 2N megakaryocyte, since many people have shown the shift in ploidy to higher levels with thrombopoietic stimulation.

Levine: But there could be increased influx, too.

Jackson: I agree that these small cells are related to cholinesterase cells. As you know, in 1974 we reported that the small cholinesterase-positive cells also fluoresced when labeled with an antibody to whole platelets in rats. Anti-rat platelet antibodies cross-react with human platelets.

Mazur: It was your work that encouraged us to look for these cells.

Evatt, Levine, and Williams, Editors
MEGAKARYOCYTE BIOLOGY AND PRECURSORS:
IN VITRO CLONING AND CELLULAR PROPERTIES

Relationship of Small Acetylcholinesterase Positive Cells to Megakaryocytes and Clonable Megakaryocytic Progenitor Cells

Michael W. Long and Neil Williams

Sloan-Kettering Institute for Cancer Research, Rye, New York

Certain indirect evidence has led to the assumption that the bone marrow contains a small cell which is restricted to the megakaryocyte compartment. In certain species these cells, along with megakaryocytes, are positive for acetylcholinesterase (ACh-E) activity (Zajicek, 1954; Jackson, 1973; Long and Henry, 1979). These small ACh-E positive cells (SACHE) are elevated following the administration of anti-platelet antisera (Jackson, 1973) and are depressed following transfusion-induced thrombocytosis (Long and Henry, 1979). Further, it has been shown that these cells have buoyant density characteristics similar to megakaryocytes (Nakeff and Floeh, 1976). Finally, small cells which are ACh-E positive have been observed in megakaryocyte colonies where a full spectrum of megakaryocyte maturation would be expected (Williams and Jackson, 1978).

To date, no evidence has been obtained to link the SACHE directly to the megakaryocyte series. In this study, cell sedimentation characteristics of SACHE were determined. Fractions enriched for SACHE were cultivated with subsequent formation of mature megakaryocytes

Methods and Materials

Velocity Sedimentation

The cells used in these experiments were obtained from the femurs of C57BL/6 mice. They were suspended in PBS and kept at 4°C throughout the separation procedure. Velocity sedimentation at unit gravity was performed according to the method of Miller and Phillips (1969). Cells (30 ml at $8.4 \pm 1.5 \times 10^6$/ml) were loaded into a cylindrical sedimentation chamber and were sedimented through a 1%–2% bovine serum albumin gradient for 2.75 hr. Cells in the collected fractions were recovered by centrifugation for 7 min at 800 × g. The pellet was resuspended in serum-free McCoy's 5A medium. The cells were counted on a hemocytometer. An aliquot was removed and stained for ACh-E activity (Jackson, 1973). In experiments where clonable megakaryocyte progenitor cells (CFU-MK) profiles were determined, cells were sedimented for approximately 4 hr before fraction collection. The assay for CFU-MK was done in the presence of WEHI-3 conditioned medium and mouse lung conditioned medium as described elsewhere (Williams and Jackson, 1978).

In Vitro Growth of Small Acetylcholinesterase Positive Cells (SACHE)

Since SACHE have a diameter of 8–12 μm and a density of approximately 1.1 g cm^{-3} (Nakeff, 1976), the fractions were pooled into three ranges of rapid sedimentation values: (I) 8.2–9.1 mm hr^{-1}; (II) 11.2–14.7 mm hr^{-1} and (III) 15.2–18.4 mm hr^{-1}. The SACHE were tested under the same culture conditions used in this laboratory for the growth of CFU-MK (Williams et al., 1978). Each pooled fraction was tested for response to Thrombopoietic Stimulatory Factor (TSF; 250 μg). Thus, each plate contained 100 μl pooled SACHE fraction, 25 μl stimulus, and 1.1 ml of agar/media (final volume 1.2 ml). Three replicate cultures were tested with and without stimulus. Cultures were incubated 2.5 days at 37°C in 7% CO_2. Additional plates were immediately dried and stained (Williams et al., 1980) to determine input megakaryocyte levels.

Results and Discussion

Velocity Sedimentation

The normalized velocity sedimentation profiles of 5 experiments are seen in Figure 1. The SACHE were found to sediment at a rate of 8.2–18.4 mm hr^{-1}. The peak value was 11.2 mm hr^{-1}, a result in good agreement with the theoretical value obtained from the diameter and density of the cells. The peak fractions contained 20–30% SACHE representing an approximate 100-fold enrichment over unfractionated marrow cells. The marrow contains approximately 0.18% SACHE. This concentration is much higher than the numbers of SACHE previously reported (Jackson, 1973; Long and Henry, 1979).

Figure 1 shows that, based on their size and density, SACHE have a high sedimentation velocity. This property allowed the separation and enrichment of these cells from the bulk of the marrow cells including CFU-MK. The sedimentation velocity characteristics are presumably highly influenced by the density component of the SACHE. An immature bone marrow cell of approximately 10 μm diameter and average buoyant density would not sediment faster than 8.0 mm hr^{-1}. This suggests that SACHE have either a highly condensed nucleus (e.g., a small cell with increased ploidy values and little cytoplasm) and/or that the beginnings of increased cytoplasmic activity (e.g., production of organelles, ACh-E activity, etc.) increases cellular density. Either explanation would imply that SACHE are the most immature of the megakaryocytes.

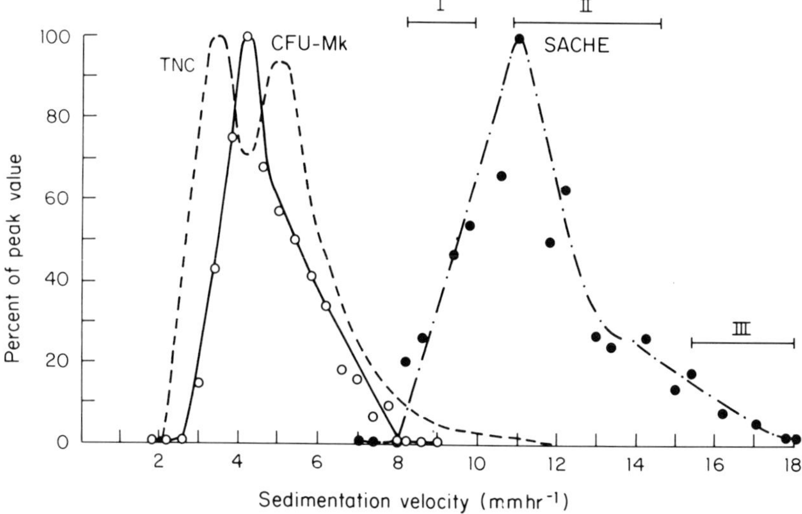

Fig.1. Separation of small acetylcholinesterase positive cells (SACHE) from normal bone marrow. SACHE (open circles) were separated from megakaryocyte progenitor cells (CFU-MK, closed circles) and the majority of total nucleated cells (TNC, dotted line). The normalized profiles of five individual experiments are shown. The yield of TNC was 85 ± 13 percent and CFU-MK 72 ± 18 percent (mean ± S.D.). The yield of SACHE was not calculated due to low input percentages (0.18% SACHE).

Response of Small Acetylcholinesterase Positive Cells to Thrombopoietic Stimulatory Factor

SACHE were observed to mature into single megakaryocyte in the presence of TSF (Fig. 2). Fractions I and II were both sensitive to the concentration of TSF tested. These fractions yielded 44.6 ± 18 and 58.3 ± 14 MK, respectively, per 10^3 SACHE plated (the input megakaryocytes were 2.0 and 4.0 megakaryocytes per plate, respectively). It should be noted that limited numbers of megakaryocytes did occur in the absence of stimulation by TSF. This response was dependent on the lot of fetal calf serum being used. However, megakaryocytes were derived in culture conditions where no spontaneous megakaryocyte maturation was observed.

It remains uncertain whether TSF influences cellular division as well as megakaryocyte maturation. Kalmaz and McDonald (Chapter 6, this volume) have shown limited proliferation of a megakaryocyte progenitor cell in the presence of TSF. However, colony size was not as great as that observed with other sources of MK-CSF. Williams et al. (1979) were unable to show that TSF could induce colony formation although TSF actively enhanced colony formation when used in conjunction with a source of megakaryocyte colony stimulating factor. The difference in the two findings may result from the relative contributions of SACHE to the two assays, since there are about 180 SACHE per 10^5 nucleated cells plated. The results

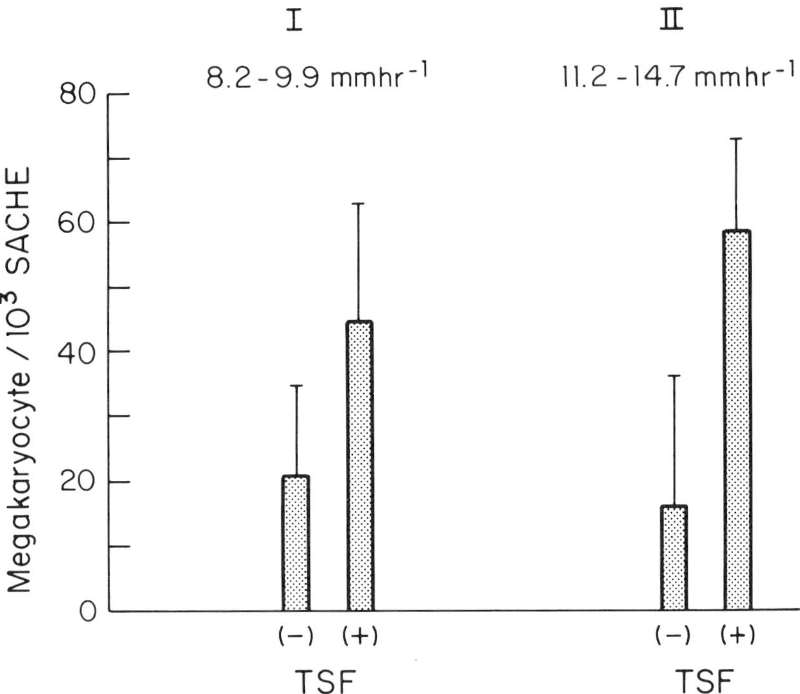

Fig. 2. Response of small acetylcholinesterase positive cells (SACHE) to Thrombopoietic Stimulatory Factor (TSF). Values are mean ± S.E.M. based on five experiments. Three replicate cultures were done in the presence and absence of TSF. Values are corrected for background (input at time zero) megakaryocytes.

in this study imply that TSF directly influenced megakaryocyte maturation from SACHE without further cell proliferation. A few two-cell aggregates were observed but the majority of the megakaryocytes were found as single cells. The potential of SACHE to undergo cellular division as opposed to nuclear endoreduplication remains to be investigated. However, preliminary data suggests that few, if any, colonies can be grown from these cell fractions.

The results presented here have demonstrated that the SACHE are part of the megakaryocyte compartment. Additionally, these cells may be biased toward further differentiation rather than cellular division.

Summary

Small acetylcholinesterase positive cells (SACHE) were grown *in vitro* into single mature megakaryocytes. This cellular maturation was stimulated by addition of TSF. The SACHE were found to be more rapidly sedimenting (8.2–18.4 mm hr^{-1}) than the majority of bone marrow cells and the megakaryocyte progenitor cell

(CFU-MK). This separation allows an approximate 100-fold enrichment of SACHE from unfractionated bone marrow cells.

ACKNOWLEDGMENTS
Thrombopoietic Stimulatory Factor (TSF) was a generous gift of Dr. T. P. McDonald. M. Long is a Leukemia Society of America Fellow. N. Williams is a Leukemia Society of America Scholar. This work was funded by Grants HL 22451 from NIH, CH-3C from the American Cancer Society and the Gar Reichman Fund.

References

Jackson, C. W. 1973. Cholinesterase as a possible marker for early cells of the megakaryocytic series. *Blood* 42:413–421.

Kalmaz, G. D. and T. P. McDonald. 1980. The effects of thrombopoietin on megakaryocytopoiesis of mouse bone marrow cells *in vitro*. In *Megakaryocytes In Vitro*. Evatt, B. L., Levine, R. F., and Williams, N., eds. New York: Elsevier North Holland, pp. 77-86.

Long, M. W. and R. L. Henry. 1979. Thrombocytosis induced suppression of small acetylcholinesterase-positive cells in bone marrow of rats. *Blood* 54:1338–1346.

Miller, R. G. and R. A. Phillips. 1969. Separation of cells by velocity sedimentation. *J. Cell. Physiol.* 73:191–201.

Nakeff, A. and D. P. Floeh. 1976. Separation of megakaryocytes from mouse bone marrow by density gradient centrifugation. *Blood* 48:133–139.

Williams, N. and H. Jackson. 1978. Regulation of the proliferation of murine megakaryocytic progenitor cells by cell cycle. *Blood* 52:163–170.

Williams, N., H. Jackson, A. P. C. Sheridan, M. J. Murphy, Jr., A. Elste, and M. A. S. Moore. 1978. Regulation of megakaryopoiesis in long–term bone marrow cultures. *Blood* 51:245–255.

Williams, N., T. P. McDonald, and E. M. Rabellino. 1979. Maturation and regulation of megakaryocytopoiesis. *Blood Cells* 5:43–55.

Williams, N., H. Jackson, R. R. Eger, and M. W. Long. 1980. The different roles of factors in murine megakaryocyte colony formation. In *Megakaryocytes In Vitro*. Evatt, B. L., Levine, R. F., and Williams, N., eds. New York: Elsevier North Holland, pp. 59-75.

Zajicek, J. 1954. Studies on the histogenesis of blood platelets. I. Histochemistry of acetylcholinesterase activity of megakaryocytes and platelets in different species. *Acta Haematol.* 12:238–244.

Discussion

Jackson: How long does this differentiation take in culture?

Long: An incubation time of 60 hr was used.

Ebbe: Why do you think the plating efficiency is so low; you plate a thousand, you get 60 megakaryocytes?

Long: Some of these cells are out of cycle and some of them are in cycle. Some would not come into cycle and might stay in a quiescent state. However, I think it is mostly a matter of proper nutritional requirements in the culture dish itself.

Hempling: These small cells with the high sedimentation velocities, did you run them through a particle size analyzer to get a measure of their mean corpuscular volumes in each of the fractions?

Long: No, we haven't done that.

Levine: There may be a stage in the megakaryocyte precursors after the proliferative phase, but before differentiation. Such an inbetween stage may be the target cells for thrombopoietin. It may act continuously on all or most differentiated megakaryocytes, as Drs. Evatt and Kellar have suggested, and/or perhaps the effect is programmed at one point in time to control the ultimate ploidization, as Dr. Paulus and Levin suggested.

Evatt, Levine, and Williams, Editors
MEGAKARYOCYTE BIOLOGY AND PRECURSORS:
IN VITRO CLONING AND CELLULAR PROPERTIES

A New Method for the Analysis of Megakaryocyte Kinetics by the Fluorescence Activated Cell Sorter

M. Mayer, W. Queisser, and A. Stöhr

Center of Clinical Onkology, Faculty of Clinical Medicine, Mannheim, and Deutsches Krebsforschungszentru, Heidelberg, Federal Republic of Germany

Study of the megakaryocytic cell system suffers from the relatively small numbers of morphologically recognizable megakaryocytes and from inability to distinguish the megakaryocyte precursor cells by light microscope criteria. Among the possible specific staining techniques the demonstration of platelet-specific antigens by immunofluorescence seems to be highly specific for megakaryocytes (Mayer and Schaefer, 1978; Pretlow and Stinson, 1976).

This study reports the use of the fluorescence activated cell sorter FACS II (Becton Dickinson) as a new tool for the separation of megakaryocytes by simultaneous membrane and DNA-fluorescence. Additionally, this method was tested for use in purification and concentration of megakaryocytes.

Methods and Materials

Specific rabbit anti-mouse platelet serum (RAMPS) was prepared by immunizing rabbits with washed mouse platelets as described earlier (Mayer and Schaefer, 1978). The specifity of the serum was tested by (1) induction of an isolated thrombocytopenia in mice after IV injection, (2) by absence of immunofluorescence with FITC labeled anti-rabbit gamma-globulin of nucleated blood cells whereas platelets showed a bright green fluorescence, and (3) by absence of positive fluorescence following adsorption of RAMPS with washed mouse platelets.

Preparation of Bone Marrow

The femora of the mice were flushed with 4 ml of cold tissue culture medium TC 199 (GIBCO) and a single cell suspension was prepared by repeated aspiration with a syringe. RNA was removed by RNAse treatment at a final concentration of 0.2

mg/10^7 cells. After washing in TC 199, the cell suspension was incubated with RAMPS for 30 min at 4°C. After removal of RAMPS, the cells were fixed with a graded concentration of ethanol. After another washing step, the cell suspension was incubated with FITC labeled anti-rabbit gammaglobulin for 30 min at 20°C. The DNA was stained specifically with propidium iodide at a concentration of 5 mg/100 ml.

The sample was applied to the FACS II with a single laser excitation of 488 nm wavelength and simultaneous red and green fluorescence detection. FITC fluorescence was measured using a bandpass filter ranging from 516 to 536 nm. The red propidium iodide fluorescence was measured by a barrier filter with a cut off at 620 nm. The signals were logarithmically amplified in both channels to compress the signal dynamics. Cells positive for FITC fluorescence were separated onto a slide for subsequent morphologic analysis.

Results and Discussion

The percentage of immunofluorescence positive cells in the bone marrow ranges from 0.1 to 0.9%. Figure 1 gives the perspective plot of simultaneous FITC and propidium iodide fluorescence and frequency distribution of the whole bone marrow suspension. On the DNA axis, peaks appear at the diploid and tetraploid stages

Fig. 1. Perspective plot with hidden line suppression of FITC – fluorescence (z-axis), propidium iodide fluorescence (x-axis), and frequency distribution (y-axis). The immunofluorescence positive cells show a DNA-distribution ranging from 2c up to 32c.

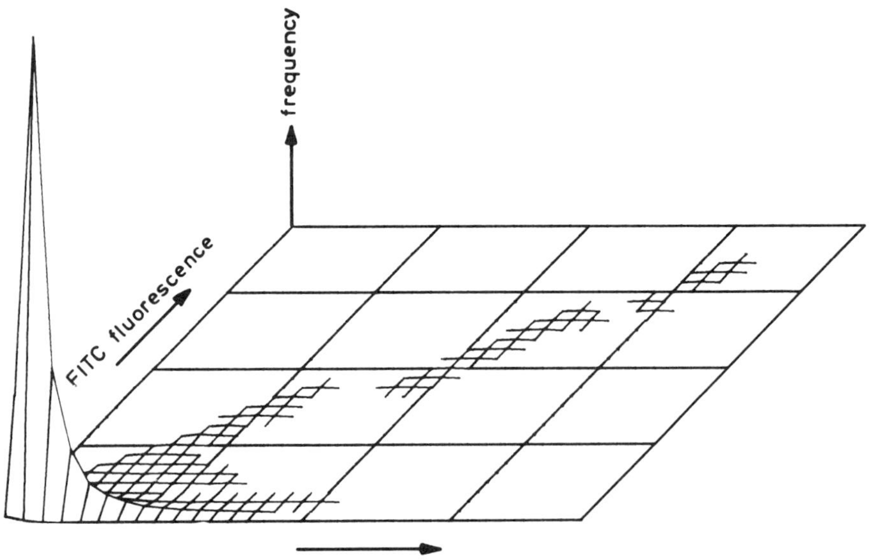

representing nonfluorescent bone marrow cells. The higher DNA values represent cells aggregated by ethanol fixation.

The cells with high FITC fluorescence show DNA values ranging from 2c up to 32c. Figure 2 gives the frequency distribution of the DNA values of immunofluorescence positive cells with peaks, especially at the 8c and 16c level. However, a considerable number of cells show DNA values between 2c and 4c. The Pappenheim morphology of these cells resembles immature cells with a small cytoplasm and a round nucleus. The cells with higher DNA values show the morphology of megakaryocytes with large nuclei. However, the morphologic characteristics of the separated cells are not as clear cut as in freshly prepared smears. Therefore, the morphologic appearance has to be interpreted with care.

The enrichment ratio of the separated cells ranges from 89%–95%. The contaminating cells are mononuclear cells, erythroblasts, and granulocytes. Some of these nonfluorescent cells are attached to the megakaryocytes. Fluorescent platelets or megakaryocyte cytoplasmic fragments are adhering to a few cells, so that they may be sorted falsely as "megakaryocytes."

Experimental data suggest that the megakaryocytic cell system is continously fed from a precursor cell compartment which itself is the progeny of a so-called pluripotent stem cell (Ebbe and Stohlman, 1965). In stained smears, these progenitor cells are not recognizable. Several attempts have been made to characterize the megakaryocytic series by specific staining techniques. The periodic acid Schiff's reagent is positive in megakaryocytes and probably in the precursor cells too, but not specific. A specific enzyme is the platelet peroxidase (Vainchenker et al., 1979) demonstrable in sections of both mature and immature megakaryocytes. A further

Fig. 2. DNA-pattern of immunofluorescence positive cells (upper part) compared to the DNA distribution of panoptically identifiable megakaryocytes (lower part). The DNA is arbitrarily chosen as "arbitrary units" (AE).

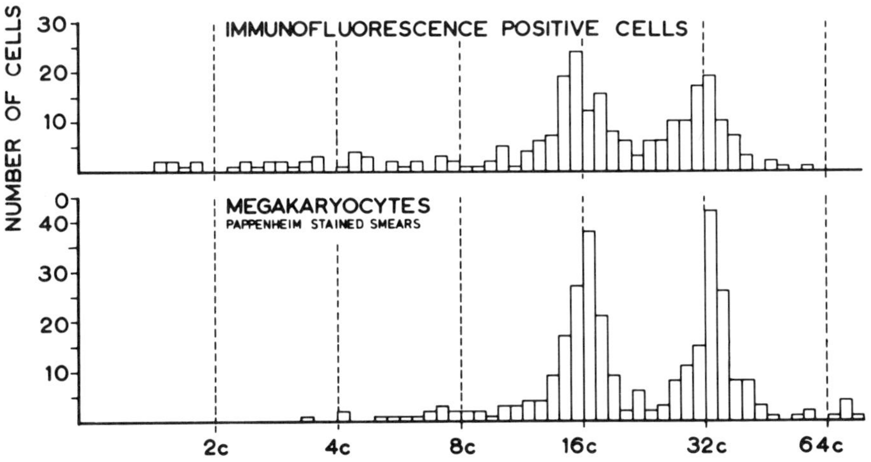

specific staining technique is the acetylcholinesterase reaction (ACH-E) in the megakaryocytes of the rat and mouse (Jackson, 1974). The smaller ACH-E positive cells are presumed to be early cells in the megakaryocytic series and their number is involved in the feedback control of the platelet count (Long and Henry, 1979). The demonstration of membrane receptors for Fc(IgG) and the immunofluorescent detection of the Ia antigen enables the recognition of megakaryocytic precursors in man (Williams et al., 1979). Using immunofluorescence, it is possible to demonstrate platelet specific antigens on megakaryocytes (Humphrey, 1955; Jackson, 1974). As reported earlier, these antigenic structures seem to be expressed in diploid and tetraploid cells that are morphologically not identifiable as megakaryocytes in Pappenheim stained smears (Mayer et al., 1978).

However, the relative low content of 0.1–0.9% of megakaryocytes makes analysis of cell kinetics difficult. Several techniques for the enrichment of megakaryocytes have been applied, i.e., velocity sedimentation (Pretlow and Stinson, 1976; Nakeff and Floeh, 1976) or discontinuous density albumin gradients (Nakeff and Floeh, 1976). The enrichment ratio obtained was approximately 1% (0.8% – 2.4%). By a two step purification technique, the results are much better (Levine and Fedorko, 1976). However, these techniques are based on physical properties of the cell, i.e., cell size or density. As the megakaryocytes are rather inhomogeneous with regard to these parameters, it cannot be ruled out that certain cell fractions will be excluded. In the present study, the FACS II was used for simultaneous determination of immunofluorescence and DNA content. Cells fulfilling certain criteria, i.e., positive immunofluorescence, can be sorted out for further evaluation. The purification and concentration up to 95% makes this method useful for research in megakaryocytopoiesis. However, further investigations have to clarify whether the preparation technique may alter certain cell functions.

The simultaneous measurement of the DNA content of the immunofluorescence positive cells that are presumably megakaryocytes and perhaps including the megakaryocytic precursor cells enables the possibility of an automatic DNA profile analysis. The DNA pattern is similar to that of Pappenheim stained megakaryocytes but further investigations are necessary to test this method under pathological conditions.

Summary

Megakaryocytic cells can be detected by demonstration of platelet specific antigens by immunofluorescence. For analysis of unfractionated mouse megakaryocytes, the bone marrow cells were labeled with highly specific antiplatelet-antibodies prepared by immunization of rabbits, repeatedly injected with mouse platelet suspension. Platelet antigen containing cells were demonstrated with FITC labeled anti-rabbit-gammaglobulin using the "sandwich" technique. After pretreatment with RNAse, the DNA was specifically stained with propidium iodide. The cell suspension was processed by a fluorescence activated cell sorter and FITC positive cells were selected. Simultaneously the DNA content was measured. The separated cells showed a DNA distribution similar to the polyploidization pattern in single cell photometry, reaching from 2c up to 32c. The morphology of the polyploid cells was that of megakaryocytes. The lower ploidy stages were immature cells, presumably megakaryocyte precursors or immature megakaryocytes.

The described method enables a rapid DNA analysis and separation of megakaryocytes in a one-step procedure.

References

Ebbe, S. and F. Stohlman, Jr. 1965. Megakaryocytopoiesis in the rat. *Blood* 26:20–35.

Humphrey, J. H. 1955. Origin of blood platelets. *Nature* 176:38.

Jackson, C. W. 1974. Some characteristics of rat megakaryocyte precursors identified using cholinesterase as a marker. In *Platelets: Production, Function, Transfusion, and Storage*. Baldini, M. G., and Ebbe, S. eds. New York: Grune and Stratton. pp. 33–40.

Levine, R. F. and M. E. Fedorko. 1976. Isolation of intact megakaryocytes from guinea pig marrow. Successful harvest made possible with inhibitors of platelet aggregation; enrichment achieved with a two-step separation technique. *J. Cell Biol.* 69:159–172.

Long, M. and R. L. Henry. 1979. Thrombocytosis-induced suppression of small acetylcholinesterase-positive cells in bone marrow of rats. *Blood* 54:1338–1346.

Mayer, M., J. Schaefer, and W. Queisser. 1978. Identification of young megakaryocytes by immunofluorescence and cytophotometry. *Blut* 37:265–270.

Nakeff, A. and D. P. Floeh. 1976. Separation of megakaryocytes from mouse bone marrow by density gradient centrifugation. *Blood* 48:133–138.

Pretlow, T. G. and A. J. Stinson. 1976. Separation of megakaryocytes from rat bone marrow cells using velocity sedimentation in an isokinetic gradient of Ficoll in tissue culture medium. *J. Cell. Physiol.* 88:317–322.

Vainchenker, W., J. Guichard, and J. Breton-Gorius. 1978. Growth of human megakaryocyte colonies in culture from fetal, neonatal, and adult peripheral blood cells: Ultrastructural analysis. *Blood Cells* 5:25–42.

Williams, N., T. P. McDonald, and E. M. Rabellino. 1979. Maturation and regulation of megakaryocytopoiesis. *Blood Cells* 5:43–55.

Discussion

Bunn: Ninety % of the cells sorted were megakaryocytes. How many cells can you obtain?

Mayer: We sort about 250 megakaryocytes per hr.

Nakeff: Is the early part of your distribution debris?

Mayer: Yes. The FITC has an overlapping spectrum with the red fluorescence of propidium iodide. Cell cytoplasm or platelets show some red fluorescence and so that background in propidium iodide measurements is present.

Evatt, Levine, and Williams, Editors
MEGAKARYOCYTE BIOLOGY AND PRECURSORS:
IN VITRO CLONING AND CELLULAR PROPERTIES

Study of *In Vitro* Megakaryocytopoiesis in a Permanent Candidate Rat Megakaryocyte Cell Line

Jerome E. Groopman, Christian C. Haudenschild, Eugene Goldwasser, Charles D. Stiles, and Charles D. Scher

Division of Hematology-Oncology, School of Medicine, University of California-Los Angeles, Los Angeles, California; Mallory Institute of Pathology, Boston University School of Medicine, Boston, Massachusetts; Department of Biochemistry, University of Chicago, Chicago, Illinois; Laboratory of Tumor Biology, Sidney Farber Cancer Institute, Department of Microbiology and Molecular Genetics, Harvard Medical School, Boston, Massachusetts; Division of Hematology-Oncology, Sidney Farber Cancer Institute and Children's Hospital Medical Center, Department of Pediatrics, Harvard Medical School, Boston, Massachusetts

The study of megakaryocytopoiesis has been hindered by difficulties in identifying, isolating, and sustaining *in vitro* megakaryocyte progenitors. Clonogenic assays in semi-solid systems of megakaryocyte precursors derived from bone marrow, fetal liver, and peripheral blood have been developed (Metcalf et al., 1975; Vainchenker et al., 1979), but provided only indirect evidence of regulation of megakaryocytopoiesis *in vitro* due to the heterogeneity of the cells present. In addition, culture in semi-solid media makes isolation of cells at progressive stages of maturation difficult. Recently, Cicoria et al. (1977) have derived a cell line (LEMP) from the bone marrow of a female Long-Evans rat that proliferates in long-term liquid suspension culture and may be the first permanent megakaryocyte progenitor cell line. The replicating cells in culture resemble small lymphocytes; however, they contain acetylcholinesterase, a cytochemical marker for probable megakaryocyte precursors in rat and murine species (Zajicek, 1954; Long and Henry, 1979).

We now present morphologic and functional evidence that LEMP cells have a number of characteristics of megakaryocytes. A serum-free medium has been developed which supports their growth in liquid suspension culture. This permanent cell line may provide a useful *in vitro* system for the study of megakaryocytopoiesis.

Methods and Materials

Cell Line

LEMP cells were a gift of Cicoria and Hempling (Cicoria et al., 1977). To assure a monoclonal population of cells for study, cells were cloned in Dulbecco's Modified Essential Medium (DME), 0.8% methylcellulose plus 30% bovine serum (Colorado Serum Co.), using a modification of the technique of Iscove et al. (1974). Clone A-1, which was used for all studies, was maintained in liquid suspension culture consisting of DME plus 30% bovine serum in 10% CO_2 at high humidity.

Culture Conditions

Cells were removed from DME plus 30% bovine serum, centrifuged at 230 × g, washed with DME, and cultured in microtiter wells (Linbro) at final volume 1 ml in (a) DME alone; (b) DME plus 1% bovine serum albumin (BSA); (c) DME plus 10% bovine serum; (d) DME plus 10% human platelet-poor plasma (Pledger et al., 1977); and (e) serum-free medium consisting of DME plus transferrin 350 g/ml (Sigma Fraction II), insulin 1.6×10^{-4} M (Sigma), BSA 10 mg/ml (Sigma), hydrocortisone succinate 5 :ml 10^{-7} M (Sigma), $FeCl_3$ 1.5×10^{-6} M, and Na_2SeO_3 1×10^{-7} M (Sigma) (Hayashi and Sata, 1976; Golde et al., 1980). In some experiments, partially purified somatomedin C (gift from Dr. J. J. VanWyk), partially purified human urinary erythropoietin, obtained from National Heart and Lung Blood Resources Division (NIH E8LSL; 40 IU/mg protein), and highly purified electrophoretically homogeneous erythropoietin (70,000 IU/mg protein) (Mayake et al., 1977) were added to serum-free cultures. Viable cells were counted at appropriate intervals by trypan blue exclusion.

LEMP cells were removed from liquid suspension, cultured, and seeded at appropriate dilutions in 1 ml of semi-solid medium consisting of DME, 0.8% methylcellulose plus 30% bovine serum. Colonies consisting of six or more cells were scored at 5 days.

Morphology

Cells in cultures were observed daily using an inverted microscope. Cytocentrifuge preparations were stained with Wright-Giemsa or for acetylcholinesterase (Karnovsky and Roots, 1964). Cells were prepared for electron microscopy by fixation *in situ* with phosphate buffered 4% glutaraldehyde — 1% formaldehyde followed by 1% aqueous osmium tetrahydroxide. Specimens were stained with uranyl acetate and lead citrate and examined in a Phillips 300 electron microscope.

Factor VIII Antigen

Cells were cytocentrifuged, fixed with 95% ethanol, washed with PBS, and incubated for 60 min with a 1:30 dilution of rabbit antiserum (Calbiochem) to human factor VIII protein. Duplicate cultures were incubated with normal rabbit serum. After washing with PBS, specimens were incubated with a 1:10 dilution of fluorescein-conjugated goat anti-rabbit immunoglobulins (Calbiochem), embedded, and examined with a fluorescent microscope.

DNA Content

LEMP cells grown in DME plus 30% bovine serum or serum-free medium were incubated for 2 hr with colcemid 0.04 g/ml (Calbiochem), transferred to 0.075 M KC1, and fixed. After staining with Wright Giemsa, chromosomes were counted. Duplicate cultures were stained supravitally with the relatively DNA-specific dye Hoechst 33342 and analyzed by fluorescence activated cell sorting (FACS II; Becton-Dickinson) according to the method of Nakeff et al. (1979).

Serotonin Uptake

1×10^5 cells/ml were incubated with L-2-^{14}C-serotonin (New England Nuclear; specific activity 53.5 mCi/mmol) at a final concentration of 0.5 μM at 37°C or 4°C in serum-supplemented medium. The technique of Fedorko (1977) was modified by centrifuging the cells at 230 × g, washing with PBS, and solubilizing the cell pellet in 1% sodium dodecyl sulfate. Samples in Aquasol (National Diagnostics) were counted on the ^{14}C channel of a liquid scintillation counter. The Jones chloroma, a rat promyelocytic leukemia cell line (Greenberger et al., 1978), was used as a control in these studies.

Results and Discussion

The growth of LEMP cells in liquid suspension culture is illustrated in Figure 1. Either human platelet-poor plasma or bovine serum support LEMP growth with a doubling time of approximately 35–45 hr. LEMP cells grew more slowly in serum-free cultures than in serum-supplemented medium, with a doubling time of approximately 80 hr. Refeeding of serum-free cultures every other day with fresh medium sustained growth and viability beyond two weeks. LEMP cells in serum-free medium were found to have an absolute requirement for insulin, transferrin, and Na_2SeO_3 for growth. A partially purified preparation of somatomedin C at a protein concentration of 3–30 ng/ml was equivalent in growth promoting activity to insulin at 0.5–1 μg/ml. Either a partially purified or a highly purified erythropoietin preparation (Mayake et al., 1977) at 2 IU/ml potentiated LEMP growth in serum-free medium (Table 1).

LEMP cells have a cloning efficiency of approximately 15–20% in 0.8% methylcellulose. Cells at low densities (10^2–10^3 cells/ml) contained exuberant colonies consisting of several hundred cells by two weeks.

LEMP cells resemble lymphocytes morphologically but shed cytoplasmic fragments in liquid suspension culture. Electron microscopy demonstrated cytoplasmic organelles resembling dense granules in both parent cells and shed fragments, yet distinct demarcation membranes and alpha granules were not found (Fig. 2).

LEMP cells take up serotonin by a temperature-dependent process, as has been described for guinea pig megakaryocytes (Fedorko, 1977). The LEMP cells concentrated approximately ten-fold more serotonin than the rat promyelocytic leukemia cell line (Fig. 3). LEMP cells had cytochemically detectable acetylcholinesterase, a marker for cells of megakaryocytic lineage in the rat; factor VIII protein was demonstrated by immunofluorescence in both serum-free and serum-containing

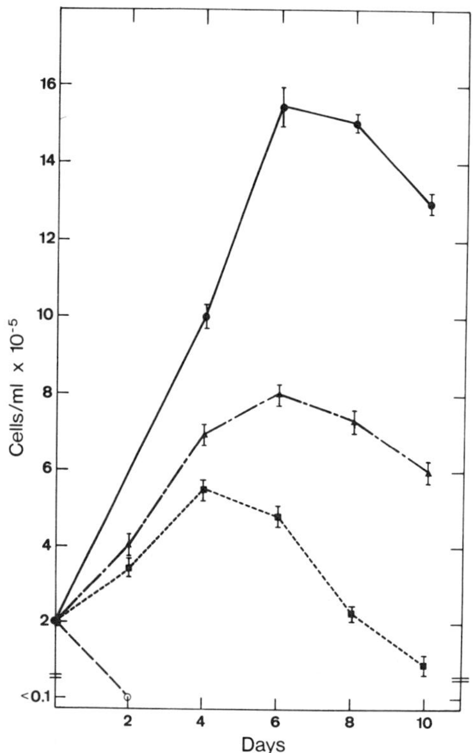

Fig. 1. Growth of LEMP cells in: 10% platelet-poor plasma (●); 10% bovine serum (▲); serum-free medium (■); DME alone (○).

Table 1. Effect of 2 IU/ml Erythropoietin on LEMP Growth in Serum-Free Cultures.[a]

Serum-free	Crude	Highly purified
$8 \pm 2 \times 10^5$ cells/ml	$14 \pm 3 \times 10^5$ cells/ml	$40 \pm 7 \times 10^5$ cells/ml

[a]Cells were plated at initial concentration of 5×10^5 /ml and counted by trypan blue exclusion 60 hours later. Mean ± SD triplicate cultures. Serum-free medium consisted of: insulin 1.6×10^{-4} *M*, transferrin 350 μg/ml, $FeCl_3$ 1.5×10^{-6} *M*, Na_2SeO_3 1×10^{-7} *M*, BSA 10 mg/ml, and hydrocortisone succinate 5×10^{-7} M.

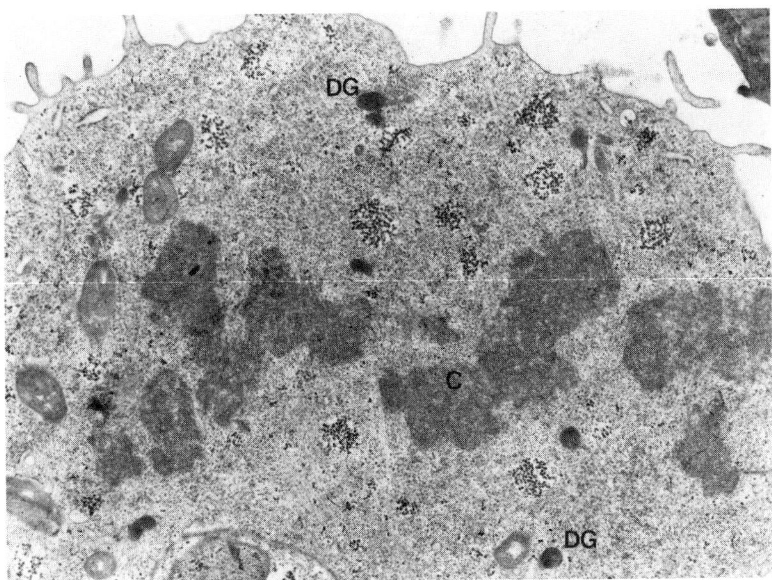

Fig. 2. Transmission electron micrograph of LEMP cell in mitosis in serum-free medium. Note dense granules (DG) and chromosomes (C). Magnification 16,000 ×.

cultures. LEMP cells have a pseudodiploid karyotype; a population of polyploid cells was not detected by fluorescence activated cell sorting of approximately 10^6 cells.

LEMP cells morphologically and cytochemically resemble the small lymphoid, acetylcholinesterase-positive bone marrow cells described by Jackson (1973) that are believed to be precursor cells of the megakaryocyte series. Like megakaryocytes, they contain dense granules, acetylcholinesterase (Zajicek, 1954; Jackson, 1973; Long and Henry, 1979), and factor VIII antigen (Nachman et al., 1977). Furthermore, they concentrate serotonin (Fedorko, 1977). However, they do not become polyploid under the culture conditions described, lack distinct demarcation membranes and alpha granules, and therefore differ from mature bone marrow megakaryocytes. They appear to be megakaryocyte progenitors.

Growth of LEMP cells in platelet-poor plasma or serum-free media allows study of the effects of various agents on both LEMP proliferation and potentially their synthesis of megakaryocyte-related proteins. Both crude and highly purified erythropoietin preparations at 2 IU/ml appear to stimulate LEMP growth. Potentiation of megakaryocyte colony formation by similar concentrations of erythropoietin has previously been reported. McLeod et al. (1976) noted a response to erythropoietin using murine bone marrow cells cultured in plasma clot. Recently, Vainchenker et al. (1979) have demonstrated similar potentiation of human megakaryocyte colony

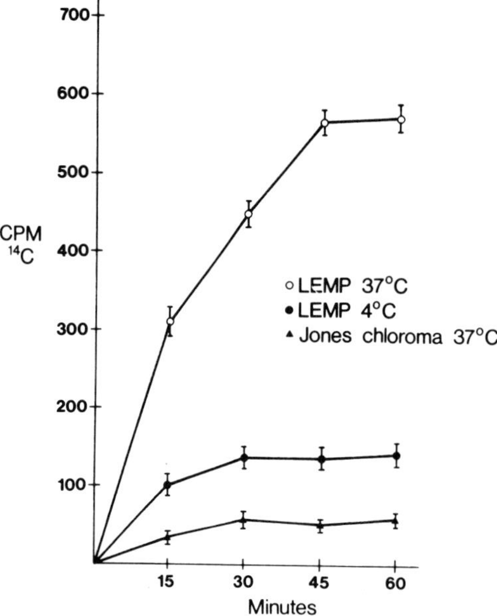

Fig. 3. ^{14}C-Serotonin uptake by LEMP cells and Jones chloroma.

formation by crude and highly purified erythropietin preparations using bone marrow, peripheral blood, and fetal liver mononuclear cells. LEMP cells should provide a useful system for the study of regulation of early stages of megakaryocytopoiesis *in vitro* by erythropoietin.

Summary

Cicoria and Hempling have established a permanent cell line (LEMP) from the bone marrow of a Long Evans rat that proliferates in liquid suspension culture and is a candidate megakaryocyte progenitor cell line. Replicating cells in culture resemble small lymphocytes but are acetylcholinesterase-positive. This enzyme is found within rat bone marrow in megakaryocytes, but not in other hematopoeitic cells. We present morphologic and functional evidence that this permanent cell line has a number of characteristics of megakaryocyte precrusors. Electron microscopy demonstrates the presence of very dense granules. LEMP in culture binds ^{14}C-serotonin in a temperature dependent fashion. Enucleated cellular particles that resemble platelets are noted in culture. However, no alpha granules have been observed. A serum-free medium has been devised to allow growth of these cells.

ACKNOWLEDGMENTS

We thank A. D. Cicoria and H. G. Hempling for the LEMP cell line, J. Greenberger for the Jones chloroma cell line and M. J. Cline, D. W. Golde, P. I. Hartzband, D. Kenney, J. M. Lipton, and D.

G. Nathan for helpful discussions. S. Latt kindly analyzed these cells by fluorescence activated cell sorting. Partially purified erythropoietin was obtained from the Division of Blood Diseases and Resources of the National Heart, Lung and Blood Institute. This investigation was supported by Grants CA 27113, CA 18662, CA 9297, CA 18375, HL 23567, HL 26176, and CA 22427. C. D. Scher is a Scholar of the Leukemia Society of America.

References

Burstein, S.A., J.W. Adamson, and L.A. Harker. Characteristics of murine megakaryocytic colonies *in vitro*. *Blood* 54:169–179, 1979.

Cicoria, A. D., A. DuPre, and H. G. Hempling. 1977. The osmotic properties and membrane permeability to water of proliferating and maturing mammalian cells. *The Physiologist* 20:17 (abstr.), 1977.

Fedorko, M. The functional capacity of guinea pig megakaryocytes. I. Uptake of ^{3}H-serotonin by megakaryocytes and their physiologic and morphologic response to stimuli for the platelet release reaction. *Lab. Invest.* 36:310–320, 1977.

Golde, D. W., N. Bersch, S. G. Quan, and A. J. Lusis. 1980. Production of erythroid potentiating activity by a human T-lymphoblast cell line. *Proc. Nat. Acad. Sci. USA* 77:593–596, 1980.

Greenberger, J. S., P. E. Newburger, A. Karpa, and W. C. Moloney. Constitutive and inducible granulocyte-macrophage functions in mouse, rat and human hyeloid leukemia-derived continuous tissue culture lines. *Cancer Res.* 38:3340–3345, 1978.

Guilbert, L. J., and N. N. Iscove. Partial replacement of serum by selenite, transferrin, albumin and lecithin in hematopoietic cell cultures. *Nature* 263:594–596, 1976.

Hayashi, I., and G. H. Sato. Replacement of serum by hormones permits growth of cells in a defined medium. *Nature (London)* 259:132–134, 1976.

Iscove, N. N., F. Sieber, and K. H. Winterhalter. Erythroid colony formation in cultures of mouse and human marrow: analysis of requirement for erythropoetin by gel filtration and affinity chromatography on an agarose concanavalia A. *J. Cell Physiol.* 83:309–320, 1974.

Jackson, C. W. Cholinesterase as a possible marker for early cells of the megakaryocytic series. *Blood* 42:413–420, 1973.

Karnovsky, M. J. and L. Roots. A direct-coloring method for cholinesterases. *J. Histochem. Cytochem.* 12:219–221, 1964.

Long, M. W. and R. L. Henry. Thrombocytosis-induced suppression of small acetylcholinesterase-positive cells in bone marrow of rats. *Blood* 54:1338–1346, 1979.

Mayake, T., C. K.-H. Kung, and E. Goldwasser. Purification of human erythropoietin. *J. Biol. Chem.* 252:5558–5564, 1977.

McLeod, D. L., M. M. Shreeve, and A. A. Axelrad. Induction of megakaryocyte colonies with platelet formation *in vitro*. *Nature* 261:492–494, 1976.

Metcalf, D., H. R. McDonald, N. Odartchenko, and B. Sordat. Growth of mouse megakaryocyte *in vitro*. *Proc. Nat. Acad. Sci. (USA)* 72:1744–1748, 1975.

Nachman, R. L., R. Levine, and E. A. Jaffe. Synthesis of Factor *VM* antigen by cultured guinea pig megakaryocytes. *J. Clin. Invest.* 60:914–921, 1977.

Nakeff, A. and S. Daniels-McQueen. *In vitro* colony assay for a new class of megakaryocyte precursor: colony-forming unit megakaryocyte (CFU-M). *Proc. Soc. Exp. Biol. Med.* 151:587–590, 1976.

Nakeff, A., F. Valeriote, J. W. Gray, and R. J. Grabske. Application of flow cytometry and cell sorting to megakaryocytopoiesis. *Blood* 53:732–745, 1979.

Pledger, W. J., C. D. Stiles, H. N. Antoniades, and C. D. Scher. Induction of DNA synthesis in Balb C3T3 cells by serum components. *Proc. Natl. Acad. Sci. (USA)* 73:4481–4485, 1977.

Vainchenker, W., J. Guichard, and J. Breton-Gorius. Growth of human megakaryocyte colonies in culture from fetal, neonatal and adult peripheral blood cells: ultrastructural analysis. *Blood Cells* 5:25–42, 1979.

Vainchenker, W., J. Bouget, J. Guichard, and J. Breton-Gorius. Megakaryocyte colony formation from human bone marrow precursors. *Blood* 54:940–946, 1979.

Williams, N., H. Jackson, A. P. C. Sheridan, M. J. Murphy, Jr., A. Elste, and M. A. S. Moore. Regulation of megakaryocytopoiesis in long-term murine bone marrow cultures. *Blood* 51:245–255, 1978.

Zajicek, J. Studies on the histogenesis of blood platelets. *Acta Haematol.* 12:238–243, 1954.

Discussion

The discussion of this paper was included with the discussion of the work presented in Chapter 28.

Evatt, Levine, and Williams, Editors
MEGAKARYOCYTE BIOLOGY AND PRECURSORS:
IN VITRO CLONING AND CELLULAR PROPERTIES

In Vivo Studies of a Rat Bone Marrow Derived Cell Line: A Bipotential Stem Cell?

Howard N. Steinberg, Martha Tracy, Robert Weinstein, Thomas Maciag, Michael B. Stemerman, and Stephen H. Robinson

Department of Medicine and Thorndike Laboratory, Beth Israel Hospital and Harvard Medical School, Boston, Massachusetts

In a recent study, A. Cicoria and H. G. Hempling described the isolation and characterization of a normal bone marrow derived cell line with presumptive megakaryocyte-like properties (Cicoria, 1979). We have been able to confirm this hypothesis and characterize the cell line as a rat promegakaryoblast (RPM) cell capable of maturation *in vitro* (Weinstein et al., unpublished observations). The RPM cell line has been shown to possess the nature of a megakaryoblast by biochemical and morphological criteria which include: (A) presence of (1) a potent smooth muscle cell mitogen, (2) fibrinogen and factor VIII antigen; (B) uptake and release of serotonin; and (C) formation of a demarcation membrane system as a function of maturation (Weinstein et al., unpublished observations). The nature of these cells as megakaryocytic precursors has been further defined by the *in vivo* diffusion chamber studies described in this preliminary report. Based on these observations and the results of studies using the spleen colony assay of Till and McCulloch (1961), the RPM cell line may actually represent a bipotential stem cell capable of differentiation into either megakaryocytic or the erythroid cell line depending upon the presence of appropriate regulators or microenvironment factors.

Methods and Materials

With the kind permission of Drs. A. Cicoria and H. G. Hempling (Medical University of South Carolina, Charleston), the rat promegakaryoblast cells (RPM) were obtained from Dr. C. Scher of the Sidney Farber Cancer Institute in Boston, Mass. Cell stocks were maintained in DME + 10% fetal calf serum (FCS) at 37°C and 5% CO_2 in air.

Both the conventional and the plasma clot diffusion chamber culture (PCDC) systems have been previously described (Benestad, 1970; Steinberg et al., 1976). Briefly, 1 × 10^3 RPM cells were seeded into diffusion chambers consisting of 2 nucleopore filters — 0.45 μ (Nucleopore Corp., Pleasanton, Ca.) glued to both sides of a lucite ring. Chambers were plugged and surgically implanted into the peritoneal cavity of normal and sublethally irradiated (450 R) 100–125g Sprague-Dawley rats (Zivic Miller Labs., Allison Park, Pa.).

In contrast to the conventional diffusion chamber assay, RPM cells seeded into PCDC cultures were suspended in a clot matrix by the addition of 20 μl bovine citrated plasma and 10 μl beef embryo extract (GIBCO, Grand Island, N.Y.). Implanted chambers were harvested at varying time intervals up to 14 days. Cells from conventional diffusion chambers were harvested by dissolving the naturally formed clot with a 0.5% pronase solution for 55 min at room temperature. Duplicate cell counts and cytocentrifuge slide preparations were made to document changes in cell number and morphology within the diffusion chamber. Colony formation by RPM cells was determined in PCDC cultures fixed with 5% gluteraldehyde immediately after chamber harvest. Using techniques described elsewhere (Hoyer et al., 1973; Karnovsky and Roots, 1964), factor VIII antigen and acetylcholinesterase enzyme activity, determined by immunofluorescence and special cytochemical staining, respectively, were analyzed on cytocentrifuge preparations of conventional diffusion chamber contents and on clots obtained from plasma clot diffusion chamber cultures.

The spleen colony assay of Till and McCulloch (1961) was used to study the RPM's ability to reconstitute the spleens and bone marrows of lethally irradiated rats. Following a lethal dose of irradiation (850 R), rats were injected with either 4 × 10^4 RPM cells or normal bone marrow cells obtained from the femur of a normal rat. Control rats received a 0.1 ml injection of cell-free alpha medium only. Spleens and bone marrow were harvested at varying intervals after reconstitution. Histological specimens were fixed in Bouin's fixative or 4% buffered formalin, embedded in paraffin, cut and stained with Wright-Giemsa or benzidine-hematoxylin stain. Total cell counts in each organ were obtained from suspensions of unfixed spleens and bone marrows. Cytocentrifuge preparations from these suspensions were prepared. In some instances, touch preparations of spleens were prepared and stained with benzidine-hematoxylin stain.

Results and Discussion

RPM cells (1 × 10^3) seeded into conventional diffusion chamber cultures and implanted into normal rat hosts grew exponentially to a maximum cell number of 3 × 10^6 cells per chamber by day 10 of culture. This was followed by a plateau in cell number through day 14. In the exponential phase, the RPM cells grew with a doubling time of 12 hr. This is significantly faster than the 30-hr doubling time observed in *in vitro* cultures (Weinstein et al., unpublished observations). Morphologically, the RPM cells inoculated into diffusion chambers were small (<

10 μm), undifferentiated blast-like cells with a high nuclear to cytoplasmic ratio, with a prominent large, well-rounded nucleus and a small strip of basophilic cytoplasm, as found with Stage I megakaryocytes (Ebbe and Stohlman, 1965). However, as early as day 4 of culture, increases in both cell size and the amount of cytoplasm were observed. As found with Stage II megakaryocytes, nuclei were more eccentric and the cytoplasm more abundant and less basophilic. The cell population, however, remained homogenous in appearance through day 10. At this point, cells began to vary in size and changes in the nuclear configuration were noted. Such large (> 50 μm) cells, having up to 8 nuclear lobes, are associated with Stage III of megakaryocyte development.

The appearance of functional markers as well as morphological changes were also observed after day 10. After day 10, during the plateau in cell number, assays for acetylcholinesterase enzyme activity and factor VIII antigen were both positive, in contrast to negative results obtained prior to that time.

In a single experiment, a significantly enhanced rate of growth and cellular maturation was observed in diffusion chambers implanted into sublethally irradiated rats (450 R). Cell numbers reached a maximum of 6×10^6 cells per diffusion chamber and large, binucleated cells appeared approximately 2 days earlier than in chambers implanted into normal host animals. Therefore, RPM cells can undergo increased growth and apparently more rapid differentiation in response to irradiation of the host animals.

In 2 experiments, RPMs grew as large and discrete colonies of cells in PCDC cultures implanted into normal hosts. The number of colonies appeared to increase linearly up to day 8, when it became impossible to determine colony number. While colonies appeared negative for acetylcholinesterase and factor VIII on day 4 of culture, most colonies appeared to be positive for both functional markers after day 8.

Based on the RPM's ability to differentiate within the physiological environment of the host rat, it was of interest to examine the ability of these cells to repopulate the spleen and bone marrow of lethally irradiated rats. Histological sections of paraffin-embedded spleens and bone marrows showed the development of discrete microscopic colonies in animals reconstituted with either normal bone marrow or RPM cells at both 7 and 10 days after reconstitution. Spleens and bone marrow of medium only-injected rats were markedly hypocellular. These observations were confirmed by findings of comparable increases in total cell counts in normal bone marrow and RPM reconstituted spleens and bone marrows, in contrast to only a small increase observed in medium-injected rats. While macrocopic surface colonies were not observed in any rat group discrete macroscopic colonies were found within the spleens of lethally irradiated (850 R) CD-1 mice injected with RPM or normal mouse marrow cells. Morphologically, cells within the spleen's colonies of RPM-injected rats resembled erythroid precursors. Large multinucleated forms or small blast-like cells, characteristic of RPM cells during early and late phases of diffusion chamber culture, were absent. Touch preparations of spleens harvested 10 and 14 days after RPM injection and stained for hemoglobin with benzidine-hematoxylin supported the erythroid nature of the observed cells. This preliminary observation suggests that RPM cells, which demonstrate properties of megakaryocytes in diffusion chambers, may in fact be bipotential stem cells since they appear to differentiate along the erythroid line in the microenvironment provided by the spleen. At the present time, the nature of cells in the RPM reconstituted bone marrow remain to be identified.

In the present study, we examined the changes that occur in a normal bone marrow-derived cell line when it is cultured in the physiological environment of a normal host rat. This cell undergoes significant changes in cell size and apparently in cell ploidy in diffusion chamber cultures. The changes in morphology and the development of acetylcholinesterase and factor VIII antigen activity demonstrate the megakaryocytic nature of this cell line. While several factors, including thrombopoietin, are postulated to govern the regulation of megakaryocyte proliferation and development (Williams et al., 1979), the nature of the diffusible factor(s) influencing RPM proliferation and differentiation in these *in vivo* cultures is as yet unknown.

The finding in diffusion chamber cultures are very similar to the *in vitro* observations of Weinstein et al. (unpublished observations). While the blast-like morphology of the RPM cells is maintained with DME + 10% fetal calf serum, lowering the amount of serum to 1% results in the maturation of these cells. Under these conditions, cells increase in size, achieve lobulated nuclei and, by immunofluorescence, become positive for both factor VIII and fibrinogen antigens. There is also preliminary evidence of platelet shedding from these cells *in vitro*. While electron microscopic studies of RPM cells isolated from diffusion chambers remain to be done, electron microscopic studies on cells maturing *in vitro* show the development of demarcation membranes, surface membrane pseudopods or "blebs," extensive rough endoplasmic reticulum, microtubules, Golgi zones, and dense granules. By several criteria, therefore, RPM cells appear to be megakaryocytic precursor cells. Taken together, the results obtained *in vitro* and *in vivo* suggest that the RPM may provide a powerful model system for the study of megakaryocytopoiesis and its regulation.

Advances in *in vitro* culture techniques have led to the identification of hemopoietic stem cells committed to a single line of differentiation (Bradley and Metcalf, 1966; McLeod et al., 1974; Nakeff and Daniels-McQueen, 1976). Recently the existence of a bipotential stem cell common to both the erythroid and megakaryocytic cell lines has been postulated (McLeod et al., 1976). While the diffusion chamber studies confirm that RPM cells are megakaryocytic precursors, the presence of erythroid precursors and the absence of megakaryocytes in the spleens of lethally irradiated rats injected with RPM cells suggest that this cell line may actually represent a bipotential progenitor cell. The line of differentiation of RPM cells in the spleen may be influenced by the microenvironment of this organ which is known to favor erythroid development. Experiments are in progress to examine these provocative findings.

Summary

Experiments were performed to determine whether a clonal cell line derived from normal bone marrow possesses characteristics of megakaryocyte precursors. When seeded in *in vivo* diffusion chamber cultures implanted into normal or sublethally irradiated (450 R) host rats, these rat promegakaryoblast-like cells (RPMs) grew exponentially for 10 days followed by a plateau phase through day 14. Cells grew to twice control levels in irradiated hosts. In the early days of culture, the RPM cells, which originally were small and undifferentiated with a high nuclear to cytoplasmic ratio, became enlarged, with a decreased nuclear to cytoplasmic ratio, and developed bleb-like surface structures. During the plateau phase very large cells with multilobed nuclei were observed. Many nuclei contained between 2 and 8 lobes. Before and during early culture, these cells gave a negative histochemical stain for acetylcholinesterase and failed to show factor VIII antigen by immunoflourescence. The cells became positive for both markers in the later days of culture. These studies suggest that RPM cells differentiate *in vivo* and take on megakaryocytic characteristics during culture in diffusion chambers, supporting the conclusion that they are megakaryocytic precursor cells. However, RPM cells injected into lethally irradiated (850 R) rats formed spleen colonies composed of erythroid precursors as shown by benzidine-hematoxylin staining. Large multinucleated cells appeared to be absent from the spleens of RPM reconstituted animals. Therefore, the RPM cell line may actually represent a bipotential stem cell line capable of directional differentiation under the influence of different stimuli.

ACKNOWLEDGMENTS

Supported by USPHS Grant AM 17148, USPHS Hematology Training Grant AM 05391, American Cancer Society Grant DH-51D, and National Institutes of Health Grant HL-25066.

References

Benestad, H. B. 1970. Formation of granulocytes and macrophages in diffusion chamber cultures of mouse blood leukocytes. *Scand. J. Haematol.* 7:279–288.

Bradley, T. R. and D. Metcalf. 1966. The growth of mouse bone marrow cells *in vitro*. *Aust. J. Exp. Biol. Med. Sci.* 44:287–299.

Cicoria, A. 1979. Properties of maturing cells: The maturation of membrane function in a normal population of differentiating megakaryocyte precursor cells. Doctoral Thesis. Medical University of South Carolina, Charleston.

Ebbe, S. and F. Stohlman, Jr. 1965. Megakaryocytopoiesis in the rat. *Blood* 26:20–35.

Hoyer, L. W., R. P. de la Santos, and J. R. Hoyer. 1973. Antihemophilic factor antigen. Localization in endothelial cells by immunoflourescent microscopy. *J. Clin. Invest.* 52:2737–2744.

Karnovsky, M. J. and L. Roots. 1964. A "direct-coloring" thiocholine method for cholinesterases. *J. Histochem. Cytochem.* 12:219–221.

McLeod, D. L., M. M. Shreeve, and A. A. Axelrad. 1974. Improved plasma culture system for production of erythrocytic colonies *in vitro*: Quantitative method for CFU-E. *Blood* 44:517–534.

McLeod, D. L., M. M. Shreeve, and A. A. Axelrad. 1976. Induction of megakaryocyte colonies with platelet formation *in vitro*. *Nature* 261:492–494.

Nakeff, A. and S. Daniels-McQueen. 1976. *In vitro* assay for a new class of megakaryocyte precursor: Colony-forming unit megakaryocyte (CFU-M). *Proc. Soc. Exp. Biol. Med.* 151:587–590.

Steinberg, H. N., E. S. Handler, and E. E. Handler. 1976. Assessment of erythrocytic and granulocytic colony formation in an *in vivo* plasma clot diffusion chamber culture system. *Blood* 47:1041–1051.

Till, J. E. and E. A. McCulloch. 1961. A direct measurement of the radiation sensitivity of normal bone marrow cells. *Rad. Res.* 14:213–222.

Williams, N., T. P. McDonald, and E. M. Rabellino. 1979. Maturation and regulation of megakaryocytopoiesis. *Blood Cells.* 5:43–55.

Discussion

Hempling: The definite reference for the cell line is Dr. Cicoria's thesis: "Osmotic properties of maturing cells, the maturation of membrane function in a normal population of differentiating megakaryocyte precursor cells."

Long: How intense was the acetylcholinesterase staining? Was there variability in the reaction?

Groopman: In my clonal isolate of the original cells, it is positive but light and uniform. In contrast to the findings of Steinberg, I have looked, but been unable to obtain benzidine-positive cells in both methylcellulose cultures and in liquid suspension cultures, even with erythropoietin.

Long: In the original findings, Cicoria found that the cell line could be stimulated and bring about differentiation of demarcation membrane formation. Have either of you been able to repeat the adult calf serum-induced differentiation of the cell line?

Groopman and Steinberg: No.

Long: We have also attempted these experiments, but could not find acetylcholinesterase cells or induce differentiation in irradiated animals.

Groopman: I think that we are working on different cells. My studies were based on a clonal isolate, and we gave the cells to Steinberg's groups.

Long: It is important that one of you finds benzidine-positive cells; the observation implies heterogeneity in the cell line.

Hoffmann: What do you think the significance of the membrane blebbing is?

Groopman: The blebbing is probably a tissue culture artifact representing cell death. I don't think that it shows anything about the formation of platelets *in vivo*. We have done careful electron microscopic studies and while blebbing occurs in normal cultures, it is markedly increased under conditions of starvation and may reflect degeneration. A large number of the fragments that are extruded are tissue culture artifact. In many cells, the nuclear membrane appears to be separating from the cytoplasm; and many of the membranous structures representing degenerating endoplasmic reticulum and might be confused with demarcation membranes.

Steinberg: We cannot be sure about that yet.

Weil: Your results show bipotentiality. Are any of the rats being rescued, which would suggest multipotency?

Steinberg: Many of these rats do live longer than normally reconstituted animals. We will repeat the experiments to look at the bone marrow.

Groopman: We have also done the irradiated rats reconstitution, and been unable to see erythroid or myeloid repopulation.

Evatt, Levine, and Williams, Editors
MEGAKARYOCYTE BIOLOGY AND PRECURSORS:
IN VITRO CLONING AND CELLULAR PROPERTIES

Serotonin Storage in Megakaryocytes

Paul K. Schick and Mitchell Weinstein

The Specialized Center for Thrombosis Research, Temple University Health Sciences Center, Philadelphia, Pennsylvania and The Department of Physiology and Biochemistry, The Medical College of Pennsylvania, Philadelphia, Pennsylvania

It is known that platelets avidly take up serotonin but the role of platelet serotonin is not understood. There is evidence that megakaryocytes can accumulate serotonin (Tanaka et al., 1967; Tranzer et al., 1972; White, 1971; Fedorko, 1977) and we have shown that serotonin storage in megakaryocytes exceeds that occurring in platelets (Schick and Weinstein, 1980). This study was designed to estimate serotonin accumulation in individual megakaryocytes. Two questions were asked: (1) Does serotonin uptake and storage occur in a specific population of megakaryocytes such as in immature megakaryocytes? (2) Can the capacity for serotonin accumulation be used as a marker for the identification of megakaryocytes?

Whole marrow cell suspensions were incubated with ^{3}H-serotonin to achieve the accumulation of radiolabeled serotonin in megakaryocytes. Guinea pig megakaryocytes were isolated to 80 to 90% purity by the method of Levine and Fedorko (1976). Autoradiography was carried out in cells, either whole marrow suspensions or purified megakaryocytes, that had been cytocentrifuged on glass slides and subsequently were stained with Wright's stain.

The examination of the unpurified whole marrow suspensions revealed that only megakaryocytes showed evidence of accumulated radioactivity and thus serotonin. The examination of the pellet of the albumin density gradient revealed that over 95% of the megakaryocytes had accumulated the radiolabeled serotonin. Therefore, the fraction of megakaryocytes present in the pellet and not recovered in the final purified megakaryocyte suspensions retained the serotonin they had accumulated.

Virtually all megakaryocytes in the purified megakaryocyte suspensions had accumulated serotonin. The grains present per cell area of megakaryocytes were equivalent regardless of the size or nuclear/cytoplasmic maturation of the megakaryocytes. It was evident that the amount of serotonin in large megakaryocytes with a multilobulated nucleus and a small megakaryocyte with less developed nuclei were similar.

In conclusion, the study demonstrates that the capacity for serotonin accumulation is well-established in all populations of megakaryocytes and most likely occurs in immature megakaryocytes. Conceivably, serotonin plays a role in megakaryocyte maturation. There is evidence that serotonin is involved in the differentiation of nervous tissue (Tissari, 1966). The study also indicates that serotonin accumulation can be used as a marker for megakaryocytes, particularly young megakaryocytes.

ACKNOWLEDGMENT
This work was supported in part by USPHS Award HL 22633.

References

Fedorko, M. E. 1977. The functional capacity of guinea pig megakaryocytes. *Lab. Invest.* 36:310–320.

Levine, R. F. and M. E. Fedorko. 1976. Isolation of intact megakaryocytes from guinea pig femoral marrow. *J. Cell Biol.* 69:159–172.

Schick, P. K. and M. Weinstein. 1980. Serotonin accumulation and metabolism in megakaryocytes. *Fed. Proc.* 39:457.

Tanaka, C, I. Kunuma, and A. Kuramato. 1967. Histochemical demonstration of 5-hydroxytryptamine in platelets and megakaryocytes. *Blood* 30:54–61.

Tissari, A. 1966. 5-hydroxytryptamine, tryptophan decarboxylase and monoamine oxidase during fetal and postnatal development of the guinea pig. *Acta Physiol. Scand.* 67 (Suppl. 265):9–80.

Tranzer, J. P., M. DaPrada, and A. Pletscher. 1972. Storage of 5-hydroxytryptamine in megakaryocytes. *J. Cell Biol.* 52:191–197.

White, J. G. 1971. Serotonin storage organelles in human platelets. *Am. J. Path.* 63:403–408.

Evatt, Levine, Williams, Editors
MEGAKARYOCYTE BIOLOGY AND PRECURSORS:
IN VITRO CLONING AND CELLULAR PROPERTIES

Abnormalities of Megakaryocytes and Megakaryocyte Progenitor Cells (CFU-M) in W/W^v Mice

Sigurdur R. Petursson and Paul A. Chervenick

University of Pittsburgh School of Medicine, Pittsburgh, Pennsylvania

W/W^v mice have a number of genetic abnormalities. In addition to sterility, these include a macrocytic anemia, decreased blood and marrow granulocytes, and decreased marrow megakaryocytes (Chervenick and Boggs, 1969; Ebbe et al., 1973). The pluripotent stem cell, the CFU-S, is abnormal and does not form macroscopic spleen colonies when injected into lethally irradiated normal (+/+) littermates (McCulloch et al., 1964). In addition, these mice also have decreased numbers of committed granulocyte-macrophage progenitor cells in their bone marrow (Benestad et al., 1975; Bennett et al., 1968). In order to determine whether megakaryocyte progenitor cells were abnormal, we studied megakaryocyte and megakaryocyte progenitor cell (CFU-M) production in W/W^v mice in the short-term soft gel (methylcellulose) *in vitro* culture system and in the continuous marrow liquid culture (CLMC) system as described by Dexter and Lajtha (1974).

Methods and Materials

Male W/W^v mice and homozygous (+/+) littermates were obtained from Jackson Laboratories (Bar Harbor, Maine). Total nucleated marrow cell counts per humerus were determined on a Coulter Counter after flushing cells from the marrow as described previously (Dexter and Lajtha, 1974). Platelet counts were performed at 1:10 dilution in 1% ammonium oxalate on a hemacytometer under phase contrast microscopy at 400×. Bone marrow megakaryocytes were determined by the technique described by Ebbe and Phalen (1973). A suspension of marrow cells was stained with 0.5% new methylene blue and megakaryocytes were counted on a hemacytometer at 40× (Ebbe et al., 1973). Megakaryocyte progenitor cells (CFU-M) were assayed by their ability to form megakaryocyte colonies in the soft gel (methylcellulose) *in vitro* culture system. Marrow cells were cultured at a concentration of 7.5×10^4 cells/ml. Colony formation was induced by pokeweed-stimulated lymphocyte conditioned medium (Nakeff and Daniels-McQueen, 1976). CFU-M colonies were scored with an inverted microscope at 40× after incubation for 9 days at 37°C in 10% CO_2. Continuous liquid marrow cultures were initiated by pooling fresh marrow cells in 25 cm^2 plastic flasks (Falcon 3013). Cultures were incubated at 33°C and 5% CO_2 in 10 ml of Fischer's medium supplemented with penicillin-streptomycin and 20% horse serum (Flow Laboratories). In cultures supplemented with hydrocortisone, hydrocortisone sodium succinate (Solu-Cortef; Upjohn) was added to produce a final concentration of 10^{-6} M. Weekly, aliquots of one half of the medium and one half of the freely suspended cells were removed from sets of three or four culture flasks, pooled, and assayed for total cell number, megakaryocytes, and CFU-M. Medium which was removed was replaced with an equal amount of fresh medium. Cultures were not recharged with fresh cells at three weeks as has been done by other investigators (Dexter and Lajtha, 1974). The percent of CFU-M in DNA synthesis was measured by determining the degree to which colony forming capacity of the cultured cells was lost after tritiated thymidine suicide (Joyce and Chervenick, 1977).

Results and Discussion

Peripheral blood platelet counts were normal, whereas marrow megakaryocytes were decreased in W/W^v mice compared to normal controls (4.06 ± 0.27 vs $7.93 \pm 0.58 \times 10^3$/humerus; $p<0.01$). (Chervenick and Boggs, 1969; Ebbe et al., 1973). The concentration of CFU-M/10^5 marrow cells was also significantly less in W/W^v mice (19.6 ± 2.4 vs 27.2 ± 2.7; $p<0.03$) as was the total number of CFU-M per humerus (1294 ± 218 vs 2477 ± 326; $p<0.01$). The percent of W/W^v CFU-M in S phase was significantly less than that found in +/+ controls (14.2 ± 3.2% vs 38.6 ± 2.1%; $p<0.001$).

The proliferation of megakaryocytes from W/W^v and +/+ marrow in continuous liquid cultures is seen in Figure 1. By 3 weeks megakaryocytes in W/W^v cultures were significantly decreased compared to controls, and after 8 weeks were undetectable. In contrast, megakaryocyte production in control cultures continued until the eleventh week. Figure 2 illustrates the proliferation of CFU-M in continuous marrow cultures. In cultures of W/W^v marrow, CFU-M failed to proliferate actively after the second week and were exhausted by the seventh week. In +/+ cultures, CFU-M proliferated at higher levels and continued for several weeks longer.

The addition of hydrocortisone at 10^{-6} M concentration to continuous marrow culture resulted in a marked increase in cell proliferation. In Table 1, the proliferation of megakaryocytes in W/W^v and +/+ cultures supplemented with hydrocortisone can be seen. Megakaryocytes in W/W^v and +/+ cultures increased slowly during the first 3 weeks. However by 6 weeks, megakaryocytes in both sets of cultures were increased, with W/W^v cultures showing a greater increase. This contrasts with the proliferation in the absence of hydrocortisone, where W/W^v megakaryocyte proliferation was always less than +/+ controls.

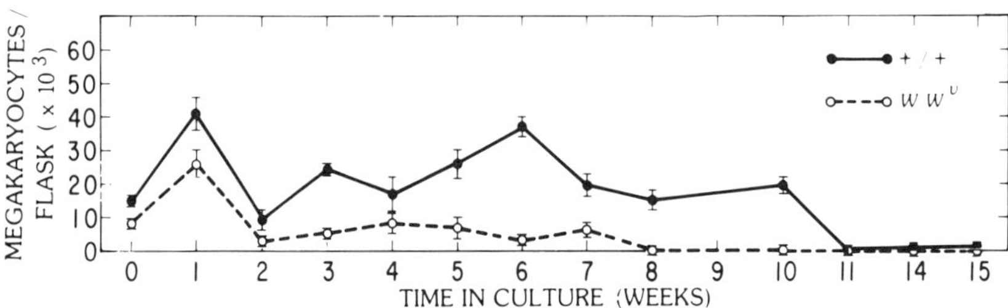

Fig. 1.

CFU-M proliferation followed a similar pattern when hydrocortisone was added to the culture system (Table 1). During the early part of the culture period, CFU-M from W/W^v mice were considerably less than controls. However, by 6–8 weeks, CFU-M from W/W^v mice increased to a greater extent than those from +/+ marrow.

Considerable evidence is available from studies in mice to indicate that pluripotent hematopoietic stem cells give rise to committed progenitor cells which in turn produce the differentiated progeny (Seller, 1970; Lajtha, 1979). That the pluripotent stem cell is abnormal in W/W^v mice has been demonstrated in transplantation experiments wherein W/W^v hematopoietic cells fail to produce spleen colonies (CFU-S) in lethally irradiated recipients. In turn, unirradiated W/W^v mice form CFU-S on their spleens following injection of marrow cells from their normal littermates, and their anemia is cured (McCulloch et al., 1964; Russell, 1970).

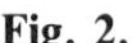
Fig. 2.

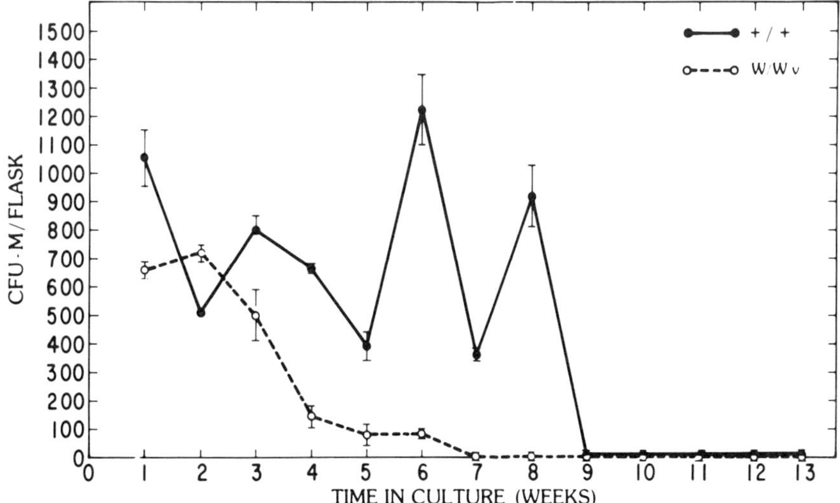

Table 1. The Effect of Hydrocortisone (10^{-6} M) on Megakaryocyte and Megakaryocyte Progenitor Cell (CFU-M) Proliferation in Liquid Culture.

	Megakaryocytes (× 10^3/flask)		CFU-M/flask	
Week in culture	**W/W^V**	**+/+**	**W/W^V**	**+/+**
0	8.1	15.9	6020	6540
3	12.0	22.7	260	980
6	81.7	51.4	1880	490
8	71.0	96.8	600	270
12	95.3	33.3	1060	320

Evidence for abnormalities in the committed progenitor cells in W/W^v mice is available from *in vitro* studies of colony forming cells. Total marrow CFU-M, BFU-E, and CFU-E have been shown to be decreased in W/W^v mice, although their concentrations are normal (Benestad et al., 1975; Bennett et al., 1968). Our study demonstrated decreased concentration (72.1% of +/+ controls) and total number of CFU-M/humerus (52.2% of +/+ controls) in W/W^v animals, the latter being similar to the observations made by Nakeff (1977).

The defect of the CFU-S in W/W^v mice is not clearly understood. A decreased feed-in from an abnormally proliferating pluripotent stem cell compartment could be invoked to explain the reduced numbers of committed progenitor cells and differentiated marrow and blood progeny. It is difficult to assess the proliferative activity of the W/W^v CFU-S since it does not form macroscopic spleen colonies. Microscopic colonies are formed, however, but it is uncertain how these should be compared to macroscopic colonies. Attempts to assess the proliferative activity of committed progenitor cells have included sizing of CFU-GM, BFU-E, and CFU-E colonies growing *in vitro*. These studies have failed to demonstrate any differences in size between CFU-GM colonies from W/W^v and +/+ mice (Iscove, 1978; Bennett et al., 1968). Our tritiated thymidine suicide studies do reveal however a marked difference in DNA synthetic activity of W/W^v CFU-M compared to +/+ CFU-M. Only about one fifth (14.2% of 1294 CFU-M/humerus vs 38.6% of 2477 CFU-M/humerus) as many CFU-M/humerus in W/W^v mice are in S phase compared to +/+ mice. Thus, under homeostatic conditions, there is markedly less megakaryopoietic activity in W/W^v mice, and this appears to be at least in part due to an intrinsic proliferative defect at the committed progenitor cell level.

Proliferation of cells in continuous marrow cultures can be utilized to study the activity of the CFU-S and committed progenitor cells. CFU-M and megakaryocytes in W/W^v cultures proliferated less and for shorter periods of time compared to +/+ cultures. However, whether this was due to CFU-S defect or a defect in the CFU-M or both, is uncertain. The addition of corticosteroids to these cell cultures has been shown by Greenberger (1978) to enhance proliferation of cells in this culture system. Hydrocortisone increased the number of cells in the adherent layer and the number and size of adipocyte colonies in the adherent layer. CFU-S and

CFU-GM were increased in number and proliferated for longer periods in hydrocortisone treated cultures. Whether enhanced proliferation of CFU-S and CFU-GM resulted from the direct effects of corticosterioids on these cells, or secondarily from an improved adherent layer (microenvironment) or both, is unknown (Greenberger, 1978). The addition of hydrocortisone, in the present studies, to W/W^v cultures resulted in a marked increase in megakaryocyte and CFU-M production which was sustained for 14+ weeks. In fact, hydrocortisone treated W/W^v cultures sustained megakaryopoietic activity at least as well as +/+ cultures treated with hydrocortisone. This suggests that hydrocortisone in some manner corrected the proliferative defect of CFU-M *in vitro*.

Summary

The present studies indicate that in addition to a numerical deficiency in megakaryocytes and the committed megakaryocyte progenitor cells in W/W^v mice, there is also a defect in DNA synthesis of CFU-M. Additionally, megakaryocyte and CFU-M production is decreased in W/W^v cells grown in continuous liquid marrow cultures. This decrease in megakaryocytopoiesis in these cultures appears to be ameliorated by culturing W/W^v cells in long term cultures *in vitro* in the presence of hydrocortisone. The mechanism of this improved proliferation is unclear at present and further studies of this change and its relationship to the defect in the CFU-S in W/W^v mice are required.

References

Benestad, H. B., A. Boyum, and S. T. Warhuus. 1975. Hematopoietic defects of W/W^v mice studied with the spleen colony, agar colony, and diffusion chamber techniques. *Scand. J. Haematol.* 15:219–227.

Bennett, M., G. Cudkowicz, and R. S. Foster, Jr. 1968. Hemopoietic progenitor cells of W anemic mice studied *in vivo* and *in vitro*. *J. Cell. Physiol.* 71:211–226.

Chervenick, P. A. and D. R. Boggs. 1969. Decreased neutrophils and megakaryocytes in anemic mice of genotype W/W^v. *J. Cell. Physiol.* 73:25–30.

Dexter, T. M. and L. G. Lajtha. 1974. Proliferation of haemopoietic stem cells *in vitro*. *Br. J. Haematol.* 28:525–530.

Ebbe, S., E. Phalen, and F. Stohlman, Jr. 1973. Abnormalities of megakaryocytes in W/W^v mice. *Blood* 42:857–864.

Greenberger, J. S. 1978. Sensitivity of corticosteriod-dependent insulin-resistant lipogenisis in marrow preadipocytes of obese-diabetic (db/db) mice. *Nature* 275:752–754.

Iscove, N. N. 1978. Committed erythroid precursor populations in genetically anemic W/W^v and $S1/S1^d$ mice. In *Aplastic Anemia*. Proceedings of 1st International Symposium on Aplastic Anemia. Japan Medical Research Foundation. Tokyo: Univ. Tokyo Press. pp. 31–36.

Joyce, R. A. and P. A. Chervenick. 1977. Corticosteriod effect on granulopoiesis in mice after cyclophosphamide. *J. Clin. Invest.* 60:277–283.

Lajtha, L. G. , 1979. Stem cell concepts. *Nouv. Rev. Fr. Hematol.* 21:59–65.

McCulloch, E. A., L. Siminovitch, and J. E. Till. 1964. Spleen-colony formation in anemic mice of genotype W/W^v. *Science* 144:844–846.

Nakeff, A. and S. Daniels-McQueen. 1976. *In vitro* colony assay for a new class of megakaryocyte precursor: Colony-forming unit megakaryocyte (CFU-M). *Proc. Soc. Exp. Biol. Med.* 151:587–590.

Nakeff, A. 1977. Megakaryocyte proliferation in W/W^v and Sl/Sl^d mice as revealed by CFU-M analysis. *Blood* 50 (Supp 1):248.

Russell, E. S. 1970. Abnormalities of erythropoiesis associated with mutant genes in mice. In *Regulation of Hematopoiesis* (Vol. 1). Gordon, A. S., ed. New York: Appleton-Century-Crofts. pp. 649–675.

Seller, M. J. 1970. The speed and order of colonization of some organs of anemic mice of the W series transplanted with allogeneic haemopoietic tissue. *Transplantation* 9:303–309.

Discussion

Burstein: Have you given hydrocortisone to any of the W/W^v mice *in vivo*?

Petursson: Not yet.

Nakeff: This study and one of our own shows a decrease in the total number of CFU-M in the marrow, but different results in the proportion of progenitor cells in cell cycle was observed. Our studies show an increase in susceptibility to S-phase drugs of about a 2-fold, while you show a decrease.

Petursson: I cannot explain it. I agree you would expect an increased turnover.

Nakeff: We obtained about the same reduction with both ara-C and hydroxyurea.

Petursson: Our studies were only with tritiated thymidine.

Evatt, Levine, and Williams, Editors
MEGAKARYOCYTE BIOLOGY AND PRECURSORS:
IN VITRO CLONING AND CELLULAR PROPERTIES

Human Megakaryocyte Derived Growth Factor(s)

Studies in 3T3 Cells and Bone Marrow Fibroblasts

Hugo Castro-Malaspina, Enrique M. Rabellino, Andrew Yen, Richard Levene, Ralph L. Nachman, and Malcolm A. S. Moore

Laboratories of Developmental Hematopoiesis and Hematopoietic Cell Kinetics, Sloan-Kettering Institute for Cancer Research, and Division of Hematology-Oncology, Department of Medicine, Cornell University Medical College, New York, New York

Human platelets store in their alpha granules (Kaplan et al., 1979) a growth factor (PDGF) that promotes *in vitro* the proliferation of a variety of human cells as well as cell lines derived from other species. Cells responding to PDGF include fibroblasts, smooth muscle cells and glial cells. PDGF is a heat stable, cationic polypeptide with a molecular weight of 10,000–30,000 daltons and an isoelectric point of 9.5–10.4 (Scher et al., 1979). In the intact organism, PDGF as well as other platelet substances are released into the tissues of injured sites during the clotting process. It is believed that PDGF plays a role in wound repair by promoting the proliferation of connective tissue cells (Balk, 1971). Ross et al. (1974) have also postulated that PDGF may have a role in the early stages of atherosclerosis by causing the proliferation of arterial wall smooth muscle cells.

As platelets do not have a biosynthetic apparatus (Hovig, 1968) it seems unlikely that platelets synthesize growth factor. It has been suggested that the liver and the central nervous system may be the source of PDGF(s). Two growth factors with molecular weights similar to growth factors isolated from platelets have been purified from perfusates of rat liver and from bovine brain (Witkoski et al., 1979). Platelets are fragments of megakaryocyte cytoplasm; megakaryocytes are nucleated cells which contain an intact biosynthetic apparatus and a full complement of genetic information. It seems more plausible that PDGF is formed by the megakaryocyte and retained in the cytoplasmic fragments which are shed as platelets. We have investigated this possibility by assessing the mitogenic activity elicited from extracts of megakaryocyte enriched bone marrow cells on 3T3 cells and on well characterized human bone marrow fibroblasts.

Methods and Materials

Bone marrow cells were obtained from rib fragments routinely removed from patients undergoing thoracic surgery at the New York Hospital and the Memorial Sloan-Kettering Cancer Center. Marrow cells were harvested in calcium- and magnesium-free Hank's balanced salt solution (HBSS-CAT), pH 7.0, containing 0.38% sodium citrate, 10^{-3} M adenosine, 2×10^{-3} M theophylline, and 30 U/ml DNAse (Rabellino et al., 1979). Megakaryocytes were enriched by density centrifugation and velocity sedimentation in protein-free Percoll gradients. Density centrifugation in a discontinuous Percoll gradient produced 3 fractions: Fraction III (cell density >1.050 g/cm^3), Fraction II (1.050 g/cm^3), and Fraction I (<1.050 g/cm^3). Most of the megakaryocytes were found in the low density fraction. Cells of Fraction I were separated by velocity sedimentation on a continuous Percoll gradient into two fractions: slowly sedimenting cells and quickly sedimenting cells which contained the majority of the megakaryocytes. Aliquots of all fractions were resuspended in alpha medium at 1×10^6 cells/ml. The number of megakaryocytes was determined by hemocytometer counts under phase-contrast microscopy. Cell extracts were then prepared by 2 cycles of freeze-thawing. Extracts were heat inactivated at 56°C for 30 min. To prepare platelet extracts, platelets were isolated from anticoagulated blood as previously described (Castro-Malaspina et al., in press). Platelets washed in HBSS-CAT, pH 7.0, were resuspended in alpha medium and disrupted by two cycles of freeze-thawing. Extracts were heat inactivated at 56°C for 30 min and Millipore filtered. Serum from platelet poor plasma (PPPS) and from whole blood (WBS) were obtained by the recalcification method of Ross et al. (1978). Bone marrow fibroblasts were obtained by culturing marrow cells in alpha medium supplemented with 20% fetal calf serum (FCS). Details of the characteristics of the marrow fibroblasts and their culture are reported elsewhere (Castro-Malaspina et al., 1980). Growth stimulation was assayed in 3T3 cells and human marrow fibroblasts by the incorporation of tritium-labeled thymidine (^{3}H-TdR) into cellular DNA. In some experiments, growth was assessed by determining the DNA content by flow microfluorometry. Details of the methods are described elsewhere (Castro-Malaspina et al., in press); Yen and Riddle, 1979). Briefly, for ^{3}H-TdR uptake, 20–30 $\times$ 10^3 trypsin dispersed 3T3 cells and early passages of bone marrow fibroblasts (BMF) were resuspended in 10% WBS alpha medium and seeded into 2cm^2 wells (Linbro plates). The cells were incubated 24 hr, washed twice with medium alone, and refed with medium supplemented with 1% PPPS. Twenty-four hr (3T3) or 48 hr (BMF) after the initial medium change the test material was added to the cultures. Cells were labeled with ^{3}H-TdR, 1 μCi/ml/well (spec. act. 20 Ci/mMol) for 4 hr (3T3) or 6 hr (BMF) prior to ending the incubation period at 24 hr (3T3) or 28 hr (BMF) after sample addition. Cells were harvested onto glass fiber filters with a Titertek automated cell harvester. The radioactivity was counted in a Packard liquid scintillation spectrometer. For microfluorometric measurements, fibroblasts were removed with trypsin-EDTA, washed in 10% FCS medium, and resuspended in propidium iodide (0.05 mg/ml in 1.1% sodium citrate containing 0.1% Triton X-100). Cellular DNA content analysis was done with a flow microfluorometer (Biophysics Systems 4800A) having a 488 nm emission line.

Results and Discussion

To assess the growth activity associated with megakaryocytes, extracts of unseparated bone marrow and of the different marrow fractions obtained by the separation procedures were added to resting 3T3 cells and BMF (Table 1). The megakaryocyte containing fraction effected the greatest ^{3}H-TdR incorporation. Cell proliferation was also measured by flow microfluorometry before and after addition of megakaryocyte-enriched marrow cell extracts (Fig. 1). Initially quiescent cells having G_1 DNA content entered S phase and proliferated after addition of marrow extracts.

To confirm that the growth activity was derived from megakaryocytes various quickly sedimenting marrow fractions containing 1×10^6 cells/ml but different proportions of megakaryocytes were tested for growth promoting activity(ies) on marrow fibroblasts. The proportions of erythroid cells of quickly sedimenting fractions used in these experiments did not vary significantly; the proportion of early myeloid precursors and some lymphoid-like cells decreased when the number of megakaryocytes increased. The growth activity of quickly sedimenting marrow cells was proportional to their megakaryocyte content. The increase of growth activity was not related to the loss of inhibitory activity(ies) associated with myeloid or lymphoid cells. Simultaneous addition of extracts of platelets and well purified peripheral blood or marrow leukocytes to quiescent fibroblasts do not inhibit the growth promoting activity of platelet extracts.

The growth activities elicited from platelets and megakaryocytes were compared. Treatment with trypsin for 30 min at 37°C abolished all growth activity in both cell extracts. Treatment with heat at 100°C for 5 min modified slightly the activity derived from both cell extracts. In addition, when dilutions of both extracts were added to cultured marrow fibroblast, the resulting dose-response curves were parallel

Table 1. Growth Promoting Activity of Megakaryocyte Enriched Bone Marrow Cell Extracts.

		3T3 cells		Marrow fibroblasts	
Test material[a]	**Mk (%)**	**CPM**[b]	**SR**[c]	**CPM**	**SR**
Media	–	1420	–	820	–
Platelet extract	–	28450	20.0	22450	27.4
Unseparated marrow extract	0.04	1560	1.1	1050	1.3
Marrow fraction extracts					
Fraction III	<0.01	1380	0.9	980	1.2
Fraction II	0.02	1550	1.1	1100	1.3
Fraction SS	0.08	2030	1.4	1350	1.6
Fraction QS	15.00	13860	9.8	8640	10.5

[a]Platelet extracts and marrow cell extracts containing 10^9 platelets/ml and 1×10^6 cells/ml respectively were tested at 10% final concentration.

[b]Values are mean cpm of triplicates from a typical experiment.

[c]SR: Stimulation ratio: cpm stimulated cultures/cpm medium control cultures.

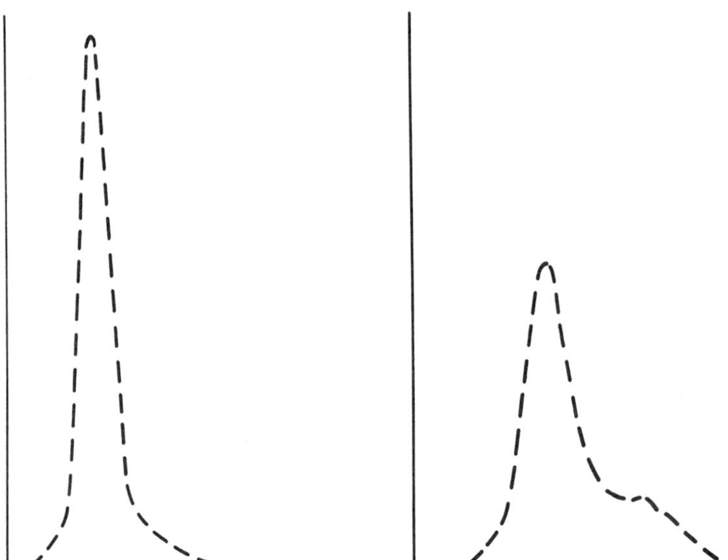

Fig. 1 Effect of megakaryocyte-enriched marrow extracts on marrow fibroblast DNA content frequency distribution histograms of fibroblasts at 0 hour (left) and 18 hours (right) after addition of marrow extracts. (10% final concentration). Ordinate: relative number of cells; Abscissa: DNA content.

indicating that the growth activities derived from both extracts have similar mitogenic effects (Castro-Malaspina et al., in press).

The results of these studies provide evidence that human megakaryocytes contain a growth factor (MkDGF) with properties similar to PDGF that stimulates the proliferation of 3T3 cells and human marrow fibroblasts. Chernoff et al. (1980) have recently reported that guinea pig megakaryocytes also contain a fibroblast growth factor. Taking into consideration the fact that very little protein synthesis occurs in platelets (Boullin et al., 1972) and that platelets originate from megakaryocytes, our findings implicate that the synthesis of PDGF may occur in megakaryocytes.

The biological role of PDGF in normal and pathological states has been outlined. MkDGF may also have some implications in human pathology. Primary Myelofibrosis (PMF) is a chronic myeloproliferative disorder characterized by an abnormal accumulation of collagen in bone marrow and an excessive proliferation of marrow myeloid cells within the bone marrow and in extramedullary sites. The pathophysiological mechanisms leading to marrow fibrosis are unknown. Cytogenetic marker analysis of hematopoietic cells and fibroblastlike cells derived from bone marrow has provided evidence indicating that the marrow fibrosis observed in the so-called Primary Myelofibrosis is a secondary phenomenon associated with a clonal proliferation of hematopoietic cells (Jacobson et al., 1978). Marrow megakaryocytosis with marked morphological abnormalities of these cells is a constant finding in

PMF. Despite marrow megakaryocytosis platelet counts are within or below normal values. Platelet kinetic studies have demonstrated that ineffective thrombopoiesis occurs in PMF (Harker and Finch, 1969). Morphological and kinetic studies suggest that the death rate of megakaryocytes in PMF is highly increased. The process leading to marrow fibrosis may be triggered by the abnormal presence in marrow intercellular space of MkDGF, which stimulates marrow fibroblast proliferation and its collagen secretion.

There is little information concerning the site of synthesis of the platelet derived growth factor (PDGF). The fact that platelets have a rudimentary biosynthetic apparatus suggests that PDGF may be synthesized by megakaryocytes, stored in alpha granules and merely transported by platelets. Human marrow cells were processed sequentially by density centrifugation and velocity sedimentation in serum-free Percoll gradients. Cell homogenates were made from the resulting fractions and tested for growth promoting activity(ies) in 3T3 cells and in well characterized human marrow fibroblasts. Growth was measured by ^{3}H-TdR incorporation and changes in DNA cell content; the latter was determined by flow microfluorometry. The highest mitogenic activity in both 3T3 cells and marrow fibroblasts was derived from marrow extracts containing the highest number of megakaryocytes. Like the growth promoting activity derived from platelet extracts, the growth promoting activity derived from megakaryocyte enriched marrow cell extracts was temperature resistant and trypsin sensitive. The dose response curves for both growth factors were parallel. These findings suggest that megakaryocytes may be the site of synthesis of PDGF.

ACKNOWLEDGMENTS

This investigation was supported by NIH CA-08748, CA-17353, CA-20194, HL-18828, Public Health Service International Research Fellowship F05-TW-02524, the Gar Reichman Foundation, and the New York Community Trust Award.

References

Balk, S. D. 1971. Calcium as a regulator of the proliferation of normal, but not transformed, chicken fibroblasts in a plasma-containing medium. *Proc. Nat. Acad. Sci. USA* 68:271–275.

Boullin, D. J., M. Votavova, and A. R. Green. 1972. Protein synthesis by human blood platelets after accumulation of leucine and arginine. *Thromb. Diathes. Haemorrh.* 28:54–64.

Castro-Malaspina H., E. M. Rabellino, A. Yen, R. L. Nachman, and M. A. S. Moore. Human megakaryocyte stimulation of proliferation of bone marrow fibroblasts. *Blood*, in press.

Castro-Malaspina, H., R. E. Gay, G. Resnick, N. Kapoor, P. Meyers, D. Chiarieri, S. McKenzie, H. E. Broxmeyer, and M. A. S. Moore. 1980. Characterization of human bone marrow fibroblast colony forming cells (CFU-F) and their progeny. *Blood* 56:289–301.

Chernoff, A., R. F. Levine, and D. S. Goodman. 1980. Origin of platelet-derived growth factor in megakaryocytes in guinea pigs. *J. Clin. Invest.* 65:926–930.

Harker, L. H. and C. A. Finch. 1969. Thrombokinetics in man. *J. Clin. Invest.* 48:963–974.

Hovig, T. 1968. The ultrastructure of blood platelets in normal and abnormal states. *Ser. Haematol.* 1:3–64.

Jacobson, R. J., A. Salo, and P. J. Fialkow. 1978. Agnogenic myeloid metaplasia: A clonal proliferation of hematopoietic stem cells with secondary myelofibrois. *Blood* 51:189–194.

Kaplan, D. R., F. C. Chao, C. D. Stiles, H. N. Antoniades, and C. D. Scher. 1979. Platelet-granules contain a growth factor for fibroblasts. *Blood* 53:1043–1052.

Rabellino, E. M., R. L. Nachman, N. Williams, R. J. Winchester, and G. D. Ross. 1979. Human Megakaryocytes. I. Characterization of the membrane and cytoplasmic components of isolated marrow megakaryocytes. *J. Exp. Med.* 149:1273–1287.

Ross, R. J., J. Glomset, B. Kariya, and L. Harker. 1974. A platelet-dependent serum factor that stimulates the proliferation of arterial smooth muscle cells *in vitro*. *Proc. Nat. Acad. Sci. USA* 71:1207–1210.

Ross, R., C. Nist, B. Kariya, M. J. Rivest, E. Raines, and J. Callis. 1978. Physiological quiescence in plasma-derived serum: Influence of platelet-derived growth factor on cell growth in culture. *J. Cell. Physiol.* 97:497–508.

Scher, C. D., R. C. Shepard, H. N. Antoniades, and C. D. Stiles. 1979. Platelet-derived growth factor and the regulation of the mammalian fibroblast cell cycle. *Biochim. Biophys. Acta* 560:217–241.

Witkoski, E., M. F. Schuler, R. C. Feldhoff, S. I. Jacob, L. S. Jefferson, and A. Lipton. 1979. Liver as a source of transformed cell growth factor. *Exp. Cell. Res.* 124:261–267.

Yen, A.G. and V. G. H. Riddle. 1979. Plasma and platelet associated factors act in G_1 to maintain proliferation and to stabilize arrested cells in a viable quiescent state. A temporal map of control points in the G_1 phase. *Exp. Cell. Res.* 120:349–357.

Concluding Remarks and Summary

Evatt, Levine, and Williams, Editors
MEGAKARYOCYTE BIOLOGY AND PRECURSORS:
IN VITRO CLONING AND CELLULAR PROPERTIES

Concluding Remarks

Jack Levin

The Johns Hopkins University, School of Medicine and Hospital, Baltimore, Maryland

This conference was exciting and stimulating and these published proceedings establish the study of megakaryocytes *in vitro*, and associated investigations of thrombopoiesis and thrombopoietin, as a vital and established field. I would like to thank Bruce Evatt, his Atlanta colleagues Kathi Kellar, Rosemary Ramsey, and Neile McGrath; Richard Levine; and Neil Williams for organizing this symposium.

Before I comment on the present and presumptuously try to look into the future, it seems appropriate to make some remarks about the past. Less than 15 years ago, two papers, one by Bradley and Metcalf (1966) and the other by Pluznik and Sachs (1965), were published. Both reported the successful culture, *in vitro*, of hematopoietic cells and thereby greatly expanded our ability to study hematopoiesis. The history of Ray Bradley's work is appropriate to recount. He was attempting to culture tumor cells, based on some techniques he had recently learned at the National Institutes of Health. His attempts to grow tumor cells were unsuccessful; but when he evaluated cultures in which he had used bone marrow cells as one of the control tissues while attempting to detect clonogenic cells in the thymus of AKR mice, colonies of hematopoietic cells were detected. The nature of many important albeit unexpected contributions to scientific progress is further exemplified by the fact that Ray Bradley was not a hematologist at that time.

Utilization of a histochemical stain for acetylcholinesterase has figured prominently in this conference and demonstrates how far we have progressed during the past few decades. There have been many comments about potential relationships between erythropoiesis and thrombopoiesis and the possible presence of colony forming cells from which either megakaryocytes or erythroid cells can be derived. Therefore, it is important to recall that well into the 20th century some believed that red blood cells produced platelets. During the early part of the twentieth century, an American physiologist climbed Pike's Peak in order to determine the effects of hypoxia on platelets (Robb-Smith, 1967). The classical study of Zajicek (1957) which described the detection of acetylcholinesterase in hematopoietic cells was carried out not primarily to study megakaryocytopoiesis, but to establish whether or not platelets were derived from red blood cell precursors. His demonstration that, in most species studied, acetylcholinesterase was primarily either in red cells or in platelets and that acetylcholinesterase activity was similar in the megakaryocytes and platelets of an animal helped establish that platelets are derived from megakaryocytes, not from erythroid precursors.

Although this meeting has focused on *in vitro* studies of megakaryocytes, many of our concepts and much of our ability to interpret current data are based on previous *in vivo* studies. It is appropriate to acknowledge that there are two people who have contributed enormously to our knowledge of megakaryocytopoiesis. They are Shirley Ebbe, currently of the Donnor Laboratory in Berkeley, California, and Theodore Odell, of the Oak Ridge National Laboratory, in Oak Ridge, Tennessee.

Technological progress makes it possible to ask new questions and to reconsider old ones. Three major developments have contributed to rapid advances in our studies. The first is a group of techniques for the culture, *in vitro*, of hematopoietic cells, with the resultant ability to evaluate and characterize colony forming cells. The second is a series of techniques for the purification and maintenance, for *in vitro* study, of recognizable megakaryocytes obtained from bone marrow. The third is the application of flow cytometry for the study of megakaryocytes, the only polyploid hematopoietic cell.

Megakaryocyte Colony-Forming Cells

A marked increase in megakaryocyte colony-forming cells (Meg-CFC) has been demonstrated following induction of acute thrombocytopenia in mice (Burstein: 127; Levin: 180; Petursson: 191).* However, the time of increase in frequency of Meg-CFC (3–5 days after thrombocytopenia) indicates that this response is not the cause of the changes that occur in bone marrow within the 48-hr period following acute thrombocytopenia (Table 1). Furthermore, the important role of the spleen in this reaction has been documented (Levin, Chapt. 15), indicating an important probable difference between the response of the mouse and the human to hematopoietic stress. The studies that have been presented indicate that we do not understand the stimulus for the increase in Meg-CFC or the feedback mechanism that regulates their frequency. At what point MEG-CFC occur in the sequence of the stages of megakaryocytopoiesis is unknown, but presumably they are more primitive than committed megakaryocyte precursors. It is important to emphasize that there is no single type of stem cell but rather that all so-called compartments are in themselves continua and in addition part of a larger continuum, ranging from

* References to data presented at this symposium are indicated in the text by the name of the senior author and the page number in this volume.

pluripotential stem cells through proliferating precursors and differentiating precursors to maturing megakaryocytes.

Analysis of Cultures

Great caution is required for the accurate interpretation of cultures. The use of plasma clots may not yield the same results as cultures of cells in agar or methylcellulose (Nakeff: 111). Culture systems also introduce the problem of the effects of various stimulators and potentiators that are currently used, as emphasized by Williams and his colleagues (Williams: 59). We know little about the reagents which currently are required for successful hematopoietic culture. For example, why is spleen conditioned medium effective in supporting the growth of megakaryocyte colonies in soft agar cultures, whereas similarly prepared supernatants from other normal organs are ineffective? Williams' work has successfully utilized a conditioned medium from WEHI-3 murine leukemic cells. However, the utilization of tumor lines is potentially unreliable because of their ability to change. Our attempts to use the parent tumor line from which Williams' cells were obtained were unsuccessful; only granulocyte and macrophage colonies were detected. Finally, it is important to emphasize that at this time, no pure megakaryocyte colony-stimulating factor (Meg-CSF) is available; and even if it were, its effects might be difficult to evaluate. It has been demonstrated recently that pure granulocyte-macrophage colony-stimulating factor (GM-CSF) can affect erythroid precursors (Metcalf and Johnson, 1979). To add to the complexity of interpreting the physiologic significance of *in vitro* stimulators are our observations that following induction of acute thrombocytopenia in mice, there was an increase in GM-CSF but no detectable increase in Meg-CSF in the blood.

In many studies, high concentrations of erythropoietin (EPO) have been required for growth of megakaryocyte colonies. I think it is reasonable to conclude that EPO is not a physiological regulator of thrombopoiesis for the following reasons: secondary polycythemia is characteristically not associated with thrombocytosis, despite elevated erythropoietin levels; nephrectomized humans do not become thrombocytopenic; studies of persons at high altitudes have indicated that platelet counts are normal; our investigations of alterations in endogenous levels of thrombopoietin

Table 1. The Response of the Bone Marrow to Acute Thrombocytopenia.

1. Increase:	Small AChE positive cells. Mitotic index of Megakaryocytes. Number of Megakaryocytes.
2. Increase:	Megakaryocyte size, ploidy, and the rate of maturation.
3. Increased rate of platelet production	

and erythropoietin have indicated that elevated levels of thrombopoietin do not stimulate erythropoiesis and similarly that elevated levels of erythropoietin do not stimulate thrombopoiesis (Evatt et al., 1976); and finally, nephrectomized animals apparently respond normally to the stress of thrombocytopenia (Levin and Evatt, 1979). However, large doses of erythropoietin could produce non-specific effects since it is well recognized that apparently specific receptors for hormones can respond to other hormones, if the latter are in sufficiently high concentration. Therefore, conservatism is required in interpreting the effects of addition of various reagents to cultures.

In addition, our own studies have indicated that administration of platelet antiserum to animals resulted in an increase not only of Meg-CFC but surprisingly also of GM-CFC. What was predicted to be a specific stimulus for one type of CFC was demonstrated to be a generalized one. Therefore, I think it is incumbent when evaluating cultures to quantify not only megakaryocyte colonies but also the other types of colonies concomitantly present if we are to establish the specificity of an effect. The well known variability of culture systems indicates a further need for caution. Different batches of fetal calf serum or horse serum can produce markedly different results and the true frequency of Meg-CFC in the bone marrow and spleen of different rodent species and man remains unknown. We obviously must develop systems with more refined components. Accurate scoring of colonies is an important requirement. Although acetylcholinesterase, a specific marker for the megakaryocytes of some rodents, does not occur in human megakaryocytes, utilization of fluorescent labeled antibodies may serve to specifically identify human megakaryocytes (Messner: 161; Mazur: 281; Rabellino: 233).

Physiologic Significance of Data Obtained from Cultures

We do not know the physiological pertinence of at least some of the data obtained from cultures. Most previous reports have not indicated the levels of cells in the peripheral blood of mice when marked changes in colony-forming cells were detected. Our investigations have indicated that despite a 10-fold increase in GM-CFC (with the potential for a tremendous amplification in numbers of mature cells), mice did not develop leukocytosis. Furthermore, an increase in Meg-CFC was not associated with thrombocytosis (Levin et al., 1980). Splenectomized mice still failed to demonstrate increased Meg-CFC in bone marrow following induction of acute thrombocytopenia, and yet their platelet counts returned to normal levels at the usual rate (Levin et al., 1980). Therefore, some of the experimental models reported at this meeting have not yet demonstrated a correlation between numbers of colony forming cells detected *in vitro* and demonstrable effects on levels of peripheral blood cells.

Measurement of DNA

Polyploidy is a unique feature of megakaryocytes. Bunn (Chapt. 19) has elegantly summarized information derived from the use of flow cytometry. He has reported the surprising observation that the proportion of 8N megakaryocytes in bone marrow is higher than previously believed. The apparent explanation is that previous investigators have selected definitely recognizable megakaryocytes for establishment of their ploidy distribution, thereby overlooking the often more difficult to identify 8N megakaryocytes. Although more data will be required to establish with certainty the distribution of ploidy levels in megakaryocytes because of potential sampling problems, the studies of Bunn (Bunn: 219) and Mayer et al. (Mayer: 299) have demonstrated the power of flow cytometry. We obviously are on the threshold of a series of important biological and clinical investigations based on measurement of ploidy levels of megakaryocytes in health and disease.

Microdensitometry remains an important technique for the evaluation of cells obtained from cultures because it provides the opportunity to identify and measure DNA in each individual cell in a colony (Levin: 180; Paulus: 171). Levin et al. (Levin: 180) and Paulus et al. (Paulus: 171) have confirmed the previous observations of Metcalf et al. (1975), Williams and Jackson (1978), and Burstein et al. (1979) that there are two types of murine megakaryocyte colonies. One is composed of essentially big, mature appearing, acetylcholinesterase positive cells and the other of larger numbers of a more heterogeneous population, many of which are acetylcholinesterase negative (Levin: 180). Studies of ploidy levels in megakaryocyte colonies, using microdensitometry and the Feulgen-Schiff stain, have indicated that there are two distinct ploidy patterns, one associated with big cell colonies and the other with heterogeneous colonies (Levin: 180; Paulus: 171). Furthermore, important biological differences exist between these colony types (Levin: 180).

Isolated Megakaryocytes

Studies of recognizable megakaryocytes, purified by the techniques described by Levine (Levine: 203) and Rabellino et al. (Rabellino: 233) already have resulted in better descriptions of megakaryocytes. Impressions have been replaced by accurate analyses of important characteristics of megakaryocytes, complementing previous elegant electron microscopic studies by Breton-Gorius (Breton-Gorius: 139), Breton-Gorius and Reyes, (1976), and Bentfield and Bainton (1975). Our ability to identify immature megakaryocytes has greatly increased. Availability of purified preparations of megakaryocytes already has led to elegant biochemical (Schick: 251) and functional (Levin: 261) studies, and I look forward to future studies describing biochemical characteristics of megakaryocytes in normal and perturbed states. These techniques will allow us to focus on factors which affect the late stages of megakaryocyte maturation, including importantly the effects of humoral regulators such as thrombopoietin.

Human Megakaryocytes

An important series of papers described colonies of human megakaryocytes *in vitro*. The reports of Vainchenker and Breton-Gorius (Breton-Gorius: 139), Messner and

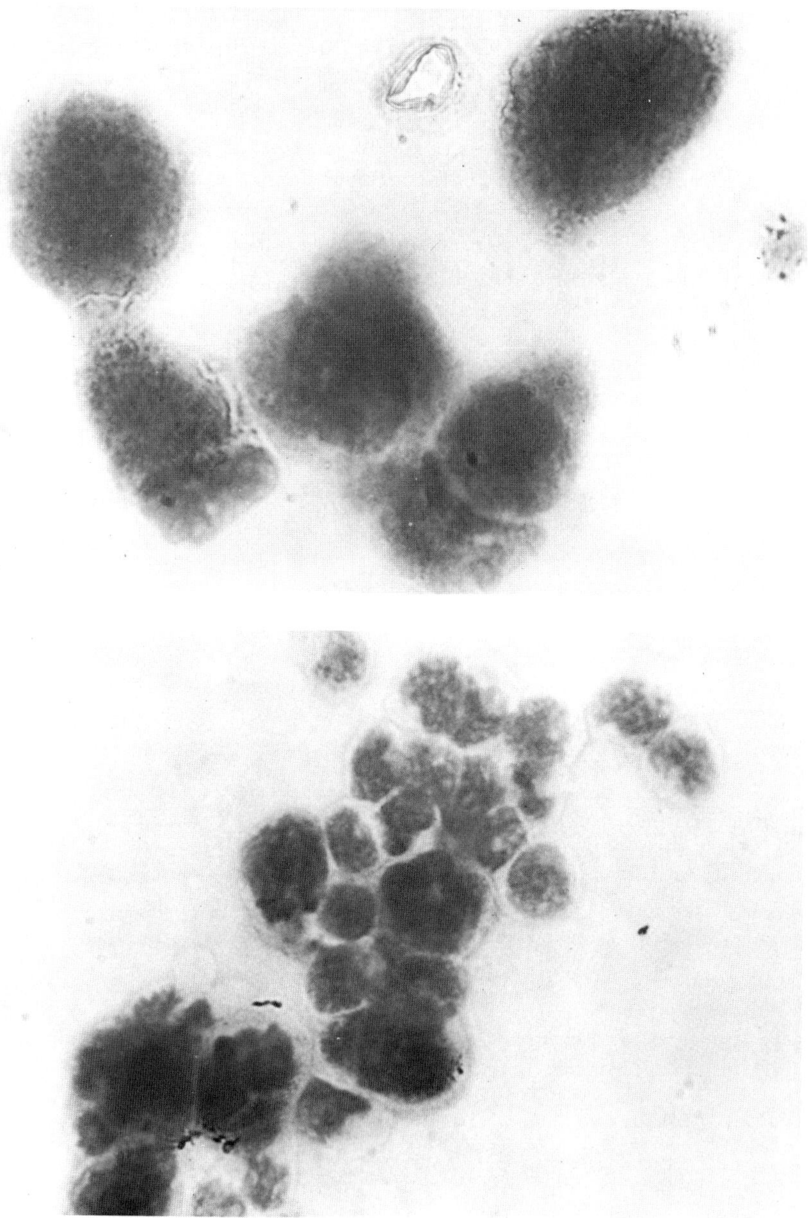

Fig. 1. Big cell colonies (top) contained large, mature appearing, acetylcholinesterase (AChE) positive cells. Heterogeneous colonies (bottom) contained large, intermediate, and small cells. Many of the small cells were AChE negative, as indicated by their clear cytoplasm. The presence of AChE is indicated by the dark, granular color in the cytoplasm. The nucleus was counterstained with methyl green (original magnification × 1,000).

Fauser (Messner and Fauser: 161) and Mazur et al. (Mazur: 281) leave no doubt that we have finally developed techniques for the growth of human megakaryocyte colonies. This has obvious importance for studies of human disease. Criteria for identification of human megakaryocytes and their precursors are critical. Unfortunately, the histochemical methods report ed by Breton-Gorius and her colleagues (Breton-Gorius and Reyes, 1976) are not easily applicable for routine use and therefore we will have to depend upon immunological techniques and biochemical characteristics (Mazur: 281; Rabellino: 233). The requirement for high concentrations of erythropoietin (1–3 units/culture) to support the growth of human megakaryocyte colonies in plasma clot systems is of concern and I believe that data will be much more valid now that we have learned to maintain these colonies without the need for erythropoietin (Mazur: 281).

Thrombopoietin

The mechanism of action of thrombopoietin (TPO) remains unknown (Levin and Evatt, 1979). Administration of thrombopoiesis-stimulating activity to rodents results in an increased incorporation of either ^{75}Se-selenomethionine or $Na_2{}^{35}SO_4$ into megakaryocytes, with subsequent appearance of labeled platelets in the circulation. The results are compatible with an increase in protein in platelets following administration of TPO. Kellar et al. (Kellar: 21) have observed a direct effect of TPO on DNA synthesis. However, the specificity of their observations remains to be established since EPO also produced an increase in DNA synthesis. She and her colleagues also observed that TPO resulted in maintenance of megakaryocytes with high ploidy levels in liquid culture, again suggesting an effect of TPO on DNA synthesis. I look forward to additional experiments confirming these observations. McDonald (McDonald: 77) has reviewed the reported effects of thrombopoietin; these are summarized in Table 2. It seems unlikely that TPO causes all of these varied

Table 2. Some Reported Effects of Thrombopoietin.

Increased incorporation of ^{75}SeM or $Na_4{}^{35}SO_4$ into megakaryocytes.
Elevation of platelet count.
Increased platelet size.
Increase in AChE positive cells in bone marrow.
Increase in Megakaryocyte colonies in the spleen colony assay with an increased number of megakaryocytes/colony.
Increased endomitotic index of bone marrow megakaryocytes.
Increased Meg-CFC, *in vitro*.

biological alterations and therefore I think that studies utilizing pure TPO in carefully characterized assay systems will be required before we can accurately establish the significant, direct biological effects of TPO on megakaryocytopoiesis. Although detection of effects of TPO upon Meg-CFC *in vitro* has been reported, other laboratories represented at this meeting have been unable to confirm these findings. It has been reported that the addition of plasma from thrombocytopenic donors to cultures increased the frequency of Meg-CFC, but this also has not been confirmed in two other laboratories. Therefore, I think that no conclusions can currently be drawn about potential direct effects of TPO on Meg-CFC.

I think it is clear that megakaryocytes and their platelet progeny have attracted a critical mass of investigators who are effectively applying newly available techniques for the study of megakaryocytopoiesis and control of platelet production. I look forward to the next megakaryocyte meeting at which time I hope to learn that platelet production has unequivocally demonstrated *in vitro* how many platelets are produced by a megakaryocyte; the primary biological effect of thrombopoietin; and the ploidy levels of small acetylcholinesterase positive cells that have been of such great interest.

ACKNOWLEDGMENT

Supported in part by Research Grant HL 01601 from the National Heart, Lung, and Blood Institute, National Institutes of Health, Bethesda, Maryland.

References

Bentfield, M. E. and D. F. Bainton. 1975. Cytochemical localization of lysosomal enzymes in rat megakaryocytes and platelets. *J. Clin. Invest.* 56:1635–1649.

Bradley, T. R. and D. Metcalf. 1966. The growth of mouse bone marrow cells *in vitro*. *Aust. J. Exp. Biol. Med. Sci.* 44:287–300.

Breton-Gorius, J. and F. Reyes. 1976. Ultrastructure of human bone marrow cell maturation. *Int. Rev. Cytol.* 46:251–321.

Burstein, S. A., J. W. Adamson, D. Thorning, and L. A. Harker. 1979. Characteristics of murine megakaryocyte colonies *in vitro*. *Blood* 54:169–179.

Evatt, B. L., J. L. Spivak, and J. Levin. 1976. Relationships between thrombopoiesis and erythropoiesis: With studies of the effects of preparations of thrombopoietin and erythropoietin. *Blood* 48:547–558.

Levin, J. and B. L. Evatt. 1979. Humoral control of thrombopoiesis. *Blood Cells* 5:105–121.

Levin, J., F. C. Levin, and D. Metcalf. 1980. The effects of acute thrombocytopenia upon megakaryocyte-CFC and granulocyte-macrophage-CFC in mice: Studies of bone marrow and spleen. *Blood* 56:274–283.

Metcalf, D. and G. R. Johnson. 1979. Interactions between purified GM-CSF, purified erythropoietin and spleen conditioned medium on hemopoietic colony formation *in vitro*. *J. Cell. Physiol.* 99:159–174.

Metcalf, D., H. R. MacDonald, N. Odartchenko, and B. Sordat. 1975. Growth of mouse megakaryocyte colonies *in vitro*. *Proc. Nat. Acad. Sci. USA* 72:1744–1748.

Pluznik, D. H. and L. Sachs. 1965. The cloning of normal "mast" cells in tissue culture. *J. Cell Comp. Physiol.* 66:319–324.

Robb-Smith, A. H. T. 1967. Why the platelets were discovered. *Br. J. Haematol.* 13:618–637.

Williams, N. and H. Jackson. 1978. Regulation of the proliferation of murine megakaryocyte progenitor cells by cell cycle. *Blood* 52:163-170.

Zajicek, J. 1957. Studies on the histogenesis of blood platelets and megakaryocytes. *Acta Physiol. Scand.* 40 (Suppl. 138):1–32.

Evatt, Levine, and Williams, Editors
MEGAKARYOCYTE BIOLOGY AND PRECURSORS:
IN VITRO CLONING AND CELLULAR PROPERTIES

Summary

The Editors

It is now clear that megakaryocytes and platelets have many substances (B. Schick) and functions (Leven, P. Schick) in common. During differentiation, megakaryocytes synthesize, assemble, and accumulate platelet components (Rabellino, P. Schick). It is not surprising that megakaryocytes have some synthetic capabilities that platelets lack; it will be interesting to learn the extent of their similarities and differences (B. Schick) and the significance of these findings.

As a corollary to the fact that megakaryocytes share platelet components, detection of these entities in nucleated cells confirms their identity as megakaryocytes. Since the earlier work establishing the specific presence of acetylcholinesterase in even the smallest megakaryocytes of certain species (Zajicek, J., *Acta Haematol.* 12:238–244, 1954; Jackson, C.W., *Blood* 42:413–421, 1973), serotonin uptake (P. Schick) and a variety of platelet surface and granular specific antigens (Rabellino) have been demonstrated in immature as well as in mature megakaryocytes. Immunofluorescent antibodies to these cell-specific antigens in humans have been shown to be an equivalent of acetylcholinesterase in identifying megakaryocytes in rodents (Rabellino, Mazur). These markers have now been used to identify the entire population of megakaryocytes, from the smallest, newly differentiated megakaryocytes to the larger, more familiar cells. Thus, a 2N cell bearing specific megakaryocyte "differentiation markers" has been demonstrated (Mayer) and human megakaryocyte progenitors have been assayed by detection of their differentiated progeny (Mayer). The younger megakaryocytes (particularly the 8N) appear to be quantitatively predominant in the continuum of megakaryocyte maturation (Bunn) but these generally escaped detection in the past (Levine). The description and classification of immature megakaryocytes should allow new insights into megakaryocytopoiesis in human physiology and pathology.

Megakaryocytes were found to be influenced *in vitro* by thrombopoietin, a substance also active *in vivo*. Plasma thrombopoietin preparations stimulated identifiable megakaryocytes, especially the megakaryoblasts, to incorporate tritiated thymidine and undergo endomitosis (Kellar). *In vivo* studies showed that in response to thrombocytopenia in mice, there was a rapid increase in the apparent number and volume of the megakaryocytes (Burstein). In response to embryonic kidney cell thrombocytopoiesis stimulating factor (TSF), early megakaryocytes formed increased numbers of acetylcholinesterase-positive colonies (Kalmaz) and small immature megakaryocytes matured into larger, more readily identifiable cells (Long). An increase in size and thus an increase in the number of readily detectable megakaryocytes appeared to occur throughout the megakaryocyte population in response to thrombopoietin and TSF preparations, but the total number of megakaryocytes may not change as rapidly.

There is little evidence that thrombopoietin influences the progenitor cell compartment. No immediate change in the number of *in vitro* colonies derived from the marrow of thrombocytopenic animals was detected (Burstein, Lewis, Petursson). Separation of the control of the proliferative and differentiated compartments was shown by *in vitro* studies, where two separate factors were required for the development of colonies containing mature megakaryocytes (Williams). The second of these factors may be thrombopoietin. Megakaryocyte progenitor cells appear to be controlled in part by their cell cycle status, being only slowly cycling in the steady state. These clonable precursor cells achieve rapid cycling in regenerating marrow. However, suppression of thrombopoietin by injection of platelets into reconstituted, irradiated mice did not influence the cell cycle characteristics, indicating independent regulation of progenitor cells and megakaryocytes (Burstein).

The effects of progenitor cells that have been attributable to embryonic kidney cell TSF (McDonald) and other thrombopoietin preparations, both *in vivo* and *in vitro*, may be due to the presence of more than one stimulatory substance or may reside as a direct effect on differentiated megakaryocytes. Since it is only the recognizable megakaryocytes that are counted in several experimental assays, many observations may be explained by changes in the level of detection of megakaryocytes. Furthermore, it remains to be established that different sources of thrombopoietin or TSF contain the same substance(s) with identical biological activities. This unproven assumption complicates interpretation of data on regulatory factors. Specificity of action of a purified hormone on particular target cells remains to be demonstrated.

Megakaryocyte colony stimulating factor has not been separated from other colony stimulators for testing of specificity *in vitro* (Williams) and indeed the specificity of various colony stimulating factors *in vivo* remains questionable. Megakaryocyte colony stimulating factor is difficult to detect in animal fluids and tissues. Although mouse bone and lung have been shown to contain very small amounts of megakaryocyte stimulating factor serum has been consistently negative, even when obtained from stressed or mitogen-stimulated animals. The development of multipotential stem cells into megakaryocyte progenitors may be constitutive and not specifically regulated. Increases in myeloid and megakaryocytic progenitors as well as pluripotential stem cells in thrombocytopenic animals was suggestive of such a generalized mechanism (Ebbe, Levin).

A novel concept presented at this symposium was that certain megakaryocyte precursor cells may have limited proliferative capacity and undergo a finite number of mitoses, the total number being the sum of the cell divisions and the endonuclear reduplications (Paulus). However, the mechanisms by which polyploidy is acquired are unknown. The DNA levels reached by particular megakaryocytes may depend upon the level of thrombopoietin, rather than on the previous history of the cell. These hypotheses suggest possible bases for the variability in the number of endomitoses in each developing megakaryocyte. The observation of two types of colonies with few or many megakaryocytes (Levin) seems to indicate the existence of different subclasses of precursor cells, differing in the number of cell divisions of which they are capable.

Because differentiated megakaryocytes are end-stage and non-proliferating cells, these states have limited the number of megakaryocytes available for study. The availability of a rat megakaryocytic cell line (Cicoria, A.D., DuPre, A. and Hempling, H.G., *The Physiologist* 20:19, 1977) may provide large quantities of megakaryocytes. New reports have further characterized this line which certainly has or acquires a number of megakaryocytic differentiation markers (Groopman, Steinberg). However, some significant differences in these cells were also found by these two groups. Clonal evolution or different growth conditions may account for their findings. Furthermore, these may well be transformed cells, so that their exact significance remains uncertain. Further development in this area will be awaited eagerly.

The importance of observations made in animal models and the need to demonstrate the same phenomena *in vitro* were well discussed in the Introduction and in the Concluding Remarks. These chapters, by two notable contributors to the field of megakaryocytopoiesis, give full perspective on the *in vitro* opportunites for, and achievements in, solving problems identified in animal studies. The new *in vitro* approaches to the study of megakaryocytes and their precursors should allow many additional investigations beyond those addressed in this volume. Is there a steady state in the absence of thrombopoietin? How are nonspecific thrombocytoses mediated? What is the structure of the progenitor cell compartment and how does it relate to the hematopoietic pluripotent stem cells?

Many *in vivo* perturbations such as irradiation and drug sensitivity can now be examined *in vitro*. A major reason for studying megakaryocytopoiesis is to understand and to quantitate platelet production. This focus particularly demands evaluation *in vitro* and has not been considered adequately.

The *in vitro* observations reported in this volume have together made a significant impact on our basic knowledge. More precise and extensive research is anticipated.

Index